Combinatorial Reciprocity Theorems

An Invitation to Enumerative Geometric Combinatorics

Matthias Beck
Raman Sanyal

GRADUATE STUDIES IN MATHEMATICS 195

Combinatorial Reciprocity Theorems

An Invitation to Enumerative Geometric Combinatorics

Matthias Beck
Raman Sanyal

AMERICAN MATHEMATICAL SOCIETY
Providence, Rhode Island

2010 *Mathematics Subject Classification.* Primary 05Axx, 05C31, 05E45, 11P21, 52B05, 52B11, 52B20, 52B45, 52C07, 68R05.

For additional information and updates on this book, visit
www.ams.org/bookpages/gsm-195

Library of Congress Cataloging-in-Publication Data

Names: Beck, Matthias, author. | Sanyal, Raman, 1978-author.
Title: Combinatorial reciprocity theorems : an invitation to enumerative geometric combinatorics / Matthias Beck, Raman Sanyal.
Description: Providence, Rhode Island : American Mathematical Society, [2018] | Series: Graduate studies in mathematics ; volume 195 | Includes bibliographical references and index.
Identifiers: LCCN 2018034294 | ISBN 9781470422004 (alk. paper)
Subjects: LCSH: Combinatorial geometry. | Combinatorial analysis. | AMS: Combinatorics – Enumerative combinatorics – Enumerative combinatorics. msc — Combinatorics – Graph theory – Graph polynomials. msc | Combinatorics – Algebraic combinatorics – Combinatorial aspects of simplicial complexes. msc | Number theory – Additive number theory; partitions – Lattice points in specified regions. msc | Convex and discrete geometry – Polytopes and polyhedra – Combinatorial properties (number of faces, shortest paths, etc.). msc | Convex and discrete geometry – Polytopes and polyhedra – n-dimensional polytopes. msc | Convex and discrete geometry – Polytopes and polyhedra – Lattice polytopes (including relations with commutative algebra and algebraic geometry). msc | Convex and discrete geometry – Polytopes and polyhedra – Dissections and valuations (Hilbert's third problem, etc.). msc | Convex and discrete geometry – Discrete geometry – Lattices and convex bodies in n dimensions. msc | Computer science – Discrete mathematics in relation to computer science – Combinatorics. msc
Classification: LCC QA167.B3545 2018 | DDC 511/.6–dc23
LC record available at https://lccn.loc.gov/2018034294

∞ The paper used in this book is acid-free and falls within the guidelines established to ensure permanence and durability.
Visit the AMS home page at `https://www.ams.org/`

10 9 8 7 6 5 4 3 2 1 23 22 21 20 19 18

Contents

Preface

Combinatorics is not a field, it's an attitude.
Anon

A *combinatorial reciprocity theorem* relates two classes of combinatorial objects via their counting functions: consider a class $\mathcal{X}$ of combinatorial objects and let $f(n)$ be the function that counts the number of objects in $\mathcal{X}$ of size n, where *size* refers to some specific quantity that is naturally associated with the objects in $\mathcal{X}$. Similar to canonization, it requires two miracles for a combinatorial reciprocity to occur:

1. the function $f(n)$ is the restriction of some reasonable function (e.g., a polynomial) to the positive integers, and
2. the evaluation $f(-n)$ is an integer of the same sign $\sigma = \pm 1$ for all $n \in \mathbb{Z}_{>0}$.

In this situation it is only human to ask if $\sigma f(-n)$ has a combinatorial meaning, that is, if there is a natural class $\mathcal{X}^\circ$ of combinatorial objects such that $\sigma f(-n)$ counts the objects of $\mathcal{X}^\circ$ of size n (where *size* again refers to some specific quantity naturally associated to $\mathcal{X}^\circ$). Combinatorial reciprocity theorems are among the most charming results in mathematics and, in contrast to canonization, can be found all over enumerative combinatorics and beyond.

As a first example we consider the class of maps $[k] \to \mathbb{Z}_{>0}$ from the finite set $[k] := \{1, 2, \dots, k\}$ into the positive integers, and so $f(n) = n^k$ counts the number of maps with codomain $[n]$. Thus $f(n)$ is the restriction of a polynomial and $(-1)^k f(-n) = n^k$ satisfies our second requirement above. This relates the number of maps $[k] \to [n]$ to itself. This relation is a genuine combinatorial reciprocity but the impression one is left with is that of being underwhelmed rather than charmed. Later in the book it will become clear that this example is not boring at all, but for now let's try again.

The term *combinatorial reciprocity theorem* was coined by Richard Stanley in his 1974 paper [**162**] of the same title, in which he developed a firm foundation of the subject. Stanley starts with an appealing reciprocity that he attributes to John Riordan: For a set S and $d \in \mathbb{Z}_{\geq 0}$, the collection of d-**subsets**[1] of S is

$$\binom{S}{d} := \{A \subseteq S : |A| = d\} .$$

For d fixed, the number of d-subsets of S depends only on the cardinality $|S|$, and the number of d-subsets of an n-set is

$$f(n) = \binom{n}{d} = \frac{1}{d!}\, n(n-1)\cdots(n-d+2)(n-d+1)\,, \tag{0.0.1}$$

which is the restriction of a polynomial in n of degree d. From the factorization we can read off that $(-1)^d f(-n)$ is a positive integer for every $n > 0$. More precisely,

$$(-1)^d f(-n) = \frac{1}{d!}\, n(n+1)\cdots(n+d-2)(n+d-1) = \binom{n+d-1}{d},$$

which is the number of *d-multisubsets* of an n-set, that is, the number of picking d elements from $[n]$ with repetition but without regard to the order in which the elements are picked. Now this is a combinatorial reciprocity! In formulas it reads

$$(-1)^d\binom{-n}{d} = \binom{n+d-1}{d}. \tag{0.0.2}$$

This is enticing in more than one way. The identity presents an intriguing connection between subsets and multisubsets via their counting functions, and its formal justification is completely within the realms of an undergraduate class in combinatorics. Equation (0.0.2) can be found in Riordan's book [**143**] on combinatorial analysis without further comment and, charmingly, Stanley states that his paper [**162**] can be considered as "further comment". That further comment is necessary is apparent from the fact that the formal proof above falls short of explaining why these two sorts of objects are related by a combinatorial reciprocity. In particular, comparing coefficients in (0.0.2) cannot be the method of choice for establishing more general reciprocity relations.

In this book we develop tools and techniques for handling combinatorial reciprocities. However, our own perspective is firmly rooted in *geometric* combinatorics and, thus, our emphasis is on the geometric nature of the

[1] All our definitions will look like that: incorporated into the text but bold-faced and so hopefully clearly visible.

combinatorial reciprocities. That is, for every class of combinatorial objects we associate a geometric object (such as a polytope or a polyhedral complex) in such a way that combinatorial features, including counting functions and reciprocity, are reflected in the geometry. In short, this book can be seen as *further comment with pictures*. At any rate, our text was written with the intention to give a comprehensive introduction to contemporary enumerative geometric combinatorics.

A Quick Tour. The book naturally comes in two parts with a special role played by the first chapter: Chapter 1 introduces four combinatorial reciprocity theorems that we set out to establish in the course of the book. Chapters 2–4 are for-the-most-part-independent introductions to three major themes of combinatorics: partially ordered sets, polyhedra, and generating functions. Chapters 5–7 treat more sophisticated topics in geometric combinatorics and are meant to be digested in order. Here is what to expect.

Chapter 1 sets the rhythm. We introduce four functions to count colorings and flows on graphs, order-preserving functions on partially ordered sets, and lattice points in dilations of lattice polygons. The definitions in this chapter are kept somewhat informal, to provide an easy entry into the themes of the later chapters. In all four cases we state a surprising combinatorial reciprocity and we point to some of the relations and connections between these examples, which will make repeated appearances later on. All in all, this chapter is a source of examples and motivation. You should revisit it from time to time to see how the various ways to view these objects shape your perspective.

Chapter 2 gives an introduction to partially ordered sets (*posets*, for short). Relating posets by means of order-preserving maps gives rise to the order polynomials from Chapter 1. One of the highlights here is a purely combinatorial proof of the reciprocity surrounding order polynomials (and only later will we see that there was geometry behind it). This gives us an opportunity to introduce important machinery, including Möbius inversion, zeta polynomials, and Eulerian posets in a hands-on and nonstandard form.

Geometry enters (quite literally) the picture in Chapter 3, in which we introduce convex polyhedra. Polyhedra are wonderful objects to study in their own right, as we hope to convey here, and much of their combinatorial structure comes in poset-theoretic terms. Our main motivation, however, is to develop a language that enables us to give the objects from Chapters 1 and 2 a geometric incarnation. The main player in Chapter 3 is the Euler characteristic, which is a powerful tool to obtain combinatorial truths from geometry. Two applications of the Euler characteristic, which we will witness in this chapter, are Zaslavsky's theorem for hyperplane arrangements and the Brianchon–Gram relation for polytopes.

Chapter 4 sets up the main algebraic machinery for our book: (rational) generating functions. We start gently with natural examples of compositions and partitions, and combinatorial reciprocity theorems appear almost instantly and just as naturally. The second half of Chapter 4 connects the world of generating functions with that of polyhedra and cones, where we develop Ehrhart and Hilbert series from first principles, including Stanley's reciprocity theorem for rational simplicial cones, which is at the heart of this book. This connection, in turn, allows us to view the first half of Chapter 4 from a new, geometric, perspective.

Chapter 5 is devoted to decomposing polyhedra into simple pieces. In particular, organizing the various pieces automatically suggests to view triangulations and, more generally, subdivisions as posets. Together with the technologies developed in the first part of the book, this culminates in a proof of our main combinatorial reciprocity theorems for polytopes and cones. The theory of subdividing polyhedra is worthy of study in its own right and we only glimpse at it by studying various ways to subdivide polytopes in a geometric, algorithmic, and, of course, combinatorial fashion. A powerful tool is that of half-open decompositions that quite remarkably help us to see some deep combinatorics in a clear way.

In Chapter 6 we give general posets life in Euclidean space as polyhedral cones. The theory of order cones allows us to utilize Chapters 2–5, often in surprisingly interconnected ways, to study posets using geometric means and, at the same time, interesting arithmetic objects derived from posets. Just as interesting are applications of this theory, which include permutation statistics, order polytopes, P-partitions, and their combinatorial reciprocity theorems.

Chapter 7 finishes the framework that was started in Chapter 1: we develop a unifying geometric approach to certain families of combinatorial polynomials. The last missing piece of the puzzle is formed by hyperplane arrangements, which constitute the main players of Chapter 7. They open a window to certain families of graph polynomials, including chromatic and flow polynomials, and we prove combinatorial reciprocity theorems for both. Hyperplane arrangements also naturally connect to two important families of polytopes, namely, alcoved polytopes and zonotopes.

The prerequisites for this book are minimal: undergraduate knowledge of linear algebra and combinatorics should suffice. The numerous exercises throughout the text are designed so that the book could easily be used for a graduate class in combinatorics or discrete geometry. The exercises that are needed for the main body of the text are marked by ⬠.

Acknowledgments. The first (and very preliminary) version of this manuscript was tried on some patient and error-forgiving students and researchers at the Mathematical Sciences Research Institute in Spring 2008 and in a course at the Freie Universität Berlin in Fall 2011. We thank them for their crucial input at the early stages of this book. In particular Lennart Claus, who took the 2011 class and did not see this book finally being finished, is vividly remembered for his keen interest, his active participation, and his *Mandelkekse*.

Since then, the book has, like its authors, matured (and aged). In particular it has expanded in breadth and depth (and, inevitably, length). We have had the fortune of receiving many valuable suggestions and corrections; we would like to thank in particular Tewodros Amdeberhan, Spencer Backman, Hélène Barcelo, Seth Chaiken, Adam Chavin, Susanna Fishel, Curtis Greene, Christian Haase, Max Hlavacek, Katharina Jochemko, Florian Kohl, Cailan Li, Sebastian Manecke, Jeremy Martin, Tyrrell McAllister, Louis Ng, Peter Paule, Bruce Sagan, Steven Sam, Paco Santos, Miriam Schlöter, Tom Schmidt, Christina Schulz, Matthias Schymura, Sam Sehayek, Richard Sieg, Christian Stump, Ngô Viêt Trung, Andrés Vindas Meléndez, Wei Wang, Russ Woodroofe, Tom Zaslavsky, and Günter Ziegler. Richard Stanley does not only also belong to this list, but he deserves special thanks: as one can see in the references throughout this text, he has been the main creative mind behind the material that forms the core of this book.

We thank the organizers and students of several classes, graduate schools, and workshops, in which we could test run various parts of the book: the 2011 Rocky Mountain Mathematics Consortium in Laramie, the 2013 Spring School in Hanoi, a Winter 2014 combinatorics class at the Freie Universität Berlin, and the 2015 Summer School at the Research Institute for Symbolic Computation in Linz.

We are grateful to the editorial staff at the American Mathematical Society, particularly Sergei Gelfand, who was relentlessly cheerful of this book project from its inception to its final polishing; his patience and wit have not only been much appreciated but needed. We thank Ed Dunne, Chris Thivierge, and the Editorial Committee and reviewers for many helpful insights, Mary Letourneau for her meticulous copy-editing, and the AMS TEX gurus, particularly Brian Bartling and Barbara Beeton, for invaluable assistance. David Austin made much of the geometry in this book come to life in the figures featured here; we are big fans of his art.

We thank the US National Science Foundation for their support, San Francisco State University for a presidential award (the resulting sabbatical allowed M.B. to give the above-mentioned lectures at MSRI), and the DFG Collaborative Research Center TRR 109 *Discretization in Geometry and*

Dynamics (sponsoring M.B.'s guest professorship at Freie Univerität in Fall 2014).

M.B. is deeply grateful to Tendai for her love, support, and patience while he tries to turn coffee into theorems, to Kumi for her energy and emotional support, and to his family *zuhause* and *kumusha* for their love. The idea for this book was conceived on numerous long trips to spend precious time with his *Papa* during the last months of his life. He dedicates this book to his memory.

R.S. is eternally grateful to Vanessa and Konstantin for their support, their patience, and, above all, for their love. When living with somebody who often times concentratedly stares at nothing (while figuring something out), all three merits are surely necessary. This book is dedicated to them. R.S. also thanks the *Villa people* at Freie Universität Berlin in the years 2011–2016, in particular Günter, for sharing the atmosphere, the freedom, and their wisdom (mathematically and otherwise).

San Francisco
Frankfurt
June 2018

Matthias Beck
Raman Sanyal

Chapter 1

Four Polynomials

To many, mathematics is a collection of theorems. For me, mathematics is a collection of examples; a theorem is a statement about a collection of examples and the purpose of proving theorems is to classify and explain the examples...
John B. Conway

In the spirit of the above quote, this chapter serves as a source of examples and motivation for the theorems to come and the tools to be developed. Each of the following four sections introduces a family of examples together with a combinatorial reciprocity statement which we will prove in later chapters.

1.1. Graph Colorings

Graphs and their colorings are all-time favorites in introductory classes on discrete mathematics, and we too succumb to the temptation to start with one of the most beautiful examples. A **graph** $G = (V, E)$ is a discrete structure composed of a finite[1] set of **nodes** V and a collection $E \subseteq \binom{V}{2}$ of unordered pairs of nodes, called **edges**. More precisely, this defines a **simple** graph as it excludes the existence of multiple edges between nodes and, in particular, edges with equal endpoints, i.e., **loops**. We will, however, need such nonsimple graphs in the sequel but we dread the formal overhead nonsimple graphs entail and will trust your discretion to make the necessary modifications. The most charming feature of graphs is that they are easy to visualize and their natural habitat is the margins of textbooks or notepads. Figure 1.1 shows some examples.

[1] *Infinite* graphs are interesting in their own right; however, they are no fun to color-count and so will play no role in this book.

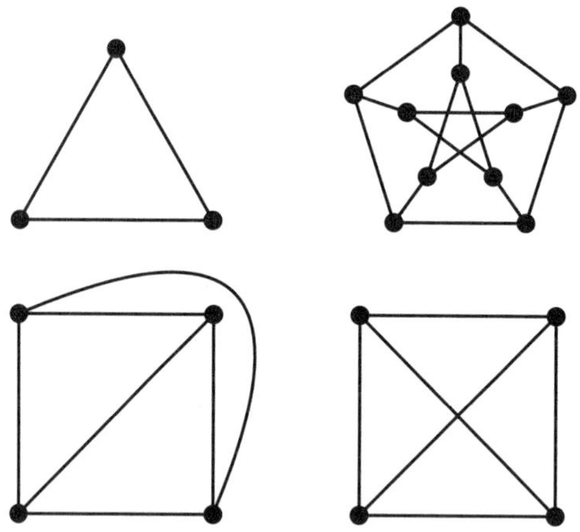

Figure 1.1. Various graphs.

An n-**coloring** of a graph G is a map $c : V \to [n] := \{1, 2, \ldots, n\}$. An n-coloring c is called **proper** if no two nodes sharing an edge get assigned the same color, that is,

$$c(u) \neq c(v) \qquad \text{whenever} \qquad uv \in E .$$

The name *coloring* comes from the natural interpretation of thinking of $c(v)$ as one of n possible colors that we use for the node v. A proper coloring is one where adjacent nodes get different colors. Here is a first indication why considering simple graphs often suffices: the existence and even the number of n-colorings is unaffected by parallel edges, and there are simply no proper colorings in the presence of loops.

Much of the fame of graph colorings stems from a question that was asked around 1852 by Francis Guthrie and answered only some 124 years later. In order to state the question in modern terms, we call a graph G **planar** if G can be drawn in the plane (or scribbled in the margin) such that edges do not cross except possibly at nodes. For example, the last row in Figure 1.1 shows a planar and nonplanar embedding of the (planar) graph K_4. Here is Guthrie's famous conjecture, now a theorem.

Four-color Theorem. *Every planar graph has a proper* 4*-coloring.*

There were several attempts at the Four-color Theorem before the first correct proof by Kenneth Appel and Wolfgang Haken. Here is one particularly interesting (but not yet successful) approach to proving the four-color theorem, due to George Birkhoff. For a (not necessarily planar) graph G, let

$$\chi_G(n) \; := \; |\{c : V \to [n] \text{ proper } n\text{-coloring}\}| .$$

The following observation, due to George Birkhoff and Hassler Whitney, is that $\chi_G(n)$ is the restriction to $\mathbb{Z}_{>0}$ of a particularly nice function.

Proposition 1.1.1. *If $G = (V, E)$ is a loopless graph, then $\chi_G(n)$ agrees with a polynomial of degree $|V|$ with integral coefficients. If G has a loop, then $\chi_G(n) = 0$.*

By a slight abuse of notation, we identify $\chi_G(n)$ with this polynomial and call it the **chromatic polynomial** of G. Nevertheless, we emphasize that, so far, only the values of $\chi_G(n)$ at positive integral arguments have an interpretation in terms of G.

Birkhoff's motivation to introduce the chromatic polynomial was that the four-color theorem is equivalent to the statement $\chi_G(4) > 0$ for all planar graphs G.

One proof of Proposition 1.1.1 is interesting in its own right, as it exemplifies *deletion–contraction* arguments which we will revisit in Chapter 7. For $e \in E$, the **deletion** of e results in the graph $G \setminus e := (V, E \setminus \{e\})$. The **contraction** G/e is the graph obtained by identifying the two nodes incident to e and removing e. An example is given in Figure 1.2.

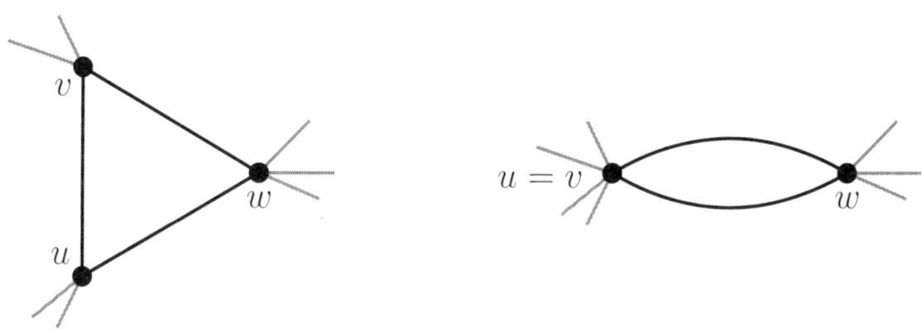

Figure 1.2. Contracting the edge $e = uv$.

Proof of Proposition 1.1.1. If G has a loop, then it admits no proper coloring by definition. For the more interesting case that G is loopless, we induct on $|E|$.

For $|E| = 0$ there are no coloring restrictions and $\chi_G(n) = n^{|V|}$. One step further, assume that G has a single edge $e = uv$. Then we can color all nodes $V \setminus \{u\}$ arbitrarily and, assuming $n \geq 2$, can color u with any color $\neq c(v)$. Thus, the chromatic polynomial is $\chi_G(n) = n^{d-1}(n-1)$, where $d = |V|$.

For the induction step, let $e = uv \in E$. We claim

$$\chi_G(n) = \chi_{G \setminus e}(n) - \chi_{G/e}(n). \tag{1.1.1}$$

Indeed, a coloring c of $G \setminus e$ fails to be a coloring of G if $c(u) = c(v)$. That is, we are over-counting by all proper colorings that assign the same color to u and v. These are precisely the proper n-colorings of G/e.

By (1.1.1) and the induction hypothesis, $\chi_G(n)$ is the difference of a polynomial of degree $d = |V|$ and a polynomial of degree $\leq d-1$, both with integer coefficients. □

Figure 1.3. A graph of Berlin.

The deletion–contraction relation (1.1.1) is a natural computing device. For example, the planar graph B in Figure 1.3 that models neighboring districts of Berlin comes with the impressive-looking chromatic polynomial

$$
\begin{aligned}
\chi_B(n) \;=\; & n^{23} - 53\,n^{22} + 1347\,n^{21} - 21845\,n^{20} + 253761\,n^{19} - 2246709\,n^{18} \\
& + 15748804\,n^{17} - 89620273\,n^{16} + 421147417\,n^{15} - 1653474650\,n^{14} \\
& + 5465562591\,n^{13} - 15279141711\,n^{12} + 36185053700\,n^{11} \\
& - 72527020873\,n^{10} + 122562249986\,n^{9} - 173392143021\,n^{8} \\
& + 203081660679\,n^{7} - 193650481777\,n^{6} + 146638574000\,n^{5} \\
& - 84870973704\,n^{4} + 35266136346\,n^{3} - 9362830392\,n^{2} \\
& + 1191566376\,n\,,
\end{aligned} \tag{1.1.2}
$$

which, nevertheless, can be easily computed on any computer. (And yes, $\chi_B(4) = 383904$ is not zero.)

Our proof of Proposition 1.1.1 and, more precisely, the deletion–contraction relation (1.1.1) reveal more about chromatic polynomials, which we invite you to show in Exercise 1.6:

Corollary 1.1.2. *Let G be a loopless graph on $d \geq 1$ nodes and $\chi_G(n) = c_d n^d + c_{d-1} n^{d-1} + \cdots + c_0$ its chromatic polynomial. Then*

(a) *the leading coefficient $c_d = 1$;*

(b) *the constant coefficient $c_0 = 0$;*

(c) $(-1)^d \chi_G(-n) > 0$ *for all integers $n \geq 1$.*

In particular the last property prompts the following natural question which we alluded to in the preface and which lies at the heart of this book.

Do the evaluations $(-1)^{|V|}\chi_G(-n)$ have combinatorial meaning?

This question was first asked (and beautifully answered) by Richard Stanley in 1973. To reproduce his answer, we need the notion of orientations on graphs. Again, to keep the formal pain level at a minimum, we denote the nodes of G by $v_1, v_2, \dots, v_d$. We define an **orientation** on G through a subset $\rho \subseteq E$; for an edge $e = v_i v_j \in E$ with $i < j$ we direct

$$v_i \stackrel{e}{\longleftarrow} v_j \ \text{ if } \ e \in \rho \qquad \text{and} \qquad v_i \stackrel{e}{\longrightarrow} v_j \ \text{ if } \ e \notin \rho\,.$$

We denote the oriented graph by ${}_\rho G$ and will sometimes write ${}_\rho G = (V, E, \rho)$. Said differently, we may think of G as canonically oriented by directing edges from small index to large, and ρ records the edges on which this orientation is reversed; see Figure 1.4 for an example.

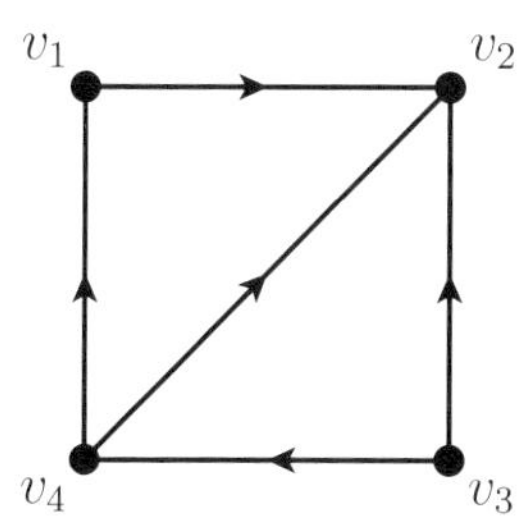

Figure 1.4. An orientation given by $\rho = \{14, 23, 24\}$.

A **directed path** in ${}_\rho G$ is a sequence $v_0, v_1, \dots, v_s$ of distinct nodes such that $v_{j-1} \to v_j$ is a directed edge in ${}_\rho G$ for all $j = 1, \dots, s$. If $v_s \to v_0$ is also a directed edge, then $v_0, v_1, \ldots, v_s, v_{s+1} := v_0$ is called a **directed cycle**. An orientation ρ of G is **acyclic** if there are no directed cycles in ${}_\rho G$.

Here is the connection between proper colorings and acyclic orientations: Given a proper coloring c, we define the orientation

$$\rho := \{v_i v_j \in E : i < j, \ c(v_i) > c(v_j)\}.$$

That is, the edge from lower index i to higher index j is directed along its **color gradient** $c(v_j) - c(v_i)$. We call this orientation ρ **induced** by the coloring c. For example, the orientation pictured in Figure 1.4 is induced by the coloring shown in Figure 1.5.

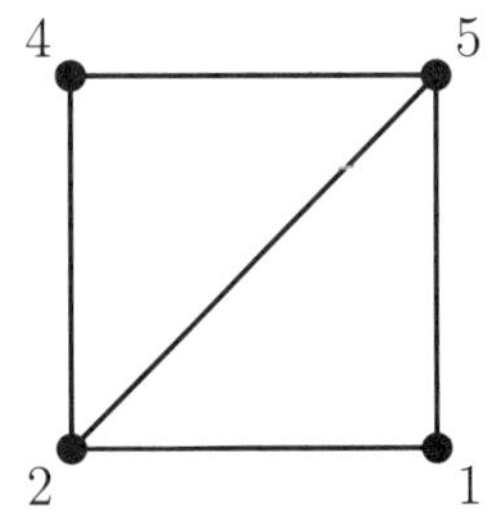

Figure 1.5. A coloring that induces the orientation in Figure 1.4.

Proposition 1.1.3. *Let $c : V \to [n]$ be a proper coloring and ρ the induced orientation on G. Then ${}_\rho G$ is acyclic.*

Proof. Assume that $v_{i_0} \to v_{i_1} \to \cdots \to v_{i_s} \to v_{i_0}$ is a directed cycle in ${}_\rho G$. Then $c(v_{i_0}) < c(v_{i_1}) < \cdots < c(v_{i_s}) < c(v_{i_0})$, which is a contradiction. □

As there are only finitely many acyclic orientations on G, we might count colorings according to the acyclic orientation they induce. An orientation ρ and an n-coloring c of G are called **compatible** if for every oriented edge $u \to v$ in ${}_\rho G$ we have $c(u) \geq c(v)$. The pair (ρ, c) is called **strictly** compatible if $c(u) > c(v)$ for every oriented edge $u \to v$.

Proposition 1.1.4. *If (ρ, c) is strictly compatible, then c is a proper coloring and ρ is an acyclic orientation on G. In particular, $\chi_G(n)$ is the number of strictly compatible pairs (ρ, c), where c is a proper n-coloring.*

Proof. If (ρ, c) are strictly compatible, then, since each edge is oriented, $c(u) > c(v)$ or $c(u) < c(v)$ whenever $uv \in E$. Hence c is a proper coloring and ρ is exactly the orientation induced by c. The acyclicity of ${}_\rho G$ now follows from Proposition 1.1.3. □

We are finally ready for our first combinatorial reciprocity theorem.

Theorem 1.1.5. *Let G be a finite graph on d nodes and $\chi_G(n)$ its chromatic polynomial. Then $(-1)^d \chi_G(-n)$ equals the number of compatible pairs (ρ, c), where c is an n-coloring and ρ is an acyclic orientation. In particular, $(-1)^d \chi_G(-1)$ equals the number of acyclic orientations of G.*

As one illustration of this theorem, consider the graph G in Figure 1.6; its chromatic polynomial is $\chi_G(n) = n(n-1)(n-2)^2$, and so Theorem 1.1.5 suggests that G should admit 18 acyclic orientations. Indeed, there are six acyclic orientations of the subgraph formed by v_1, v_2, and v_4, and for the remaining two edges, one of the four possible combined orientations of v_2v_3 and v_3v_4 produces a cycle with v_2v_4, so there are a total of $6 \cdot 3 = 18$ acyclic orientations.

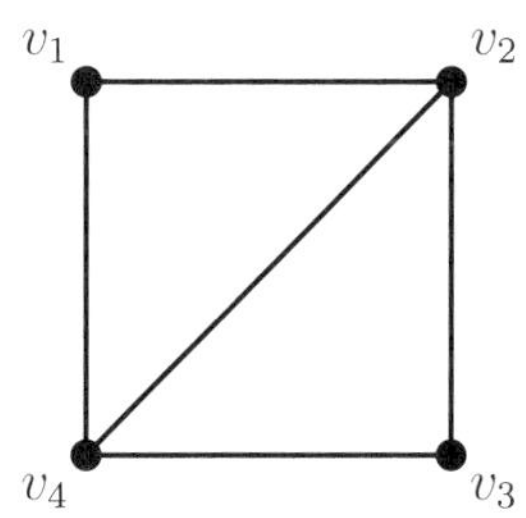

Figure 1.6. This graph has 18 acyclic orientations.

A deletion–contraction proof of Theorem 1.1.5 is outlined in Exercise 1.9; we will give a geometric proof in Section 7.1.

1.2. Flows on Graphs

Given a graph $G = (V, E)$ together with an orientation ρ and the finite Abelian group $\mathbb{Z}_n = \mathbb{Z}/n\mathbb{Z}$, a **$\mathbb{Z}_n$-flow** is a map $f : E \to \mathbb{Z}_n$ that assigns a value $f(e) \in \mathbb{Z}_n$ to each edge $e \in E$ such that there is conservation of flow at every node v:

$$\sum_{\overset{e}{\to} v} f(e) = \sum_{v \overset{e}{\to}} f(e),$$

that is, what "flows" into the node v is precisely what "flows" out of v. This physical interpretation is a bit shaky as the commodity flowing along edges are elements of $\mathbb{Z}_n$, and the flow conservation is with respect to the group structure. The set

$$\operatorname{supp}(f) := \{e \in E : f(e) \neq 0\}$$

is the **support** of f, and a $\mathbb{Z}_n$-flow f is **nowhere zero** if $\operatorname{supp}(f) = E$. In this section we will be concerned with counting nowhere-zero $\mathbb{Z}_n$-flows, and so we define

$$\varphi_G(n) := \left|\{f \text{ nowhere-zero } \mathbb{Z}_n\text{-flow on } {}_\rho G\}\right|.$$

A priori, the counting function $\varphi_G(n)$ depends on our chosen orientation ρ, but our language suggests that this is not the case, which we invite you to verify in Exercise 1.11:

Proposition 1.2.1. *The flow-counting function $\varphi_G(n)$ is independent of the orientation ρ of G.*

A **connected component** of the graph G is a maximal subgraph of G in which any two nodes are connected by a path. A graph G is **connected** if it has only one connected component.[2] As you will discover (at the latest when working on Exercises 1.13 and 1.14), G will not have any nowhere-zero flow if G has a **bridge** (also known as an **isthmus**), that is, an edge whose removal increases the number of connected components of G.

To motivate why we care about counting nowhere-zero flows, we assume that G is a *planar* bridgeless graph with a given embedding into the plane. The drawing of G subdivides the plane into connected regions in which two points lie in the same region whenever they can be joined by a path in $\mathbb{R}^2$ that does not meet G. Two such regions are neighboring if their topological closures share a proper (i.e., 1-dimensional) part of their boundaries. This induces a graph structure on the subdivision of the plane: for the given embedding of G, we define the **dual graph** G^* as the graph with nodes corresponding to the regions and two regions C_1, C_2 share an edge e^* if an original edge e is properly contained in both their boundaries. As we can see in the example pictured in Figure 1.7, the dual graph G^* is typically not simple with parallel edges. If G had bridges, G^* would have loops.

Given an orientation of G, an orientation on G^* is induced by, for example, rotating the edge clockwise. That is, the dual edge will "point" east assuming that the primal edge "points" north:

By carefully adding G^* to the picture we can see that dualizing G^* recovers G, i.e., $(G^*)^* = G$.

Our interest in flows lies in the connection to colorings: let c be an n-coloring of G, and for a change we assume that c takes on colors in $\mathbb{Z}_n$. After giving G an orientation, we can record the color gradient $t(uv) = c(v) - c(u)$ for each oriented edge $u \to v$, as shown in Figure 1.8.

[2]These notions refer to an *unoriented* graph.

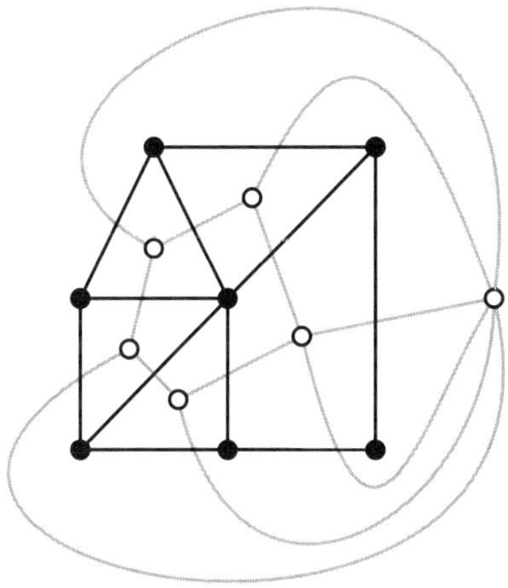

Figure 1.7. A graph and its dual.

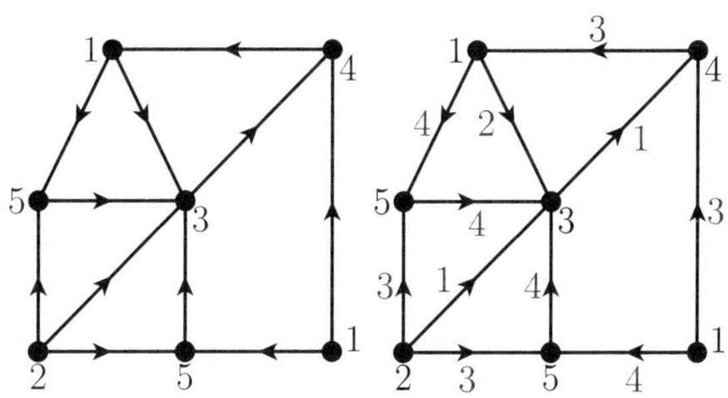

Figure 1.8. Recording color gradients, in $\mathbb{Z}_6$.

Conversely, knowing the color of a single node v_0, we can recover the coloring from $t : E \to \mathbb{Z}_n$: for a node $v \in V$ simply choose an undirected path $v_0 = p_0 p_1 p_2 \cdots p_k = v$ from v_0 to v. Then while walking along this path we can color each node p_i by adding or subtracting $t(p_{i-1}p_i)$ to $c(p_{i-1})$ depending on whether we walked the edge $p_{i-1}p_i$ with or against its orientation.

The color $c(v)$ is independent of the chosen path and thus, walking along a cycle in G the sum of the values $t(e)$ of edges along their orientation minus those against their orientation has to be zero; this is illustrated in Figure 1.9. Now, via the correspondence of primal and dual edges, t induces a map $f : E^* \to \mathbb{Z}_n$ on the dual graph G^*, shown in Figure 1.10. Each node of G^* represents a region that is bounded by a cycle in G, and the orientation on G^* is such that walking around this cycle clockwise, each edge traversed along its orientation corresponds to a dual edge into the

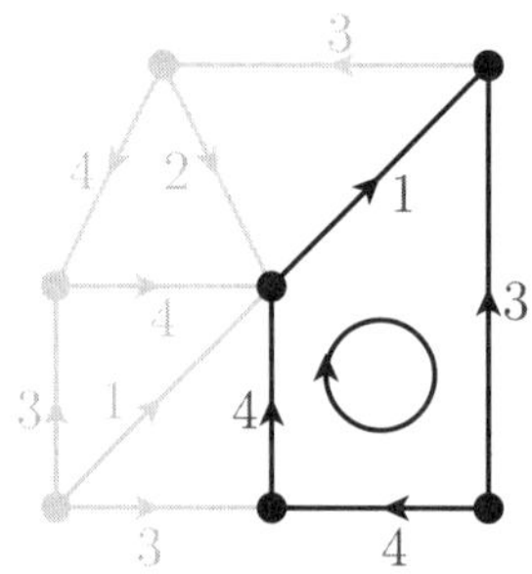

$$4 + 1 - 3 + 4 \equiv 0 \pmod 6$$

Figure 1.9. A cycle of flows.

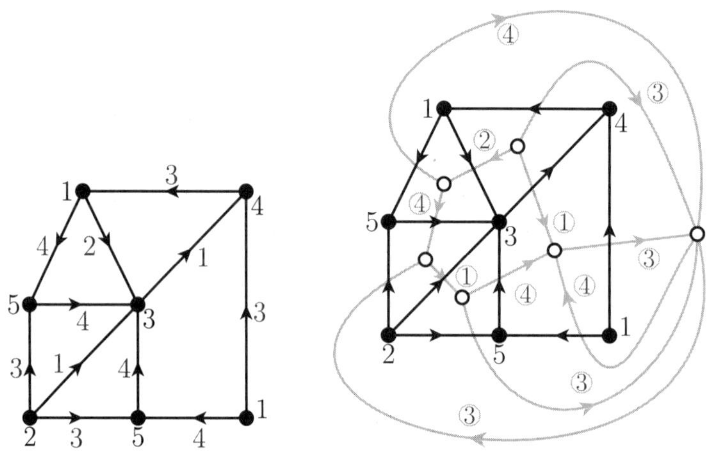

Figure 1.10. A flow and its dual.

region while counter-clockwise edges dually point out of the region. The cycle condition, illustrated in Figure 1.11, then proves:

Proposition 1.2.2. *Let G be a connected planar graph with dual G^*. For every n-coloring c of G, the induced map f is a $\mathbb{Z}_n$-flow on G^*, and every such flow arises this way. Moreover, the coloring c is proper if and only if f is nowhere zero.*

Conversely, for a given (nowhere-zero) flow f on G^* one can construct a (proper) coloring on G (see Exercise 1.12). In light of all this, we can rephrase the Four-color Theorem as follows.

Corollary 1.2.3 (Dual Four-color Theorem). *If G is a planar bridgeless graph, then $\varphi_G(4) > 0$.*

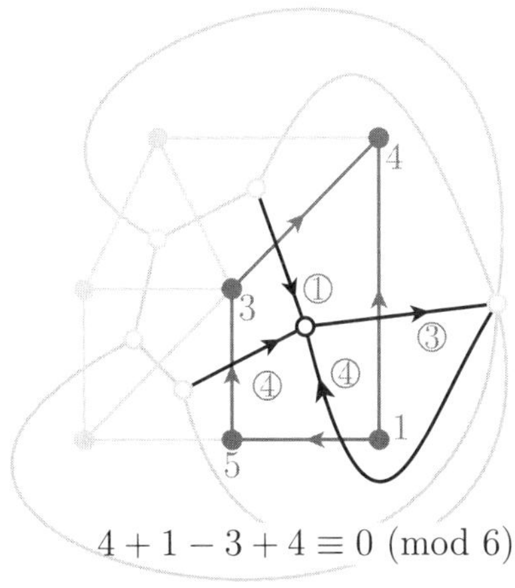

Figure 1.11. Proposition 1.2.2 illustrated.

This perspective on colorings of planar graphs was pioneered by William Tutte who initiated the study of $\varphi_G(n)$ for all (not necessarily planar) graphs. To see how much flows differ from colorings, we observe that there is no universal constant n_0 such that every graph has a proper n_0-coloring. The analogous statement for flows is not so clear and, in fact, Tutte conjectured the following:

Five-flow Conjecture. *Every bridgeless graph has a nowhere-zero $\mathbb{Z}_5$-flow.*

This sounds like a rather daring conjecture, as it is not even clear that there is *any* n such that every bridgeless graph has a nowhere-zero $\mathbb{Z}_n$-flow. However, it was shown by Paul Seymour that $n \leq 6$ works. In Exercise 1.17 you will show that there exist graphs that do not admit a nowhere-zero $\mathbb{Z}_4$-flow.

On the enumerative side, we have the following.

Proposition 1.2.4. *If G is a bridgeless connected graph, then $\varphi_G(n)$ agrees with a polynomial with integer coefficients of degree $|E|-|V|+1$ and leading coefficient* 1.

Again, we will abuse notation and refer to $\varphi_G(n)$ as the **flow polynomial** of G. The proof of the polynomiality is a deletion–contraction argument which is deferred to Exercise 1.13.

Towards a reciprocity statement, we need a notion dual to acyclic orientations: an orientation ρ on G is **totally cyclic** if every edge in $_\rho G$ is contained in a directed cycle. We quickly define the **cyclotomic number** of G as $\xi(G) := |E| - |V| + c$, where $c = c(G)$ is the number of connected components of G.

Theorem 1.2.5. *Let G be a bridgeless graph. For every positive integer n, the evaluation $(-1)^{\xi(G)}\varphi_G(-n)$ counts the number of pairs (f, ρ), where f is a $\mathbb{Z}_n$-flow and ρ is a totally-cyclic reorientation of $G/\operatorname{supp}(f)$. In particular, $(-1)^{\xi(G)}\varphi_G(-1)$ equals the number of totally-cyclic orientations of G.*

We will prove this theorem in Section 7.6.

1.3. Order Polynomials

A **partially ordered set**, or **poset** for short, is a set Π together with a binary relation $\preceq_\Pi$ that is

reflexive:	$a \preceq_\Pi a$,
transitive:	$a \preceq_\Pi b \preceq_\Pi c$ implies $a \preceq_\Pi c$, and
antisymmetric:	$a \preceq_\Pi b$ and $b \preceq_\Pi a$ implies $a = b$

for all $a, b, c \in \Pi$. We write $\preceq$ if the poset is clear from the context.

Partially ordered sets are ubiquitous structures in combinatorics and, as we will amply demonstrate soon, are indispensable in enumerative and geometric combinatorics. Most posets that we will encounter in this book are finite and when we say *poset*, we will always mean a finite poset unless stated otherwise.

The essence of a poset is encoded by its **cover relations**: an element $a \in \Pi$ is covered by an element b if

$$[a, b] \;:=\; \{z \in \Pi \,:\, a \preceq z \preceq b\} \;=\; \{a, b\}\,;$$

in plain English: $a \prec b$ and there is nothing between a and b. We write $a \prec\!\cdot\, b$ when a is covered by b. From its cover relations we can recover the poset by taking the transitive closure and adding in the reflexive relations. The cover relations can be thought of as a directed graph, and this gives an effective way to picture a poset: The **Hasse diagram** of Π is a drawing of the directed graph of cover relations in Π as an (undirected) graph where the node a is drawn lower than the node b whenever $a \prec b$. Here is an example: for $n \in \mathbb{Z}_{>0}$ we define D_n as the set $[n] = \{1, 2, \dots, n\}$ ordered by divisibility, that is, $a \preceq b$ if a divides b. The Hasse diagram of D_{10} is given in Figure 1.12.

This example truly is a *partial* order as, for example, 2 and 7 are not comparable. A poset in which each element is comparable to every other element is a **chain**. To be more precise: the poset Π is a chain if we have either $a \preceq b$ or $b \preceq a$ for any two elements $a, b \in \Pi$. The elements of a chain are **totally** or **linearly ordered**.

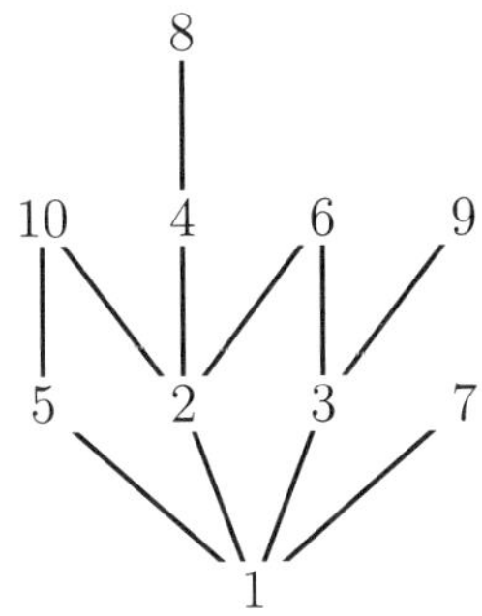

Figure 1.12. D_{10}: the set [10], partially ordered by divisibility.

A map $\phi : \Pi \to \Pi'$ is **(weakly) order preserving** if for all $a, b \in \Pi$

$$a \preceq_\Pi b \quad \Longrightarrow \quad \phi(a) \preceq_{\Pi'} \phi(b)$$

and **strictly order preserving** if

$$a \prec_\Pi b \quad \Longrightarrow \quad \phi(a) \prec_{\Pi'} \phi(b) .$$

For example, we can label the elements of a chain Π such that

$$\Pi = \{a_1 \prec a_2 \prec \cdots \prec a_n\} ,$$

which makes Π **isomorphic** to $[n] = \{1 < 2 < \cdots < n\}$, in the sense that there is a bijection $\phi : \Pi \to [n]$ such that ϕ and ϕ^{-1} are strictly order preserving.

Order-preserving maps are the natural morphisms (even in a categorical sense) between posets, and in this section we will be concerned with counting (strictly) order-preserving maps from a poset into chains.

A strictly order-preserving map ϕ from one chain $[d]$ into another $[n]$ exists only if $d \le n$ and is then determined by

$$1 \le \phi(1) < \phi(2) < \cdots < \phi(d) \le n .$$

Thus, the number of such maps equals $\binom{n}{d}$, the number of d-subsets of an n-set. In the case of a general poset Π, we define the **strict order polynomial**

$$\Omega^\circ_\Pi(n) := |\{\phi : \Pi \to [n] \text{ strictly order preserving}\}| .$$

As we have just seen, $\Omega^\circ_\Pi(n)$ is indeed a polynomial when $\Pi = [d]$. We now show that polynomiality holds for all posets Π.

Proposition 1.3.1. *For a finite poset Π, the function $\Omega^\circ_\Pi(n)$ agrees with a polynomial of degree $|\Pi|$ with rational coefficients.*

Proof. Let $d := |\Pi|$ and $\phi : \Pi \to [n]$ be a strictly order-preserving map. Now ϕ factors uniquely into a surjective map σ onto $\phi(\Pi)$ followed by an injection ι:

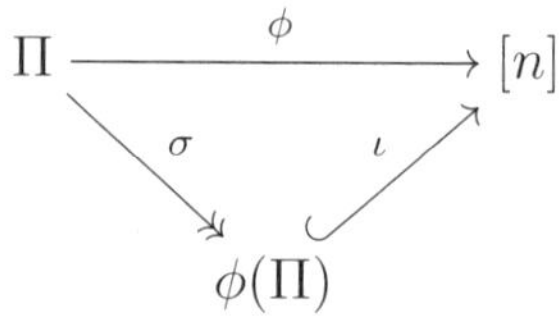

(Use the functions $\sigma(a) := \phi(a)$ and $\iota(a) := a$, defined with domains and codomains pictured above.) The image $\phi(\Pi)$ is a subposet of a chain and so is itself a chain. Thus $\Omega^\circ_\Pi(n)$ counts the number of pairs (σ, ι) of strictly order-preserving maps $\Pi \twoheadrightarrow [r] \rightarrowtail [n]$ for $r = 1, 2, \dots, d$. For fixed r, there are only finitely many order-preserving surjections $\sigma : \Pi \to [r]$, say, s_r many. As we discussed earlier, the number of strictly order-preserving maps $[r] \to [n]$ is exactly $\binom{n}{r}$, which is a rational polynomial in n of degree r. Hence, for fixed r, there are $s_r \binom{n}{r}$ many pairs (σ, ι) and we obtain

$$\Omega^\circ_\Pi(n) \;=\; s_d\binom{n}{d} + s_{d-1}\binom{n}{d-1} + \cdots + s_1\binom{n}{1},$$

which finishes our proof. □

As an aside, Proposition 1.3.1 proves that $\Omega^\circ_\Pi(n)$ is a polynomial with *integral* coefficients if we use $\left\{\binom{n}{r} : r \in \mathbb{Z}_{\geq 0}\right\}$ as a basis for the polynomial ring $\mathbb{R}[n]$. That the binomial coefficients indeed form a basis for the univariate polynomials follows from Proposition 1.3.1: if Π is an **antichain** on d elements, i.e., a poset in which no elements are related, then

$$\Omega^\circ_\Pi(n) \;=\; n^d \;=\; s_d\binom{n}{d} + s_{d-1}\binom{n}{d-1} + \cdots + s_1\binom{n}{1}. \tag{1.3.1}$$

In this case, the coefficients $s_r = S(d, r)$ are the **Stirling numbers of the second kind** which count the number of surjective maps $[d] \twoheadrightarrow [r]$. (The Stirling numbers might come in handy in Exercise 1.10.)

For the case that Π is a d-chain, the reciprocity statement (0.0.2) says that $(-1)^d\,\Omega^\circ_\Pi(-n)$ gives the number of d-multisubsets of an n-set, which equals, in turn, the number of (weak) order-preserving maps from a d-chain to an n-chain. Our next combinatorial reciprocity theorem expresses this duality between weak and strict order-preserving maps from a general poset into chains. You can already guess what is coming. We define the **order polynomial**

$$\Omega_\Pi(n) \;:=\; |\{\phi : \Pi \to [n] \text{ order preserving}\}|.$$

A slight modification (which we invite you to check in Exercise 1.20) of our proof of Proposition 1.3.1 implies that $\Omega_\Pi(n)$ indeed agrees with a polynomial in n of degree $|\Pi|$, and the following reciprocity theorem gives the relationship between the two polynomials $\Omega_\Pi(n)$ and $\Omega_\Pi^\circ(n)$.

Theorem 1.3.2. *Let Π be a finite poset. Then*

$$(-1)^{|\Pi|}\,\Omega_\Pi^\circ(-n) \;=\; \Omega_\Pi(n)\,.$$

We will prove this theorem in Chapter 2. To further motivate the study of order polynomials, we remark that a poset Π gives rise to an oriented graph by way of the cover relations of Π. Conversely, the binary relation given by an oriented graph G can be completed to a partial order $\Pi(G)$ by adding the necessary transitive and reflexive relations if and only if G is acyclic. Figure 1.13 shows an example, for the orientation pictured in Figure 1.4. The following result will be the subject of Exercise 1.18.

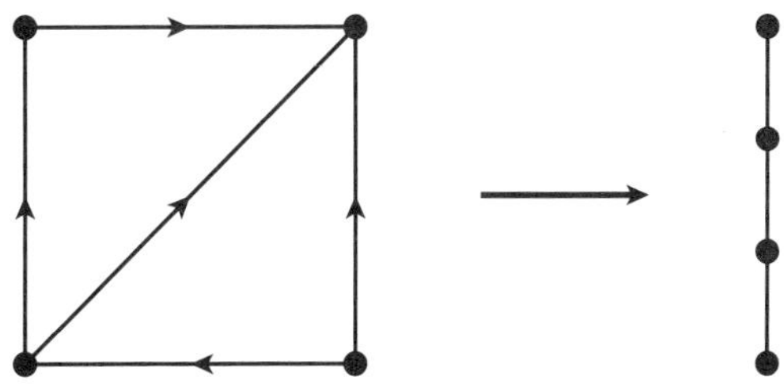

Figure 1.13. From an acyclic orientation to a poset.

Proposition 1.3.3. *Let ${}_\rho G = (V, E, \rho)$ be an acyclic graph and $\Pi({}_\rho G)$ the induced poset. A map $c : V \to [n]$ is strictly compatible with the orientation ρ of G if and only if c is a strictly order-preserving map $\Pi({}_\rho G) \to [n]$.*

In Proposition 1.1.4 we identified the number of n-colorings $\chi_G(n)$ of G as the number of colorings c strictly compatible with some acyclic orientation ρ of G, and so this proves:

Corollary 1.3.4. *The chromatic polynomial $\chi_G(n)$ of a graph G is the sum of the order polynomials $\Omega_{\Pi({}_\rho G)}^\circ(n)$ for all acyclic orientations ρ of G.*

1.4. Ehrhart Polynomials

The formulation of (0.0.1) in terms of d-subsets of an n-set has a straightforward geometric interpretation that will fuel much of what is about to come: the d-subsets of $[n]$ correspond precisely to the points in $\mathbb{R}^d$ with integral coordinates in the set

$$(1.4.1) \qquad (n+1)\triangle_d^\circ \;=\; \left\{\mathbf{x} \in \mathbb{R}^d \,:\, 0 < x_1 < x_2 < \cdots < x_d < n+1\right\}.$$

Next we explain the notation on the left-hand side: we define

$$\triangle_d^\circ := \left\{ \mathbf{x} \in \mathbb{R}^d : 0 < x_1 < x_2 < \cdots < x_d < 1 \right\},$$

and for a set $S \subseteq \mathbb{R}^d$ and a positive integer n, we set

$$n\,S := \{ n\,\mathbf{x} : \mathbf{x} \in S \},$$

the **n-th dilate** of S. (We hope the notation in (1.4.1) now makes sense.) For example, when $d = 2$,

$$\triangle_2^\circ = \{ (x_1, x_2) \in \mathbb{R}^2 : 0 < x_1 < x_2 < 1 \}$$

is the interior of a triangle, and every integer point (x_1, x_2) in the $(n+1)$-st dilate of $\triangle_2^\circ$ satisfies $0 < x_1 < x_2 < n+1$ or, equivalently, $1 \le x_1 < x_2 \le n$. We illustrate these integer points for the case $n = 5$ in Figure 1.14.

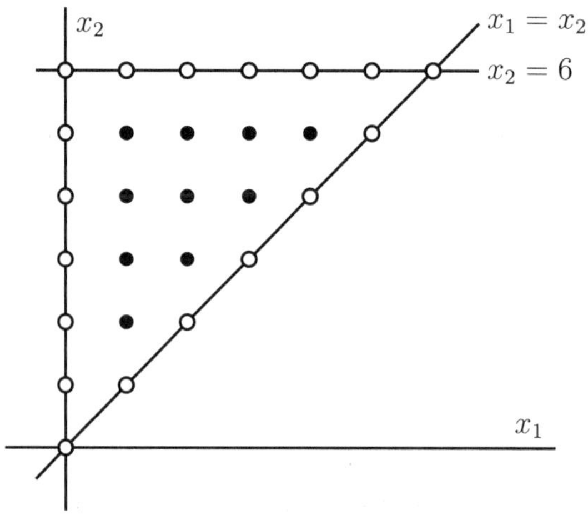

Figure 1.14. The integer points in $6\triangle_2^\circ$.

A **convex lattice polygon** $\mathsf{P} \subset \mathbb{R}^2$ is the smallest convex set containing a given finite set of noncollinear integer points in the plane. The **interior** of P is denoted by P°. Convex polygons are 2-dimensional instances of **convex polytopes**, which live in any dimension and whose properties we will study in detail in Chapter 3. For now, we count on your intuition about terms like *convex* and objects such as *vertices* and *edges* of a polygon, which will be defined rigorously in Chapter 3.

For a bounded set $S \subset \mathbb{R}^2$, we write $E(S) := \left| S \cap \mathbb{Z}^2 \right|$ for the number of integer lattice points in S. Our example above motivates the definitions of the counting functions

$$\mathrm{ehr}_{\mathsf{P}^\circ}(n) := E(n\,\mathsf{P}^\circ) = \left| n\,\mathsf{P}^\circ \cap \mathbb{Z}^2 \right|$$

and

$$\operatorname{ehr}_{\mathsf{P}}(n) := E(n\,\mathsf{P}) = \left|n\,\mathsf{P} \cap \mathbb{Z}^2\right|,$$

the **Ehrhart functions** of P. The historical reasons for this naming convention will be given in Chapters 4 and 5.

As we know from (0.0.1), the number of integer lattice points in the $(n+1)$-st dilate of $\triangle_2^\circ$ is given by the polynomial

$$\operatorname{ehr}_{\triangle_2^\circ}(n+1) = \binom{n}{2}.$$

To make the combinatorial reciprocity statement given by (0.0.1) geometric, we observe that the number of weak order-preserving maps from $[n]$ into $[2]$ is given by the integer points in the $(n-1)$-st dilate of

$$\triangle_2 = \left\{(x_1, x_2) \in \mathbb{R}^2 \,:\, 0 \le x_1 \le x_2 \le 1\right\},$$

the closure of $\triangle_2^\circ$. The combinatorial reciprocity statement given by (0.0.1) now reads $(-1)^2\binom{-n}{2}$ *equals the number of integer points in* $(n-1)\triangle_2$. Unraveling the parameters (and making appropriate shifts), we can rephrase this as: $(-1)^2 \operatorname{ehr}_{\triangle_2^\circ}(-n)$ equals the number of integer points in $n\triangle_2$. The reciprocity theorem featured in this section states that this holds for all convex lattice polygons; in Chapter 5 we will prove an analogue in all dimensions.

Theorem 1.4.1. *Let* $\mathsf{P} \subset \mathbb{R}^2$ *be a lattice polygon. Then* $\operatorname{ehr}_{\mathsf{P}}(n)$ *agrees with a polynomial of degree* 2 *with rational coefficients, and* $(-1)^2 \operatorname{ehr}_{\mathsf{P}}(-n)$ *equals the number of integer points in* $n\mathsf{P}^\circ$.

In the remainder of this section we will prove this theorem. The proof will be a series of simplifying steps that are similar in spirit to those that we will employ for the general result in Section 5.2.

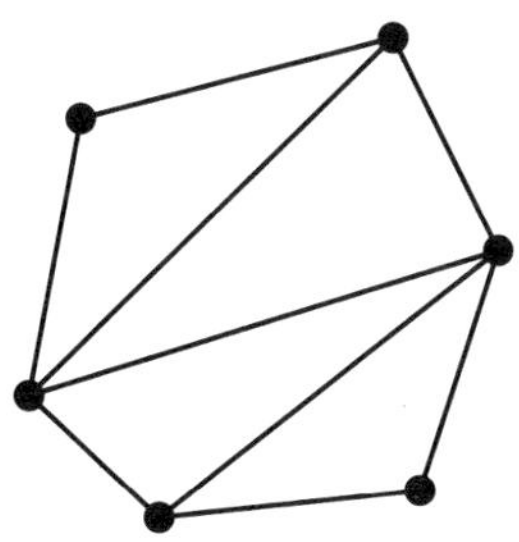

Figure 1.15. A triangulation of a hexagon.

As a first step, we reduce the problem of showing polynomiality for Ehrhart functions of arbitrary lattice polygons to that of lattice *triangles*.

Let P be a lattice polygon in the plane with n vertices. We can **triangulate** P by cutting the polygon along sufficiently many (exactly $n-3$) nonintersecting diagonals, as in Figure 1.15. The result is a set of $n-2$ lattice triangles that cover P. We denote by $\mathcal{T}$ the collection of **faces** of all these triangles, that is, $\mathcal{T}$ consists of n zero-dimensional polytopes (vertices), $2n-3$ one-dimensional polytopes (edges), and $n-2$ two-dimensional polytopes (triangles).

Our triangulation is a well-behaved collection of polytopes in the plane in the sense that they intersect nicely: if two elements of $\mathcal{T}$ intersect, then they intersect in a common face of both. This is useful, as counting lattice points is a *valuation*.[3] Namely, for $S, T \subset \mathbb{R}^2$,

$$E(S \cup T) = E(S) + E(T) - E(S \cap T), \tag{1.4.2}$$

and applying the inclusion–exclusion relation (1.4.2) repeatedly to the elements in our triangulation of P yields

$$\operatorname{ehr}_{\mathsf{P}}(n) = \sum_{\mathsf{F} \in \mathcal{T}} \mu(\mathsf{F}) \operatorname{ehr}_{\mathsf{F}}(n), \tag{1.4.3}$$

where the $\mu(\mathsf{F})$ are some coefficients that correct for over-counting. If F is a triangle, then $\mu(\mathsf{F}) = 1$—after all, we want to count the lattice points in P that are covered by the triangles. For an edge F of the triangulation, we have to make the following distinction: F is an **interior edge** of $\mathcal{T}$ if it is contained in two triangles. In this case the lattice points in F get counted twice, and in order to compensate for this, we set $\mu(\mathsf{F}) = -1$. In the case that F is a **boundary edge**, i.e., F lies in only one triangle of $\mathcal{T}$, there is no over-counting and we can set $\mu(\mathsf{F}) = 0$. To generalize this to all faces of $\mathcal{T}$, we call a face $\mathsf{F} \in \mathcal{T}$ a **boundary face** of $\mathcal{T}$ if F is contained in the boundary of P, and an **interior face** otherwise. We can give the coefficients $\mu(\mathsf{F})$ explicitly as follows.

Proposition 1.4.2. *Let $\mathcal{T}$ be a triangulation of a lattice polygon $\mathsf{P} \subset \mathbb{R}^2$. Then the coefficients $\mu(\mathsf{F})$ in (1.4.3) are given by*

$$\mu(\mathsf{F}) = \begin{cases} (-1)^{2-\dim \mathsf{F}} & \text{if F is interior,} \\ 0 & \text{otherwise.} \end{cases}$$

For boundary vertices $\mathsf{F} = \{\mathbf{v}\}$, we can check that $\mu(\mathsf{F}) = 0$ is correct: the vertex is counted positively as a lattice point by every incident triangle and negatively by every incident interior edge. As there are exactly one interior edge less than incident triangles, we do not count the vertex more than once. For an interior vertex, the number of incident triangles and incident (interior) edges are equal and hence $\mu(\mathsf{F}) = 1$. (In triangulations

[3] We'll have more to say about valuations in Section 3.4.

of P obtained by cutting along diagonals we never encounter *interior* vertices, however, they will appear soon when we consider a different type of triangulation.)

The coefficient $\mu(\mathsf{F})$ for a triangulation of a polygon was easy to argue and to verify in the plane. For higher-dimensional polytopes we will have to resort to more algebraic and geometric means. The right algebraic setup will be discussed in Chapter 2 where we will make use of the fact that a triangulation $\mathcal{T}$ constitutes a partially ordered set. In the language of posets, $\mu(\mathsf{F})$ is an evaluation of the *Möbius function* for the poset $\mathcal{T}$. Möbius functions are esthetically satisfying but are in general difficult to compute. However, we are dealing with situations with plenty of geometry involved, and we will make use of that in Chapter 5 to give a statement analogous to Proposition 1.4.2 in general dimension.

Returning to our 2-dimensional setting, showing that $\mathrm{ehr}_{\mathsf{F}}(n)$ is a polynomial whenever F is a lattice point, a lattice segment, or a lattice triangle gives us the first half of Theorem 1.4.1. If F is a vertex, then $\mathrm{ehr}_{\mathsf{F}}(n) = 1$. If $\mathsf{F} \in \mathcal{T}$ is an edge of one of the triangles and thus a lattice segment, verifying that $\mathrm{ehr}_{\mathsf{F}}(n)$ is a polynomial is the content of Exercise 1.21.

The remaining challenge now is the polynomiality and reciprocity for lattice triangles. For the rest of this section, let $\triangle \subset \mathbb{R}^2$ be a fixed lattice triangle in the plane. The idea that we will use is to triangulate the dilates $n\triangle$ for $n \geq 1$, but the triangulation will change with n. Figure 1.16 gives the picture for $n = 1, 2, 3$.

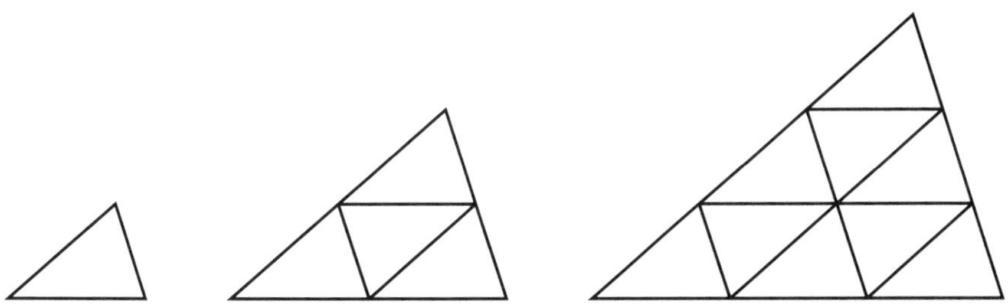

Figure 1.16. Special triangulations of dilates of a lattice triangle.

We trust that you can imagine the triangulation for all values of n. The special property of this triangulation is that up to lattice translations, there are only a few different pieces. In fact, there are only two different lattice triangles used in the triangulation of $n\triangle$: there is $\triangle$ itself and (lattice translates of) the reflection of $\triangle$ with respect to the origin, which we will denote by $\triangledown$. As for edges, we have three different kinds of edges, namely, the edges $\diagdown$, —, and $\diagup$. Up to lattice translation, there is only one vertex $\bullet$.

Now we count how many copies of each tile occur in these special triangulations; let $t(\mathsf{Q}, n)$ denote the number of times Q appears in our triangulation of $n\triangle$. For triangles, we count

$$t(\triangle, n) \;=\; \binom{n+1}{2} \qquad \text{and} \qquad t(\triangledown, n) \;=\; \binom{n}{2}.$$

For the interior edges, we observe that each interior edge is incident to a unique upside-down triangle $\triangledown$ and consequently

$$t(\diagdown, n) \;=\; t(-, n) \;=\; t(\diagup, n) \;=\; \binom{n}{2}.$$

Similarly, for interior vertices,

$$t(\bullet, n) \;=\; \binom{n-1}{2}.$$

Thus with (1.4.3), the Ehrhart function for the triangle $\triangle$ is

$$\begin{aligned} \operatorname{ehr}_{\triangle}(n) \;=\; & \binom{n+1}{2} E(\triangle) + \binom{n}{2} E(\triangledown) \\ & - \binom{n}{2} \left(E(\diagdown) + E(-) + E(\diagup) \right) \\ & + \binom{n-1}{2} E(\bullet) . \end{aligned} \tag{1.4.4}$$

This proves that $\operatorname{ehr}_{\triangle}(n)$ agrees with a polynomial of degree 2, and together with (1.4.3) this establishes the first half of Theorem 1.4.1.

To prove the combinatorial reciprocity of Ehrhart polynomials in the plane, we make the following useful observation.

Proposition 1.4.3. *If for every lattice polygon* $\mathsf{P} \subset \mathbb{R}^2$ *we have that* $\operatorname{ehr}_{\mathsf{P}}(-1)$ *equals the number of lattice points in the interior of* P, *then* $\operatorname{ehr}_{\mathsf{P}}(-n) = E(n\mathsf{P}^\circ)$ *for all* $n \geq 1$.

Proof. For fixed $n \geq 1$, we denote by Q the lattice polygon $n\mathsf{P}$. We see that $\operatorname{ehr}_{\mathsf{Q}}(m) = E(m(n\mathsf{P}))$ for all $m \geq 1$. Hence the Ehrhart polynomial of Q is given by $\operatorname{ehr}_{\mathsf{P}}(mn)$ and for $m = -1$ we conclude

$$\operatorname{ehr}_{\mathsf{P}}(-n) \;=\; \operatorname{ehr}_{\mathsf{Q}}(-1) \;=\; E(\mathsf{Q}^\circ) \;=\; E(n\mathsf{P}^\circ),$$

which finishes our proof. □

To establish the combinatorial reciprocity of Theorem 1.4.1 for triangles, we can simply substitute $n = -1$ into (1.4.4) and use (0.0.2) to obtain

$$\operatorname{ehr}_{\triangle}(-1) \;=\; E(\triangledown) - E(\diagdown) - E(-) - E(\diagup) + 3E(\bullet),$$

which equals the number of interior lattice points of $\triangledown$. Observing that $\triangle$ and $\triangledown$ have the same number of lattice points finishes the argument.

For the general case, Exercise 1.21 gives

$$\mathrm{ehr}_{\mathsf{P}}(-1) \;=\; \sum_{\mathsf{F}\in\mathcal{T}} E(\mathsf{F}^\circ) \;=\; E(\mathsf{P}^\circ)$$

and this (finally!) concludes our proof of Theorem 1.4.1.

Exercises 1.21 and 1.23 also answer the question of why we carefully triangulate P along diagonals (as opposed to cutting it up arbitrarily to obtain triangles): Theorem 1.4.1 is only true for lattice polygons. There are versions for polygons with rational and irrational coordinates but they become increasingly complicated. By cutting along diagonals we can decompose a lattice polygon into lattice segments and lattice triangles. This part becomes nontrivial already in dimension 3, and we will worry about this in Chapter 5.

In Exercise 1.25 we will look into the question as to what the coefficients of $\mathrm{ehr}_{\mathsf{P}}(n)$, for a lattice polygon P, tell us. We finish this chapter by considering the constant coefficient $c_0 = \mathrm{ehr}_{\mathsf{P}}(0)$. This is the most tricky one, as we could argue that $\mathrm{ehr}_{\mathsf{P}}(0) = E(0\mathsf{P})$ and since $0\mathsf{P}$ is just a single point, we get $c_0 = 1$. This argument is flawed: we defined $\mathrm{ehr}_{\mathsf{P}}(n)$ only for $n \geq 1$. To see that this argument is, in fact, plainly wrong, we consider $S = \mathsf{P}_1 \cup \mathsf{P}_2 \subset \mathbb{R}^2$, where P_1 and P_2 are disjoint lattice polygons. Since they are disjoint, $\mathrm{ehr}_S(n) = \mathrm{ehr}_{\mathsf{P}_1}(n) + \mathrm{ehr}_{\mathsf{P}_2}(n)$. Now $0S$ is also just a point and therefore

$$1 \;=\; \mathrm{ehr}_S(0) \;=\; \mathrm{ehr}_{\mathsf{P}_1}(0) + \mathrm{ehr}_{\mathsf{P}_2}(0) \;=\; 2\,.$$

It turns out that $c_0 = 1$ is still correct but the justification will have to wait until Theorem 5.1.8. In Exercise 1.26, you will prove a more general version for Theorem 1.4.1 that dispenses of convexity.

Notes

Graph-coloring problems started in the form of coloring maps such that countries sharing a proper part of their boundaries get colored with different colors. The graphs associated to such map-coloring problems are planar as is illustrated in Figure 1.3. So the fact that the chromatic polynomial is indeed a polynomial was proved for maps (in 1912 by George Birkhoff [**32**]) before Hassler Whitney proved it for general graphs in 1932 [**184**]. The deletion–contraction argument that we used in the proof of Proposition 1.1.1 gives an algorithm that we used, for example, for the chromatic polynomial (1.1.2) of Berlin. Complexity-theory-savvy readers might want to ponder the (exponential) complexity of this algorithm but it can be implemented with little effort (we used SAGE [**55**]) and for small graphs it works well. As we mentioned, the first proof of the Four-color Theorem is due to Kenneth Appel and Wolfgang Haken [**7,8**]. Theorem 1.1.5 is due to

Richard Stanley [**161**]. We will give a proof from a geometric point of view in Section 7.1.

As already mentioned, the approach of studying colorings of planar graphs through flows on their duals was pioneered by William Tutte [**179**], who also conceived the Five-flow Conjecture. This conjecture becomes a theorem when "5" is replaced by "6", due to Paul Seymour [**154**]; the 8-flow theorem had previously been shown by François Jaeger [**93, 94**]. Theorem 1.2.5 was proved in [**37**]. We will give a proof in Section 7.6.

The number of proper n-colorings, of nowhere-zero $\mathbb{Z}_n$-flows, and of acyclic or totally cyclic orientations can all be computed by using deletions and contractions. More generally, let f be a function that assigns any graph G a number $f(G) \in \mathbb{R}$ such that $f(G) = f(G')$ if G and G' are isomorphic. Then f is called a *generalized Tutte–Grothendieck invariant* if there are constants α, β such that for any $e \in E(G)$

$$f(G) = \begin{cases} \alpha\, f(G \setminus e) + \beta\, f(G/e) & \text{if } e \text{ is neither a loop nor a bridge,} \\ f(e)\, f(G \setminus e) & \text{otherwise.} \end{cases}$$

Here $f(e)$ is the value on the graph that consists of the edge e alone. It is not difficult to show that there is a *universal* Tutte–Grothendieck invariant in the following sense: for every graph G there is a polynomial $T_G(x, y) \in \mathbb{Z}[x, y]$ such that $f(G)$ is an evaluation of $T_G(x, y)$ in terms of α, β, and the values of f on a loop and bridge; see [**44**] for much more on this. The polynomial $T_G(x, y)$ is called the *Tutte polynomial* of G. Its evaluations, its coefficients, as well as the many mathematical contexts in which they occur are quite remarkable, and that area of geometric and algebraic combinatorics is very active. We will see the notion of deletion–contraction in a more geometric context in Chapter 7.

Order polynomials were introduced by Richard Stanley [**160, 166**] as "chromatic-like polynomials for posets" (this is reflected in Corollary 1.3.4); Theorem 1.3.2 is due to him. We will study order polynomials in depth in Chapters 2 and 6.

Theorem 1.4.1 is essentially due to Georg Pick [**136**], whose famous formula is the subject of Exercise 1.25. In some sense, this formula marks the beginning of the study of integer-point enumeration in polytopes. Our phrasing of Theorem 1.4.1 suggests that it has an analogue in higher dimensions, and we will study this analogue in Chapters 4 and 5.

Herbert Wilf [**185**] raised the question of characterizing which polynomials can occur as chromatic polynomials of graphs. This question has spawned a lot of work in algebraic combinatorics. For example, a recent theorem of June Huh [**89**] says that the absolute values of the coefficients of every chromatic polynomial form a *unimodal* sequence, that is, the sequence increases up to some point, after which it decreases. Huh's theorem had been conjectured by Ronald Read [**140**] almost 50 years earlier. In fact, Huh proved much more. In Chapter 7 we will study arrangements of hyperplanes and their associated characteristic polynomials. Huh and later Huh and Eric Katz [**90**] proved that, up to sign, the coefficients of characteristic polynomials of hyperplane arrangements (defined over any field) form a *log-concave* sequence. We will see the relation between chromatic and characteristic polynomials in Chapter 7.

Exercises

1.1 Two graphs $G_1 = (V_1, E_1)$ and $G_2 = (V_2, E_2)$ are **isomorphic** if there is a bijection $\phi : V_1 \to V_2$ such that for all $u, v \in V_1$

$$uv \in E_1 \quad \text{if and only if} \quad \phi(u)\,\phi(v) \in E_2.$$

Let G be a planar graph and let G_1 and G_2 be the dual graphs for two distinct planar embeddings of G. Is it true that G_1 and G_2 are isomorphic?

If not, can you give a sufficient condition on G such that the above claim is true? (*Hint:* A precise characterization is rather difficult, but for a sufficient condition you might want to contemplate Steinitz's theorem [**176**]; see [**190**, Ch. 4] for a modern treatment.)

1.2 Find two simple nonisomorphic graphs G and H with $\chi_G(n) = \chi_H(n)$. Can you find many (polynomial, exponential) such examples in the number of nodes? Can you make your examples arbitrarily high connected?

1.3 Find the chromatic polynomials of
   (a) the path on d nodes;
   (b) the cycle on d nodes;
   (c) the wheel with d spokes (and $d+1$ nodes); for example, the wheel with six spokes is this:

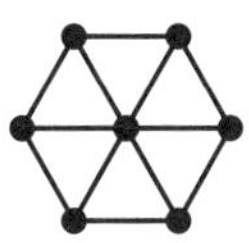

1.4 Verify that the graph of Berlin in Figure 1.3 cannot be colored with three colors. (*Hint:* Instead of evaluating the chromatic polynomial, try to find a simple subgraph that is not 3-colorable.)

1.5 Show that if G has c connected components, then n^c divides the polynomial $\chi_G(n)$.

1.6 ○ Complete the proof of Corollary 1.1.2: Let G be a loopless nonempty graph on d nodes and $\chi_G(n) = c_d n^d + c_{d-1} n^{d-1} + \cdots + c_0$ its chromatic polynomial. Then
(a) the leading coefficient $c_d = 1$;
(b) the constant coefficient $c_0 = 0$;
(c) $(-1)^d \chi_G(-n) > 0$.

1.7 Prove that every **complete graph** K_d (a graph with d nodes and all possible edges between them) has exactly $d!$ acyclic orientations.

1.8 Using a construction similar to the one in our proof of Proposition 1.3.1, show that the chromatic polynomial of a given graph G can be written as

$$\chi_G(n) = a_d \binom{n}{d} + a_{d-1} \binom{n}{d-1} + \cdots + a_1 \binom{n}{1}$$

for some (explicitly describable) nonnegative integers $a_1, a_2, \ldots, a_d$. (This gives yet another proof of Proposition 1.1.1.)

1.9 In this exercise you will give a deletion–contraction proof of Theorem 1.1.5.
(a) Verify that the deletion–contraction relation (1.1.1) implies for the function $\overline{\chi}_G(n) := (-1)^d \chi_G(-n)$ that

$$\overline{\chi}_G(n) = \overline{\chi}_{G \setminus e}(n) + \overline{\chi}_{G/e}(n) .$$

(b) Define $\mathcal{X}_G(n)$ as the number of compatible pairs of an acyclic orientation ρ and an n-coloring c. Show $\mathcal{X}_G(n)$ satisfies the same deletion–contraction relation as $\overline{\chi}_G(n)$.
(c) Infer that $\overline{\chi}_G(n) = \mathcal{X}_G(n)$ by induction on $|E|$.

1.10 The **complete bipartite graph** $K_{r,s}$ is the graph on the node set $V = \{1, 2, \ldots, r, 1', 2', \ldots, s'\}$ and edges

$$E = \{ij' : 1 \le i \le r,\ 1 \le j \le s\} .$$

Determine the chromatic polynomial $\chi_{K_{r,s}}(n)$ for $m, k \ge 1$. (*Hint:* Proper n-colorings of $K_{r,s}$ correspond to pairs (f, g) of maps $f : [r] \to [n]$ and $g : [s] \to [n]$ with disjoint ranges.)

1.11 ○ Prove Proposition 1.2.1: The flow-counting function $\varphi_G(n)$ is independent on the orientation of G.

1.12 ○ Let G be a connected planar graph with dual G^*. By reversing the steps in our proof before Proposition 1.2.2, show that every (nowhere-zero) $\mathbb{Z}_n$-flow f on G^* naturally gives rise to n different (proper) n-colorings on G.

1.13 ⬠ Prove Proposition 1.2.4: If G is a bridgeless connected graph, then $\varphi_G(n)$ agrees with a monic polynomial of degree $|E| - |V| + 1$ with integer coefficients.

1.14 ⬠ Let $G = (V, E)$ be a graph, and let n be a positive integer. An ***n*-flow** is a function $g : E \to \mathbb{Z}$ with $-n < g(e) < n$ such that conservation of flow holds at every node of G. The n-flow is **nowhere zero** if $g(e) \neq 0$ for all $e \in E$.

(a) Show that if G has a nowhere-zero n-flow, then G has a nowhere-zero $\mathbb{Z}_n$-flow.

(b) For a nowhere-zero $\mathbb{Z}_n$-flow f, define $g : E \to [-(n-1), n-1]$ such that $g(e)$ is congruent to $f(e)$ modulo n. The conservation of flow of g is not necessarily satisfied at each node. The absolute value of the different between incoming and outgoing flow at v is called the **excess**.

An **augmenting path** from a node u to a node v is a path $u = u_0 u_1 \dots u_r = v$ in the undirected graph G such that $u_{i-1} \to u_i$ is a directed edge in ${}_\rho G$ if and only if $g(u_{i-1}u_i) > 0$. Let $h : E \to \{-1, 0, 1\}$ be the function such that $h(e)g(e) > 0$ if e is on the path and $h(e) = 0$ otherwise. Show that $g + nh : E \to \mathbb{Z}$ still takes values in the interval $[-(n-1), n-1]$ and reduces the excess at some node.

(c) Prove that if G has a nowhere-zero $\mathbb{Z}_n$-flow, then G has a nowhere-zero n-flow.

(d) Prove that

$$\varphi_G(n) \neq 0 \qquad \text{implies} \qquad \varphi_G(n+1) \neq 0\,.$$

(e) Even stronger, prove that

$$\varphi_G(n) \leq \varphi_G(n+1)\,.$$

(This is nontrivial. But you will easily prove this after having read Chapter 7.)

1.15 Let ${}_\rho G = (V, E, \rho)$ be an oriented graph and $n \geq 2$.

(a) Let $f : E \to \mathbb{Z}_n$ be a nowhere-zero $\mathbb{Z}_n$-flow and let $e \in E$. Show that f naturally yields a nowhere-zero $\mathbb{Z}_n$-flow on the contraction ${}_\rho G/e$.

(b) For $S \subseteq V$ let $E^{\text{in}}(S)$ be the **in-coming** edges, i.e., $u \to v$ with $v \in S$ and $u \in V \setminus S$, and let $E^{\text{out}}(S)$ be the **out-going** edges. Show that $f : E \to \mathbb{Z}_n$ is a nowhere-zero $\mathbb{Z}_n$-flow if and only if

$$\sum_{e \in E^{\text{in}}(S)} f(e) \;=\; \sum_{e \in E^{\text{out}}(S)} f(e)$$

for all $S \subseteq V$. (*Hint:* For the sufficiency, contract all edges in S and $V \setminus S$.)

(c) Infer that $\varphi_G \equiv 0$ if G has a bridge.

1.16 Discover the notion of tensions.

1.17 Consider the **Petersen graph** G pictured in Figure 1.17.

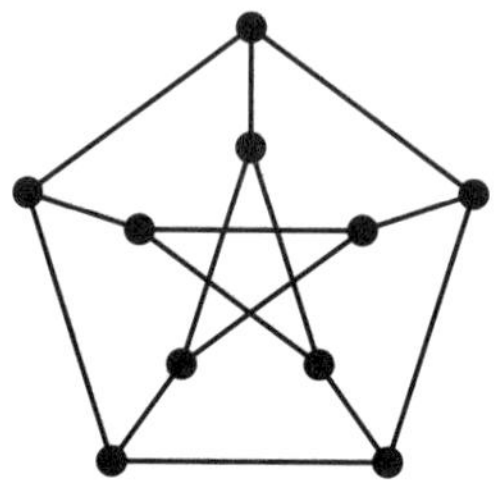

Figure 1.17. The Petersen graph.

(a) Show that $\varphi_G(4) = 0$.
(b) Show that the polynomial $\varphi_G(n)$ has nonreal roots.
(c) Construct a planar[4] graph whose flow polynomial has nonreal roots. (*Hint:* Think of the dual coloring question.)

1.18 ⬠ Prove Proposition 1.3.3: Let ${}_\rho G = (V, E, \rho)$ be an acyclic graph and $\Pi = \Pi({}_\rho G)$ the induced poset. A map $c : V \to [n]$ is strictly compatible with the orientation ρ of G if and only if c is a strictly order-preserving map $\Pi \to [n]$.

1.19 Compute $\Omega^\circ_{D_{10}}(n)$.

1.20 ⬠ Show that $\Omega_\Pi(n)$ is a polynomial in n.

1.21 ⬠ Let $\mathcal{S} = \operatorname{conv}\left\{ \binom{a_1}{b_1}, \binom{a_2}{b_2} \right\}$, with $a_1, a_2, b_1, b_2 \in \mathbb{Z}$, be a **lattice segment**.[5] Show that

$$\operatorname{ehr}_{\mathcal{S}}(n) = L\,n + 1\,,$$

where $L = |\gcd(a_2 - a_1, b_2 - b_1)|$, the **lattice length** of $\mathcal{S}$. Conclude further that $-\operatorname{ehr}_{\mathcal{S}}(-n)$ equals the number of lattice points of $n\mathcal{S}$ other than the endpoints, in other words,

$$(-1)^{\dim \mathcal{S}} \operatorname{ehr}_{\mathcal{S}}(-n) = \operatorname{ehr}_{\mathcal{S}^\circ}(n)\,.$$

Can you find an explicit formula for $\operatorname{ehr}_{\mathcal{S}}(n)$ when $\mathcal{S}$ is a segment with *rational* endpoints?

[4]The Petersen graph is a (famous) example of a nonplanar graph.

[5]We use the notation $\operatorname{conv}(V)$ to denote the convex hull of a set V of vectors.

1.22 Let O be a closed polygonal lattice path, i.e., the union of lattice segments, such that each vertex on O lies on precisely two such segments, and that topologically O is a closed curve. Show that

$$\mathrm{ehr}_O(n) \;=\; L\,n\,,$$

where L is the sum of the lattice lengths of the lattice segments that make up O or, equivalently, the number of lattice points on O.

1.23 ⬠ Let $\mathbf{v}_1, \mathbf{v}_2 \in \mathbb{Z}^2$, and let Q be the half-open parallelogram

$$\mathsf{Q} := \{\lambda \mathbf{v}_1 + \mu \mathbf{v}_2 \,:\, 0 \le \lambda, \mu < 1\}\,.$$

Show (for example, by tiling the plane by translates of Q) that

$$\mathrm{ehr}_{\mathsf{Q}}(n) \;=\; A\,n^2,$$

where $A = \left|\det\left(\begin{smallmatrix} a & c \\ b & d \end{smallmatrix}\right)\right|$.

1.24 A lattice triangle $\mathrm{conv}\{\mathbf{v}_1, \mathbf{v}_2, \mathbf{v}_3\}$ is **unimodular** if $\mathbf{v}_2 - \mathbf{v}_1$ and $\mathbf{v}_3 - \mathbf{v}_1$ form a lattice basis of $\mathbb{Z}^2$.

(a) Prove that a lattice triangle is unimodular if and only if it has area $\frac{1}{2}$.

(b) Conclude that for any two unimodular triangles $\triangle_1$ and $\triangle_2$, there exist $T \in \mathrm{GL}_2(\mathbb{Z})$ and $\mathbf{x} \in \mathbb{Z}^2$ such that $\triangle_2 = T(\triangle_1) + \mathbf{x}$.

(c) Compute the Ehrhart polynomials of all unimodular triangles.

(d) Show that every lattice polygon can be triangulated into unimodular triangles.

(e) Use the above facts to give an alternative proof of Theorem 1.4.1.

1.25 Let $\mathsf{P} \subset \mathbb{R}^2$ be a lattice polygon, and denote the area of P by A, the number of integer points inside the polygon P by I, and the number of integer points on the boundary of P by B. Prove that

$$A \;=\; I + \tfrac{1}{2}B - 1$$

(a famous formula due to Georg Alexander Pick). Deduce from this formulas for the coefficients of the Ehrhart polynomial of P.

1.26 Let $\mathsf{P}, \mathsf{Q} \subset \mathbb{R}^2$ be lattice polygons, such that Q is contained in the interior of P. Generalize Exercise 1.25 (i.e., both a version of Pick's theorem and the accompanying Ehrhart polynomial) to the "polygon with a hole" $\mathsf{P} - \mathsf{Q}$. Generalize your formulas to a lattice polygon with n "holes" (instead of one).

1.27 Let $f(t) \;=\; a_d t^d + a_{d-1} t^{d-1} + \cdots + a_0 \in \mathbb{R}[t]$ be a polynomial such that $f(n)$ is an integer for every integer $n > 0$. Give a proof or a counterexample for the following statements:

(a) All coefficients a_j are integers.

(b) $f(n)$ is an integer for all $n \in \mathbb{Z}$.

(c) If $(-1)^k f(-n) \ge 0$ for all $n > 0$, then $k = \deg(f)$.

1.28 Suppose $f(t) = a_d t^d + a_{d-1} t^{d-1} + \cdots + a_0 \in \mathbb{R}[t]$ is a polynomial with $a_d > 0$. Prove that, if all roots of $f(t)$ have negative real parts, then each $a_j > 0$.

Chapter 2

Partially Ordered Sets

Life is the twofold internal movement of composition and decomposition at once general and continuous.
Henri de Blainville

Partially ordered sets, *posets* for short, made an appearance twice so far. First (in Section 1.3) as a class of interesting combinatorial objects with a rich counting theory intimately related to graph colorings and, second (in Section 1.4), as a natural book-keeping structure for geometric subdivisions of polygons. In particular, the stage for the principle of overcounting-and-correcting, more commonly referred to as inclusion–exclusion, is naturally set in the theory of posets. Our agenda in this chapter is twofold: we need to introduce machinery that will be crucial tools in later chapters, but we will also prove our first combinatorial reciprocity theorems in a general setting, from first principles; later on we will put these theorems in a geometric context. We recall that a poset Π is a finite set with a binary relation $\preceq_\Pi$ that is reflexive, transitive, and antisymmetric.

2.1. Order Ideals and the Incidence Algebra

We now return to Section 1.3 and the problem of counting (via $\Omega_\Pi(n)$) order-preserving maps $\phi : \Pi \to [n]$ which satisfy

$$a \preceq_\Pi b \quad \Longrightarrow \quad \phi(a) \le \phi(b)$$

for all $a, b \in \Pi$. The preimages $\phi^{-1}(j)$, for $j = 1, 2, \ldots, n$, partition Π and uniquely identify ϕ, but from a poset point of view they do not have enough structure. A better perspective comes from the following observation: let $\phi : \Pi \to [2]$ be an order-preserving map into the 2-chain, and let $I := \phi^{-1}(1)$.

Now

$$y \in I \quad \text{and} \quad x \preceq_\Pi y \quad \Longrightarrow \quad x \in I\,.$$

A subset $I \subseteq \Pi$ with this property is called an **order ideal** of Π. Conversely, if $I \subseteq \Pi$ is an order ideal, then $\phi : \Pi \to [2]$ with $\phi^{-1}(1) = I$ defines an order preserving map. Thus, order-preserving maps $\phi : \Pi \to [2]$ are in bijection with the order ideals of Π. Dually, the complement $F = \Pi \setminus I$ of an order ideal I is characterized by the property that $x \succeq y \in F$ implies $x \in F$. Such a set is called a **dual order ideal** or **filter** of Π. This reasoning proves the following observation.

Proposition 2.1.1. *Let Π be a finite poset. Then $\Omega_\Pi(2)$ is the number of order ideals (or, equivalently, filters) of Π.*

To characterize general order-preserving maps into chains in terms of Π, we note that every order ideal of $[n]$ is **principal**, that is, every order ideal $I \subseteq [n]$ is of the form

$$I \;=\; \{j \in [n] \,:\, j \le k\} \;=\; [k]$$

for some k. In particular, if $\phi : \Pi \to [n]$ is order preserving, then the preimage $\phi^{-1}([k])$ of an order ideal $[k] \subseteq [n]$ is an order ideal of Π, and this gives us the following bijection.

Proposition 2.1.2. *order-preserving maps $\phi : \Pi \to [n]$ are in bijection with multichains*[1] *of order ideals*

$$\varnothing \;=\; I_0 \;\subseteq\; I_1 \;\subseteq\; \cdots \;\subseteq\; I_n \;=\; \Pi$$

of length n. The map ϕ is strictly order preserving if and only if $I_j \setminus I_{j-1}$ is an antichain for all $j = 1, 2, \ldots, n$.

Proof. We need to argue only the second part. We observe that ϕ is strictly order preserving if and only if there are no elements $x \prec y$ with $\phi(x) = \phi(y)$. Hence, ϕ is strictly order preserving if and only if $\phi^{-1}(j) = I_j \setminus I_{j-1}$ does not contain a pair of comparable elements. □

The collection $\mathcal{J}(\Pi)$ of order ideals of Π is itself a poset under set inclusion, which we call the **lattice of order ideals** or the **Birkhoff lattice**[2] of Π. What we just showed is that $\Omega_\Pi(n)$ counts the number of multichains of length n in $\mathcal{J}(\Pi) \setminus \{\varnothing, \Pi\}$. The next problem we address is counting multichains in general posets. To that end, we introduce an algebraic gadget: the **incidence algebra** $I(\Pi)$ is a $\mathbb{C}$-vector space spanned by those functions $\alpha : \Pi \times \Pi \to \mathbb{C}$ that satisfy

$$\alpha(x, y) = 0 \quad \text{whenever} \quad x \npreceq y\,.$$

[1]A **multichain** is a sequence of comparable elements, where we allow repetition.

[2]The reason for this terminology will become clear shortly.

We define the (convolution) product of $\alpha, \beta : \Pi \times \Pi \to \mathbb{C}$ as

$$(\alpha * \beta)(r,t) \ := \ \sum_{r \preceq s \preceq t} \alpha(r,s)\,\beta(s,t)\,,$$

and together with $\delta \in I(\Pi)$ defined by

$$\delta(x,y) \ := \ \begin{cases} 1 & \text{if } x = y, \\ 0 & \text{if } x \neq y, \end{cases} \tag{2.1.1}$$

this gives $I(\Pi)$ the structure of an associative $\mathbb{C}$-algebra with unit δ. (If this is starting to feel like linear algebra, you are on the right track.) A distinguished role is played by the **zeta function** $\zeta \in I(\Pi)$ defined by

$$\zeta(x,y) \ := \ \begin{cases} 1 & \text{if } x \preceq y, \\ 0 & \text{otherwise.} \end{cases}$$

For the time being, the power of zeta functions lies in their powers.

Proposition 2.1.3. *Let Π be a finite poset and $x, y \in \Pi$. Then $\zeta^n(x,y)$ equals the number of multichains*

$$x \ = \ x_0 \ \preceq \ x_1 \ \preceq \ \cdots \ \preceq \ x_n \ = \ y$$

of length n.

Proof. For $n = 1$, we have $\zeta(x,y) = 1$ if and only if $x = x_0 \preceq x_1 = y$. Arguing by induction, we assume that $\zeta^{n-1}(x,y)$ is the number of multichains of length $n-1$ for all $x, y \in \Pi$, and we calculate

$$\zeta^n(x,z) \ = \ (\zeta^{n-1} * \zeta)(x,z) \ = \ \sum_{x \preceq y \preceq z} \zeta^{n-1}(x,y)\,\zeta(y,z)\,.$$

Each summand on the right is the number of multichains of length $n-1$ ending in y that can be extended to z. □

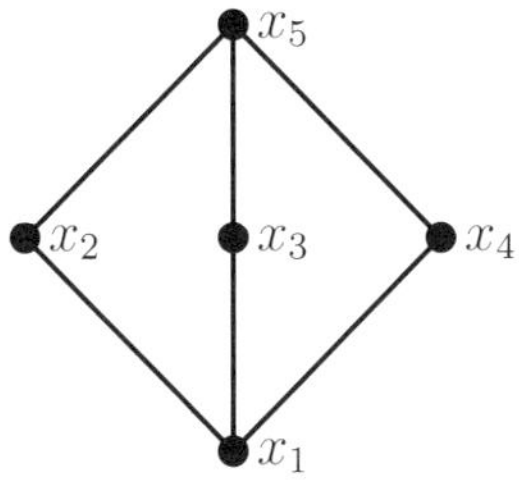

Figure 2.1. A sample poset.

As an example, the zeta function for the poset in Figure 2.1 is given in matrix form as

$$\begin{pmatrix} 1 & 1 & 1 & 1 & 1 \\ 0 & 1 & 0 & 0 & 1 \\ 0 & 0 & 1 & 0 & 1 \\ 0 & 0 & 0 & 1 & 1 \\ 0 & 0 & 0 & 0 & 1 \end{pmatrix}.$$

We encourage you to see Proposition 2.1.3 in action by computing powers of this matrix.

As a first milestone, Proposition 2.1.3 implies the following representation of the order polynomial of Π which we introduced in Section 1.3.

Corollary 2.1.4. *For a finite poset Π, let ζ be the zeta function of $\mathcal{J}(\Pi)$, the lattice of order ideals in Π. The order polynomial associated with Π is given by*

$$\Omega_\Pi(n) \;=\; \zeta^n(\varnothing, \Pi)\,.$$

Identifying $\Omega_\Pi(n)$ with the evaluation of a power of ζ does not stipulate that $\Omega_\Pi(n)$ is the restriction of a polynomial (which we know to be true from Exercise 1.20) but this impression is misleading: let $\eta \in I(\Pi)$ be defined by

(2.1.2) $$\eta(x,y) \;:=\; \begin{cases} 1 & \text{if } x \prec y, \\ 0 & \text{otherwise.} \end{cases}$$

Then $\zeta = \delta + \eta$ and hence, using the binomial theorem (Exercise 2.1),

(2.1.3) $$\zeta^n(x,y) \;=\; (\delta+\eta)^n(x,y) \;=\; \sum_{k=0}^{n} \binom{n}{k} \eta^k(x,y)\,.$$

Exercise 2.5 asserts that the sum on the right stops at the index $k = |\Pi|$ and is thus a polynomial in n of degree $\le |\Pi|$.

The arguments in the preceding paragraph are not restricted to posets formed by order ideals, but hold more generally for every poset Π that has a **minimum** $\hat{0}$ and a **maximum** $\hat{1}$, i.e., $\hat{0}$ and $\hat{1}$ are elements in Π that satisfy $\hat{0} \preceq x \preceq \hat{1}$ for all $x \in \Pi$. (For example, the Birkhoff lattice $\mathcal{J}(\Pi)$ has minimum $\varnothing$ and maximum Π.) This gives the following.

Proposition 2.1.5. *Let Π be a finite poset with minimum $\hat{0}$, maximum $\hat{1}$, and zeta function ζ. Then $\zeta^n(\hat{0},\hat{1})$ is a polynomial in n.*

To establish the reciprocity theorem for $\Omega_\Pi(n)$ (Theorem 1.3.2), we would like to evaluate $\zeta^n(\varnothing,\Pi)$ at negative integers n, so we first need to understand when an element $\alpha \in I(\Pi)$ is invertible. To this end, we pause and make the incidence algebra a bit more tangible.

Choose a **linear extension** of Π, that is, we label the $d = |\Pi|$ elements of Π by $p_1, p_2, \ldots, p_d$ such that $p_i \preceq p_j$ implies $i \le j$. (That such a labeling exists is the content of Exercise 2.2.) This allows us to identify $I(\Pi)$ with a subalgebra of the upper triangular $(d \times d)$-matrices by setting

$$\alpha \ := \ \left(\alpha(p_i, p_j)\right)_{1 \le i,j \le d} \, .$$

For example, for the poset D_{10} given in Figure 1.12, a linear extension is given by $(p_1, p_2, \ldots, p_{10}) = (1, 5, 2, 3, 7, 10, 4, 6, 9, 8)$ and the incidence algebra consists of matrices of the form

$$\begin{array}{c} \\ 1 \\ 5 \\ 2 \\ 3 \\ 7 \\ 10 \\ 4 \\ 6 \\ 9 \\ 8 \end{array}
\begin{array}{c}
\begin{array}{cccccccccc} 1 & 5 & 2 & 3 & 7 & 10 & 4 & 6 & 9 & 8 \end{array} \\
\begin{pmatrix}
* & * & * & * & * & * & * & * & * & * \\
 & * & & & & * & & & & \\
 & & * & & & * & * & * & & * \\
 & & & * & & & & * & * & \\
 & & & & * & & & & & \\
 & & & & & * & & & & \\
 & & & & & & * & & & * \\
 & & & & & & & * & & \\
 & & & & & & & & * & \\
 & & & & & & & & & *
\end{pmatrix}
\end{array} ,$$

where the stars are the possible nonzero entries for the elements in $I(\Pi)$. This linear-algebra perspective affords a simple criterion for when α is invertible; see Exercise 2.4.

Proposition 2.1.6. *An element $\alpha \in I(\Pi)$ is invertible if and only if*

$$\alpha(x, x) \neq 0 \qquad \text{for all} \ \ x \in \Pi \, .$$

2.2. The Möbius Function and Order Polynomial Reciprocity

We now return to the stage set up by Corollary 2.1.4, namely,

$$\Omega_\Pi(n) \ = \ \zeta^n_{\mathcal{J}(\Pi)}(\varnothing, \Pi) \, .$$

We would like to use this identity to compute $\Omega_\Pi(-n)$; thus we need to invert the zeta function of $\mathcal{J}(\Pi)$. Such an inverse exists by Proposition 2.1.6, and we call $\mu := \zeta^{-1}$ the **Möbius function**. For example, the Möbius function of the poset in Figure 2.1 is given in matrix form as

$$\begin{pmatrix}
1 & -1 & -1 & -1 & 2 \\
0 & 1 & 0 & 0 & -1 \\
0 & 0 & 1 & 0 & -1 \\
0 & 0 & 0 & 1 & -1 \\
0 & 0 & 0 & 0 & 1
\end{pmatrix} .$$

It is apparent that one can compute the Möbius function recursively, and in fact, unravelling the condition that $(\mu * \zeta)(x,z) = \delta(x,z)$ for all $x, z \in \Pi$ gives

$$\begin{aligned} \mu(x,z) &= -\sum_{x \prec y \preceq z} \mu(y,z) = -\sum_{x \preceq y \prec z} \mu(x,y) \quad \text{for } x \prec z, \text{ and} \\ \mu(x,x) &= 1. \end{aligned} \tag{2.2.1}$$

As a notational remark, the functions ζ, δ, μ, and η depend on the underlying poset Π, so we will sometimes write ζ_Π, δ_Π, etc., to make this dependence clear. For an example, we consider the Möbius function of the **Boolean lattice** B_d, the partially ordered set of all subsets of $[d]$ ordered by inclusion. For two subsets $S \subseteq T \subseteq [d]$, we have $\mu_{B_d}(S,T) = 1$ whenever $S = T$ and $\mu_{B_d}(S,T) = -1$ whenever $|T \setminus S| = 1$. Although this provides little data, we venture that

$$\mu_{B_d}(S,T) = (-1)^{|T \setminus S|}. \tag{2.2.2}$$

We dare you to prove this from first principles, or to appeal to the results in Exercise 2.6 after realizing that B_d is the d-fold product of a 2-chain. Towards proving the combinatorial reciprocity theorem for order polynomials (Theorem 1.3.2) we note the following.

Proposition 2.2.1.

$$\Omega_\Pi(-n) = \zeta_{\mathcal{J}}^{-n}(\varnothing, \Pi) = \mu_{\mathcal{J}}^{n}(\varnothing, \Pi),$$

where $\mathcal{J} = \mathcal{J}(\Pi)$ is the Birkhoff lattice of Π.

This proposition is strongly suggested by our notation but nevertheless requires a proof.

Proof. Let $d = |\Pi|$. By Exercise 2.5,

$$\zeta^{-1} = (\delta + \eta)^{-1} = \delta - \eta + \eta^2 - \cdots + (-1)^d \eta^d.$$

If we now take powers of ζ^{-1} and again appeal to Exercise 2.5, we calculate

$$\zeta^{-n} = \sum_{k=0}^{d} (-1)^k \binom{n+k-1}{k} \eta^k.$$

Thus the expression of ζ^n as a polynomial given in (2.1.3), together with the fundamental combinatorial reciprocity for binomial coefficients (0.0.2) given in the very beginning of this book, proves the claim. □

Expanding $\mu_{\mathcal{J}}^n$ into the n-fold product of $\mu_{\mathcal{J}}$ with itself, the right-hand side of the identity in Proposition 2.2.1 is

$$\mu_{\mathcal{J}}^n(\varnothing, \Pi) = \sum \mu_{\mathcal{J}}(I_0, I_1)\, \mu_{\mathcal{J}}(I_1, I_2) \cdots \mu_{\mathcal{J}}(I_{n-1}, I_n), \tag{2.2.3}$$

where the sum is over all multichains of order ideals

$$\varnothing = I_0 \subseteq I_1 \subseteq \cdots \subseteq I_n = \Pi$$

of length n. Our next goal is thus to understand the evaluation $\mu_{\mathcal{J}}(K, M)$, where $K \subseteq M \subseteq \Pi$ are order ideals. This evaluation depends only on

$$[K, M] := \{L \in \mathcal{J} : K \subseteq L \subseteq M\},$$

the **interval** from K to M in the Birkhoff lattice $\mathcal{J}$. Moreover, we call two posets Π and Π' **isomorphic** (and write $\Pi \cong \Pi'$) if there is a bijection $\phi : \Pi \to \Pi'$ that satisfies $x \preceq_\Pi y \iff \phi(x) \preceq_{\Pi'} \phi(y)$.

Theorem 2.2.2. *Let Π be a finite poset and $K \subseteq M$ order ideals in $\mathcal{J} = \mathcal{J}(\Pi)$. Then*

$$\mu_{\mathcal{J}}(K, M) = \begin{cases} (-1)^{|M \setminus K|} & \text{if } M \setminus K \text{ is an antichain,} \\ 0 & \text{otherwise.} \end{cases}$$

In the proof we want to use induction—not on the number of elements but on the *length* of Π. A **chain** in Π is a collection of elements $C = \{c_0, c_1, \dots, c_k\}$ such that $c_0 \prec_\Pi c_1 \prec_\Pi \cdots \prec_\Pi c_k$. The **length** of the chain C is $k - 1$. The chain C is **saturated** or **unrefineable** if $c_{i-1} \prec c_i$ is a cover relation for all $i = 1, \dots, k$. The chain is **maximal** if c_0 and c_k are minimal and maximal elements of Π. The **length** of a poset is the maximal length of a (maximal) chain in Π.

Proof. We first consider the (easier) case that $M \setminus K$ is an antichain. In this case $K \cup A$ is an order ideal for all $A \subseteq M \setminus K$. In other words, the interval $[K, M]$ is isomorphic to the Boolean lattice B_r for $r = |M \setminus K|$ and hence, with (2.2.2), we conclude $\mu_{\mathcal{J}}(K, M) = (-1)^r$.

The case that $M \setminus K$ contains comparable elements is a bit more tricky. We argue by induction on the length of the interval $[K, M]$. The base case is given by the situation that $M \setminus K$ consists of exactly two comparable elements $a \prec b$. Hence, $[K, M] = \{K \prec K \cup \{a\} \prec M\}$ and we compute

$$\mu_{\mathcal{J}}(K, M) = -\mu_{\mathcal{J}}(K, K) - \mu_{\mathcal{J}}(K, K \cup \{a\}) = -1 - (-1) = 0.$$

For the induction step we use (2.2.1), i.e.,

$$\mu_{\mathcal{J}}(K, M) = -\sum \mu_{\mathcal{J}}(K, L),$$

where the sum is over all order ideals L such that $K \subseteq L \subset M$. By the induction hypothesis, $\mu_{\mathcal{J}}(K, L)$ is zero unless $L \setminus K$ is an antichain and thus

$$\mu_{\mathcal{J}}(K, M) = -\sum \left\{ (-1)^{|L \setminus K|} : \begin{array}{l} K \subseteq L \subset M \text{ order ideal,} \\ L \setminus K \text{ is an antichain} \end{array} \right\},$$

where we have used the already-proven part of the theorem. Now let $m \in M \setminus K$ be a minimal element. The order ideals L in the above sum can be partitioned into those containing m and those that do not. Both parts of this partition have the same size: if $m \notin L$, then $L \cup \{m\}$ is also an order ideal; if $m \in L$, then $L \setminus \{m\}$ is an admissible order ideal as well. (You should check this.) Hence, the positive and negative terms cancel each other and $\mu_{\mathcal{J}}(K, M) = 0$. □

With this we can give a (purely combinatorial) proof of Theorem 1.3.2, the reciprocity theorem for order polynomials.

Proof of Theorem 1.3.2. By Theorem 2.2.2, the right-hand side of (2.2.3) equals $(-1)^{|\Pi|}$ times the number of multichains

$$\varnothing \;=\; I_0 \;\subseteq\; I_1 \;\subseteq\; \cdots \;\subseteq\; I_n \;=\; \Pi$$

of order ideals such that $I_j \setminus I_{j-1}$ is an antichain for $j \in [n]$. By Proposition 2.1.2 this is exactly $(-1)^{|\Pi|}\,\Omega_\Pi^\circ(n)$, and this proves Theorem 1.3.2. □

Our proof gives us some additional insights into the structure of $\Omega_\Pi(n)$.

Corollary 2.2.3. *Let Π be a finite poset. Then $\Omega_\Pi(-k) = 0$ for all $0 < k < m$ if and only if Π contains an m-chain.*

Proof. Since $\Omega_\Pi^\circ(k)$ is weakly increasing, it suffices to assume that m is the length of Π. Let $C = \{c_1 \prec \cdots \prec c_{m+1}\}$ be a chain in Π of maximal length. Then $i \mapsto c_i$ defines a strictly order-preserving injection $[m+1] \hookrightarrow \Pi$. We can compose this injection with a given strictly order-preserving map $\Pi \to [k]$ to create a strictly order-preserving injection $[m+1] \hookrightarrow [k]$; however, such an injection exists only if $k \geq m+1$.

To show that $\Omega_\Pi^\circ(m+1) > 0$, set $I_0 := \varnothing$ and define a sequence $I_0 \subset I_1 \subset \cdots \subset I_r = \Pi$ by the following rule. Define $I_j := I_{j-1} \cup M_j$, where M_j consists of the minimal elements of $\Pi \setminus I_{j-1}$. You should convince yourself that this is a sequence of order ideals and that $I_j \setminus I_{j-1}$ is a nonempty antichain for all $j \geq 1$. In particular $C \cap (I_j \setminus I_{j-1}) = \{c_j\}$ and hence $r = m+1$. Proposition 2.1.2 now implies that there is a strictly order-preserving map $\Pi \to [m+1]$, which completes the proof. □

2.3. Zeta Polynomials, Distributive Lattices, and Eulerian Posets

We now take a breath and see how far we can generalize Theorem 1.3.2 (by weakening the assumptions). Our starting point is Proposition 2.1.5: for a poset Π that has a minimum $\hat{0}$ and maximum $\hat{1}$, the evaluation

$$Z_\Pi(n) \;:=\; \zeta^n(\hat{0}, \hat{1})$$

is a polynomial in n, the **zeta polynomial** of Π. For example, if we augment the poset D_{10} in Figure 1.12 by a maximal element (think of the number 0, which is divisible by all positive integers), Exercise 2.10 gives the accompanying zeta polynomial as

$$Z_\Pi(n) \;=\; \tfrac{1}{24}n^4 + \tfrac{13}{12}n^3 + \tfrac{23}{24}n^2 - \tfrac{13}{12}n\,. \tag{2.3.1}$$

In analogy with the combinatorial reciprocity theorem for order polynomials (Theorem 1.3.2)—which are, after all, zeta polynomials of posets formed by order ideals—we now seek interpretations for evaluations of zeta polynomials at negative integers. Analogous to Proposition 2.2.1,

$$Z_\Pi(-n) \;=\; \zeta^{-n}(\hat{0},\hat{1}) \;=\; \mu^n(\hat{0},\hat{1})\,,$$

where μ is the Möbius function of Π. Our sample zeta function (2.3.1) illustrates that the quest for interpretations at negative evaluations is nontrivial: here we compute

$$Z_\Pi(-1) = 1 \qquad \text{and} \qquad Z_\Pi(-2) = -2$$

and so any hope of a simple counting interpretation of $Z_\Pi(-n)$ or $-Z_\Pi(-n)$ is shattered. On a more optimistic note, we can repeat the argument behind (2.2.3) for a general poset Π:

$$Z_\Pi(-n) \;=\; \mu^n(\hat{0},\hat{1}) \;=\; \sum \mu(x_0,x_1)\,\mu(x_1,x_2)\,\cdots\,\mu(x_{n-1},x_n)\,, \tag{2.3.2}$$

where the sum is over all multichains

$$\hat{0} \;=\; x_0 \;\preceq\; x_1 \;\preceq\; \cdots \;\preceq\; x_n \;=\; \hat{1}$$

of length n. The key property that put (2.2.3) to work in our proof of Theorem 2.2.2 (and subsequently, our proof of Theorem 1.3.2) was that each summand on the right-hand side of (2.2.3) was either 0 or the same constant. We thus seek a class of posets where a similar property holds in (2.3.2).

For two elements x and y in a poset Π, consider all least upper bounds of x and y, i.e., all $z \in \Pi$ such that $x \preceq z$ and $y \preceq z$ and there is no $w \prec z$ with the same property. If such a least upper bound of x and y exists and is unique, we call it the **join** of x and y and denote it by $x \vee y$. Dually, if a greatest lower bound of x and y exists and is unique, we call it the **meet** of x and y and denote it by $x \wedge y$.

A **lattice**[3] is a poset in which meets and joins exist for any pair of elements. Note that every finite lattice will necessarily have a minimum $\hat{0}$ and a maximum $\hat{1}$. A lattice Π is **distributive** if meets and joins satisfy the distributive laws

$$(x \wedge y) \vee z \;=\; (x \vee z) \wedge (y \vee z) \qquad \text{and} \qquad (x \vee y) \wedge z \;=\; (x \wedge z) \vee (y \wedge z)$$

[3]This *lattice* is not to be confused with the integer lattice $\mathbb{Z}^2$ that made an appearance in Section 1.4 and whose higher-dimensional cousins will play a central role in later chapters. Both meanings of *lattice* are well furnished in the mathematical literature; we hope that they will not be confused in this book.

for all $x, y, z \in \Pi$. The reason we are interested in distributive lattices is the following famous result, often called *Birkhoff's theorem*, whose proof is subject to Exercise 2.13.

Theorem 2.3.1. *Every finite distributive lattice is isomorphic to the poset of order ideals of some poset.*

Next we consider some of the consequences of this theorem. Given a finite distributive lattice Π, we now know that the Möbius-function values on the right-hand side of (2.3.2) can be interpreted as stemming from the poset of order ideals of some other poset. But this means that we can apply Theorem 2.2.2 in precisely the same way we used it in our proof of Theorem 1.3.2: the right-hand side of (2.3.2) becomes $(-1)^{|\Pi|}$ times the number of multichains $\hat{0} = x_0 \preceq x_1 \preceq \cdots \preceq x_n = \hat{1}$ such that the corresponding differences of order ideals are all antichains. A moment's thought reveals that this last condition is equivalent to the fact that each interval $[x_j, x_{j+1}]$ is a Boolean lattice. What we have just proved is a combinatorial reciprocity theorem which, in a sense, generalizes that of order polynomials (Theorem 1.3.2).

Theorem 2.3.2. *Let Π be a finite distributive lattice. Then $(-1)^{|\Pi|} Z_\Pi(-n)$ equals the number of multichains $\hat{0} = x_0 \preceq x_1 \preceq \cdots \preceq x_n = \hat{1}$ such that each interval $[x_j, x_{j+1}]$ is a Boolean lattice.*

There is another class of posets that comes with a combinatorial reciprocity theorem stemming from (2.3.2). To introduce it, we need a few more definitions. A finite poset Π is **graded** if every maximal chain in Π has the same length r, which we call the **rank** of Π. The **length** $l_\Pi(x, y)$ of two elements $x, y \in \Pi$ is the length of a maximal chain in $[x, y]$. A graded poset that has a minimal and a maximal element is **Eulerian** if its Möbius function is

$$\mu(x, y) = (-1)^{l_\Pi(x,y)} .$$

We have seen examples of Eulerian posets earlier, for instance, Boolean lattices; another important class of Eulerian posets are formed by faces of polyhedra, which we will study in the next chapter. What happens with (2.3.2) when the underlying poset Π is Eulerian? In this case, the Möbius-function values on the right-hand side are determined by the interval length, and so each summand on the right is simply $(-1)^r$, where r is the rank of Π. But then (2.3.2) says that $Z_\Pi(-n)$ equals $(-1)^r$ times the number of multichains of length n, which is $\zeta^n(\hat{0}, \hat{1}) = Z_\Pi(n)$. This argument yields a reciprocity theorem that relates the zeta polynomial of Π to itself.

Theorem 2.3.3. *Let Π be a finite Eulerian poset of rank r. Then*

$$Z_\Pi(-n) = (-1)^r Z_\Pi(n) .$$

We will return to this result in connection with the combinatorial structure of polytopes.

2.4. Inclusion–Exclusion and Möbius Inversion

Our approach to Möbius functions in the proof of Theorem 1.3.2 is a bit uncommon. Usually, Möbius functions are introduced as a sophisticated version of overcounting and correcting. We too follow this approach by starting with what is known as the *principle of inclusion–exclusion.*

Let A and B be two finite sets. Then

$$|A \cup B| \;=\; |A| + |B| - |A \cap B| .$$

If we have n finite sets $A_1, A_2, \ldots, A_n$, then using the fact that $\cup$ is associative and $\cap$ distributes over $\cup$ gives

$$|A_1 \cup A_2 \cup \cdots \cup A_n| \;=\; \sum_{\varnothing \neq J \subseteq [n]} (-1)^{|J|-1} \, |A_J| , \tag{2.4.1}$$

where we set $A_J := \bigcap_{j \in J} A_j$; see Exercise 2.16. If $A_1, \ldots, A_n$ are subsets of some ground set A, then we can go one step further:

$$|A \setminus (A_1 \cup \cdots \cup A_n)| \;=\; |A| - |A_1 \cup \cdots \cup A_n| \;=\; \sum_{J \subseteq [n]} (-1)^{|J|} \, |A_J| , \tag{2.4.2}$$

where $A_\varnothing := A$. The simple identities (2.4.1) and (2.4.2) form the basis for the principle of inclusion–exclusion and the theory of Möbius functions.

The typical scenario in which we use (2.4.1) or (2.4.2) is the following: Imagine that A is a set of combinatorial objects and $A_j \subseteq A$ is the collection of objects having the property j. Then (2.4.2) gives a formula for the number of objects having none of the properties $1, 2, \ldots, n$. This is of value provided we can write down $|A_J|$ explicitly.

Here is an example involving the chromatic polynomial $\chi_G(n)$ of a graph $G = (V, E)$, which we introduced in Section 1.1. For a finite set T, let $s_G(T)$ be the number of proper colorings of G with colors in T such that each color is used at least once. In other words, $s_G(T)$ is the number of surjective maps $c : V \twoheadrightarrow T$ such that $c(u) \neq c(v)$ for all $uv \in E$. Now, for $t \in T$, let A_t be the collection of proper colorings $c : V \to T$ that miss at least the color t, i.e., $t \notin c(V)$. Since the number of colorings with colors in T depends only on the number of colors, $|A_J| = \chi_G(|T| - |J|)$ for all $J \subseteq T$, and so

$$\begin{aligned} s_G(T) \;&=\; \sum_{J \subseteq T} (-1)^{|J|} |A_J| \;=\; \sum_{J \subseteq T} (-1)^{|J|} \chi_G(|T| - |J|) \\ &=\; \sum_{r=0}^{|T|} \binom{|T|}{r} (-1)^r \chi_G(|T| - r) . \end{aligned}$$

Here is another example: Let A be the collection of all d-multisubsets of $[n]$ and let A_i be the collection of multisubsets such that the element i appears at least twice. Then $|A_\varnothing| = |A| = \binom{n+d-1}{d}$ and, since we can remove two copies of i from each multisubset in A_i,

$$|A_i| \;=\; \binom{n+(d-2)-1}{d-2}.$$

Consequently $|A_J| = \binom{n+d-2|J|-1}{d-2|J|}$ for each $J \subseteq [n]$. The principle of inclusion–exclusion (2.4.2) leads to

$$\binom{n}{d} \;=\; |A \setminus (A_1 \cup \cdots \cup A_n)| \;=\; \sum_{i=0}^{n} (-1)^i \binom{n+d-2i-1}{d-2i}\binom{n}{i}.$$

Depending on your personal preference, this might not be a pretty formula, but it certainly gives a nontrivial relation among binomial coefficients which is tricky to prove by other means. See Exercise 2.18 for one further prominent application of (2.4.2).

In both examples, the key was that it was easy enough to write down the number of objects having *at least* property j whereas the number of objects having *only* property j is hard. Moreover, if we are only concerned about counting objects, we can abstract away the collection of objects A_J and only work with numbers. We recall that B_n is the Boolean lattice consisting of all subsets of $[n]$ partially ordered by inclusion. Let $f_= : B_n \to \mathbb{C}$ be a function. We can then define a new function $f_\geq : B_n \to \mathbb{C}$ by

$$f_\geq(J) \;:=\; \sum_{K \supseteq J} f_=(K)$$

for all $J \subseteq [n]$. Can we recover $f_=$ from the knowledge of $f_\geq$? To relate to the setting above, we had $f_\geq(I) := |A_I|$ and (2.4.2) gave us a way to compute $f_=(\varnothing) = |A \setminus (A_1 \cup \cdots \cup A_n)|$.

Proposition 2.4.1. *For* $I \subseteq [n]$,

$$f_=(I) \;=\; \sum_{J \supseteq I} (-1)^{|J \setminus I|} f_\geq(J).$$

Proof. The proof is a simple computation. We insert the definition of $f_\geq(K)$ into the right-hand side of the sought-after identity:

$$\sum_{J \supseteq I} (-1)^{|J \setminus I|} f_\geq(J) \;=\; \sum_{J \supseteq I} (-1)^{|J \setminus I|} \sum_{K \supseteq J} f_=(K).$$

The sets K in the second sum are supersets of I, and so we may change the order of summation to obtain

$$\sum_{J \supseteq I} (-1)^{|J \setminus I|} f_\geq(J) \;=\; \sum_{K \supseteq I} f_=(K) \sum_{I \subseteq J \subseteq K} (-1)^{|J \setminus I|}.$$

In Exercise 2.19 you will show that the interior sum on the right-hand side equals 1 if $I = K$ and 0 if $I \subsetneq K$. □

The natural next step is to generalize the setup to an arbitrary finite poset Π. For a function $f_= : \Pi \to \mathbb{C}$, we define $f_\geq : \Pi \to \mathbb{C}$ by

$$f_\geq(b) := \sum_{c \succeq b} f_=(c) . \tag{2.4.3}$$

Analogously to our question in the Boolean lattice case, we can now ask if we can recover $f_=$ from the knowledge of $f_\geq$. The right context in which to ask this question is that of the incidence algebra.

Let $\mathbb{C}^\Pi = \{f : \Pi \to \mathbb{C}\}$ be the $\mathbb{C}$-vector space of functions on Π. The incidence algebra $I(\Pi)$ operates on $\mathbb{C}^\Pi$ as follows: for $f \in \mathbb{C}^\Pi$ and $\alpha \in I(\Pi)$, we define a new function $\alpha f \in \mathbb{C}^\Pi$ by

$$(\alpha f)(b) := \sum_{c \succeq b} \alpha(b, c) f(c) . \tag{2.4.4}$$

That is, $I(\Pi)$ is a ring of operators on $\mathbb{C}^\Pi$, and

$$f_\geq = \zeta_\Pi \, f_= ,$$

where $\zeta_\Pi \in I(\Pi)$ is the zeta function of Π. Hence, the question of recoverability of $f_=$ from $f_\geq$ is that of the invertibility of ζ_Π. That, however, we sorted out in Proposition 2.1.6, and we obtain what is referred to as *Möbius inversion*.

Theorem 2.4.2. *Let Π be a poset with Möbius function μ_Π. Then for any two functions $f_=, f_\geq \in \mathbb{C}^\Pi$,*

$$f_\geq(b) = \sum_{c \succeq b} f_=(c) \qquad \text{if and only if} \qquad f_=(a) = \sum_{b \succeq a} \mu_\Pi(a, b) f_\geq(b) .$$

Likewise,

$$f_\leq(b) = \sum_{a \preceq b} f_=(a) \qquad \text{if and only if} \qquad f_=(c) = \sum_{b \preceq c} f_\leq(b) \mu_\Pi(b, c) .$$

Proof. It is instructive to do the yoga of Möbius inversion at least once: if $f_\geq$ satisfies the left-hand side of the first statement, then

$$\sum_{b \succeq a} \mu_\Pi(a, b) f_\geq(b) = \sum_{b \succeq a} \mu_\Pi(a, b) \sum_{c \succeq b} f_=(c) = \sum_{c \succeq a} f_=(c) \sum_{a \preceq b \preceq c} \mu_\Pi(a, b) .$$

By (2.2.1), the last inner sum equals 1 if $a = c$ and 0 otherwise. □

In a nutshell, Möbius inversion is what we implicitly used in our treatment of Ehrhart theory for lattice polygons in Section 1.4. The subdivision of a lattice polygon P into triangles, edges, and vertices is a genuine poset under inclusion. The function $f_=(\mathsf{P})$ is the number of lattice points in $\mathsf{P} \subseteq \mathbb{R}^2$ and we were interested in the evaluation of $f_=(\mathsf{P}) = (\mu f_\leq)(\mathsf{P})$.[4]

We conclude this section with a nontrivial application of Möbius inversion. Let $G = (V, E)$ be a simple graph. A **flat** of G is a set of edges $F \subseteq E$ such that for any $e \in E \setminus F$, the number of connected components of the graph $G[F] := (V, F)$ is strictly larger than that of $G[F \cup \{e\}]$. Let $\mathcal{L}(G)$ be the collection of flats of G ordered by inclusion; Figure 2.2 shows an example.

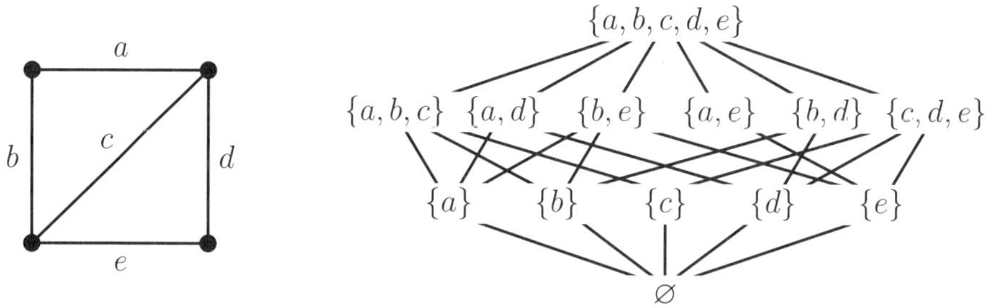

Figure 2.2. A sample graph and its flats.

Let $c : V \to [n]$ be a (not necessarily proper) coloring of G, and define

$$F_G(c) \ := \ \{uv \in E \, : \, c(u) = c(v)\} . \tag{2.4.5}$$

Then $F_G(c)$ is a flat and, as you will prove in Exercise 2.20, every flat $F \in \mathcal{L}(G)$ arises that way. We observe that c is a proper coloring precisely when $F_G(c) = \varnothing$.

For $n \geq 1$, we define the function $f_=^n : \mathcal{L}(G) \to \mathbb{Z}$ such that $f_=^n(F)$ is the number of n-colorings c with $F_G(c) = F$. In particular, the chromatic polynomial of G is $\chi_G(n) = f_=^n(\varnothing)$. This number is not so easy to determine. However, by defining

$$f_\geq^n(F) \ := \ \left|\left\{c \in [n]^V \, : \, F \subseteq F_G(c)\right\}\right| ,$$

we can say more.

Proposition 2.4.3. *Let $G = (V, E)$ be a simple graph and $F \in \mathcal{L}(G)$. Then*

$$f_\geq^n(F) \ = \ n^{\kappa(G[F])} ,$$

where $\kappa(G[F])$ is the number of connected components of $G[F]$.

[4] You might notice, upon re-reading Section 1.4, that we used a μ with only one argument, but that was simply for ease of notation: we give the full picture in our proof of Theorem 5.2.3 in Chapter 5.

Proof. Let $c : V \to [n]$ be a coloring. Then $F \subseteq F_G(c)$ if and only if c is constant on every connected component of $G[F]$. Hence, $f^n_{\geq}(F)$ equals the number of choices of one color per connected component. □

By Theorem 2.4.2,

$$\chi_G(n) \;=\; f^n_{=}(\varnothing) \;=\; \sum_{F \in \mathcal{L}(G)} \mu_{\mathcal{L}(G)}(\varnothing, F)\, n^{\kappa(G[F])}.$$

This shows again that $\chi_G(n)$ is a polynomial and, since $F = \varnothing$ maximizes the number of connected components of $G[F]$, the polynomial is of degree $|V|$, and so we obtain another (completely different) proof of Proposition 1.1.1. We will return to the Möbius function of $\mathcal{L}(G)$ in Section 7.2.

The original setup for inclusion–exclusion furnishes a natural class of posets. For a collection $A_1, A_2, \dots, A_n \subseteq A$ of sets, the **intersection poset** is

$$\mathcal{L} \;=\; \mathcal{L}(A_1, \dots, A_n; A) \;:=\; \left\{ A_I := \bigcap_{i \in I} A_i \;:\; I \subseteq [n] \right\},$$

partially ordered by *reverse* containment. So the maximum of $\mathcal{L}$ is $\hat{1} = A_{[n]} = A_1 \cap \dots \cap A_n$ whereas the minimum is $\hat{0} = A_\varnothing = A$. Of course, it would be natural to order the subsets $A_I \subseteq A$ by inclusion. However, by using reverse inclusion we have that for $I, J \subseteq [n]$,

$$I \subseteq J \quad \text{implies} \quad A_I \preceq_{\mathcal{L}} A_J$$

and hence the map $I \mapsto A_I$ is an order-preserving map from B_n to $\mathcal{L}$. The reverse implication does not hold in general—the collection of sets I that give rise to the same $A_I = S \in \mathcal{L}$ can be quite complicated. But there is always a canonical set: for $S \in \mathcal{L}$, we define

$$J_S \;:=\; \{ i \in [n] \;:\; S \subseteq A_i \}.$$

We call a set $J \subseteq [n]$ **closed** if $J = J_S$ for some $S \in \mathcal{L}$. The proof of the following simple but useful properties is outsourced to Exercise 2.22.

Lemma 2.4.4. *Let $\mathcal{L} = \mathcal{L}(A_1, \dots, A_n; A)$ and $S \in \mathcal{L}$. If $A_I = S$ for $I \subseteq [n]$, then $I \subseteq J_S$. Moreover, for $T \in \mathcal{L}$,*

$$S \preceq T \quad \Longleftrightarrow \quad J_S \subseteq J_T.$$

In particular, Lemma 2.4.4 implies that $\mathcal{L}$ is isomorphic to the subposet of B_n given by closed subsets.

For intersection posets, the principle of inclusion–exclusion gives a pedestrian way to compute the Möbius function.

Theorem 2.4.5. *Let $\mathcal{L} = \mathcal{L}(A_1, \dots, A_n; A)$ be an intersection poset. Then*

$$\mu_{\mathcal{L}}(S,T) \;=\; \sum_{\substack{J \subseteq [n] \\ A_J = T}} (-1)^{|J \setminus J_S|}$$

for all $S \preceq_{\mathcal{L}} T$.

Proof. Let $f_= : \mathcal{L} \to \mathbb{C}$ be an arbitrary function on $\mathcal{L}$ and let $f_\geq : \mathcal{L} \to \mathbb{C}$ be given by

$$f_\geq(S) \;=\; \sum_{T \succeq S} f_=(T)\,.$$

We can use $f_=$ to define a map $F_= : B_n \to \mathbb{C}$ by setting

$$F_=(I) \;:=\; \begin{cases} f_=(S) & \text{if } I = I_S \text{ for } S \in \mathcal{L}, \\ 0 & \text{if } I \text{ is not a closed set.} \end{cases}$$

Now define $F_\geq : B_n \to \mathbb{C}$ by $F_\geq(I) := \sum_{J \supseteq I} F_=(J)$. The second part of Lemma 2.4.4 implies that $f_\geq(S) = F_\geq(J_S)$ for any $S \in \mathcal{L}$. We can use Proposition 2.4.1 to compute

$$F_=(I) \;=\; \sum_{J \supseteq I} (-1)^{|J \setminus I|} F_\geq(J)\,.$$

The first part of Lemma 2.4.4 implies that $F_\geq(J) = f_\geq(A_J)$ for all $J \subseteq [n]$. For $S \in \mathcal{L}$,

$$f_=(S) \;=\; F_=(J_S) \;=\; \sum_{J \supseteq J_S} (-1)^{|J \setminus J_S|} f_\geq(A_I) \;=\; \sum_{T \succeq_{\mathcal{L}} S} f_\geq(T) \sum_{\substack{J \subseteq [n] \\ A_J = T}} (-1)^{|J \setminus J_S|}\,.$$

In contrast, we can also apply Theorem 2.4.2 to $f_\geq$ to obtain

$$\sum_{T \succeq_{\mathcal{L}} S} f_\geq(T) \mu_{\mathcal{L}}(S,T) \;=\; f_=(S) \;=\; \sum_{T \succeq_{\mathcal{L}} S} f_\geq(T) \sum_{\substack{J \subseteq [n] \\ A_J = T}} (-1)^{|J \setminus J_S|}\,.$$

Since $f_=$ was chosen arbitrarily and since the Möbius function of a poset is unique, this establishes the claim. □

We finish this section with one more way to compute the Möbius function of a poset Π, *Philip Hall's theorem.* Recall that the length of the chain $a_0 \prec a_1 \prec \cdots \prec a_k$ is k, the number of links. We denote by $c_k(a,b)$ the number of chains of length k of the form $a = a_0 \prec a_1 \prec \cdots \prec a_k = b$.

Theorem 2.4.6. *Let Π be a finite poset and $a \prec_\Pi b$. Then*

$$\mu_\Pi(a,b) \;=\; -\,c_1(a,b) + c_2(a,b) - c_3(a,b) + \cdots\,. \qquad (2.4.6)$$

Proof. A short and elegant proof can be given via the incidence algebra. Recall that $\eta \in I(\Pi)$ is given by $\eta(a,b) = 1$ if $a \prec b$ and 0 otherwise. Then $\zeta = \delta + \eta$ and we compute

$$\mu(a,b) \;=\; \zeta^{-1}(a,b) \;=\; (\delta+\eta)^{-1}(a,b) \;=\; \delta(a,b) - \eta^1(a,b) + \eta^2(a,b) - \cdots .$$

By Exercise 2.5, the sum on the right-hand side is finite and $c_k(a,b) = \eta^k(a,b)$. □

Notes

Posets and lattices originated in the nineteenth century and became subjects in their own right with the work of Garrett Birkhoff, who proved Theorem 2.3.1 [**31**], and Philip Hall [**80**], whose Theorem 2.4.6 concluded our chapter.

The oldest type of Möbius function is the one studied in number theory, which is the Möbius function (in a combinatorial sense) of the divisor lattice (see Exercise 2.7). The systematic study of Möbius functions of general posets was initiated by Gian–Carlo Rota's famous paper [**146**] which arguably started modern combinatorics. Rota's paper also put the idea of incidence algebras on firm ground, but it can be traced back much further to Richard Dedekind and Eric Temple Bell [**170**, Chapter 3]. Our proof of Theorem 2.2.2 implicitly makes use of *Rota's crosscut theorem*. Determining the Möbius function of a poset is difficult in general. Many techniques (including Rota's crosscut theorem) are explained in [**107**, Chapter 3] or [**170**].

As we already mentioned in Chapter 1, order polynomials were introduced by Richard Stanley [**160**, **166**] as chromatic-like polynomials for posets and we will see them again in Chapter 6 in geometric guise. Stanley introduced the zeta polynomial of a poset in [**162**], the paper that inspired the title of our book, and Theorem 2.3.3 appears as a side remark. Stanley also initiated the study of Eulerian posets in [**165**], though, in his own words, "they had certainly been considered earlier".

For (much) more on posets, lattices, and Möbius functions, we recommend [**159**] and [**170**, Chapter 3], which contains numerous open problems; we mention one representative: let Π_n be the set of all partitions (whose definition is given in (4.4.1) in Chapter 4) of a fixed positive integer n. We order the elements of Π_n by *refinement*, i.e., given two partitions $(a_1, a_2, \dots, a_j)$ and $(b_1, b_2, \dots, b_k)$ of n, we say that

$$(a_1, a_2, \dots, a_j) \;\preceq\; (b_1, b_2, \dots, b_k)$$

if the parts $a_1, a_2, \dots, a_j$ can be partitioned into blocks whose sums are $b_1, b_2, \dots, b_k$. Find the Möbius function of Π_n.

Exercises

2.1 ⬠ Let $(R, +, \cdot)$ be a ring with unit 1. For every $r \in R$ and $d \geq 0$ verify the **binomial theorem**

$$(1 + r)^d = \sum_{j=0}^{d} \binom{d}{j} r^j .$$

Show how this, in particular, implies (2.1.3).

2.2 ⬠ Show that every finite poset Π has a linear extension. (*Hint:* You can argue graphically by reading the Hasse diagram or, more formally, by induction on $|\Pi|$.)

2.3 A finite poset Π in which the meet of any two elements exists is called a **meet semilattice**. Show that if a meet semilattice Π has a maximal element, then Π is even a lattice.

2.4 ⬠ For $\alpha \in I(\Pi)$ with $\alpha(x, x) \neq 0$ for all $x \in \Pi$, explicitly construct the inverse $\alpha^{-1} \in I(\Pi)$.

2.5 ⬠ Let Π be a finite poset and recall that $\zeta = \delta + \eta$, where δ and η are defined by (2.1.1) and (2.1.2), respectively.

(a) Show that for $x \preceq y$,

$$\eta^k(x, y) = |\{x = x_0 \prec x_1 \prec x_2 \prec \cdots \prec x_{k-1} \prec x_k = y\}| ,$$

the number of *strict* chains of length k in the interval $[x, y]$.

(b) Infer that η is **nilpotent**, that is, $\eta^{k+1} \equiv 0$ for k the length of Π.

(c) For $x, y \in \Pi$, do you know what $(2\delta - \zeta)^{-1}(x, y)$ counts?

(d) Show that $\eta^n_{\mathcal{J}(\Pi)}(\varnothing, \Pi)$ equals the number of surjective order-preserving maps $\Pi \to [n]$.

2.6 ⬠ For posets $(\Pi_1, \preceq_1)$ and $(\Pi_2, \preceq_2)$, we define their **(direct) product** with underlying set $\Pi_1 \times \Pi_2$ and partial order

$$(x_1, x_2) \preceq (y_1, y_2) \quad :\Longleftrightarrow \quad x_1 \preceq_1 y_1 \text{ and } x_2 \preceq_2 y_2 .$$

(a) Show that every interval $[(x_1, x_2), (y_1, y_2)]$ of $\Pi_1 \times \Pi_2$ is of the form $[x_1, y_1] \times [x_2, y_2]$.

(b) Show that $\mu_{\Pi_1 \times \Pi_2}((x_1, x_2), (y_1, y_2)) = \mu_{\Pi_1}(x_1, y_1)\, \mu_{\Pi_2}(x_2, y_2)$.

(c) Show that the Boolean lattice B_n is isomorphic to the n-fold product of the chain $[2]$, and conclude that for $S \subseteq T \subseteq [n]$

$$\mu_{B_n}(S, T) = (-1)^{|T \setminus S|} .$$

2.7 (a) Let $\Pi = [d]$, the d-chain. Show that for $1 \le i < j \le d$

$$\mu_{[d]}(i,j) = \begin{cases} 1 & \text{if } i = j, \\ -1 & \text{if } i+1 = j, \\ 0 & \text{otherwise.} \end{cases}$$

(b) Write out the statement that Möbius inversion gives in this explicit case and interpret it along the lines of the Fundamental Theorem of Calculus.

(c) The Möbius function in number theory is the function $\mu : \mathbb{Z}_{>0} \to \mathbb{Z}$ defined for $n \in \mathbb{Z}_{>0}$ through $\mu(1) = 1$, $\mu(n) = 0$ if n is not *squarefree*, that is, if n is divisible by a proper prime power, and $\mu(n) = (-1)^r$ if n is the product of r distinct primes. Show that for given $n \in \mathbb{Z}_{>0}$, the partially ordered set D_n of divisors of n is isomorphic to a direct product of chains and use Exercise 2.6 to verify that $\mu(n) = \mu_{D_n}(1, n)$.

2.8 Consider the poset Π_d on $2d$ elements $a_1, a_2, \dots, a_d, b_1, b_2, \dots, b_d$, defined by the relations

$$a_1 \prec a_2 \prec \cdots \prec a_d \qquad \text{and} \qquad a_j \succ b_j \ \text{ for } 1 \le j \le d\,,$$

depicted in Figure 2.3.

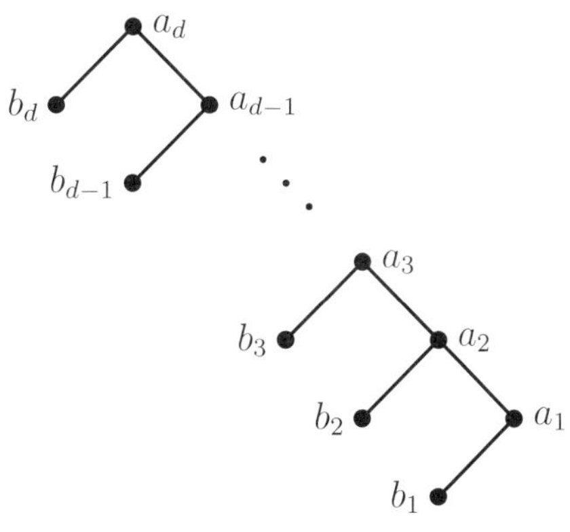

Figure 2.3. The poset of Exercise 2.8.

(a) Show that the number of linear extensions of Π_d is

$$(2d-1)!! \ := \ (2d-1)(2d-3)\cdots 3 \cdot 1\,.$$

(b) Show that the order polynomial satisfies the relation

$$\Omega_{\Pi_{d+1}}(n+1) \ = \ \Omega_{\Pi_d}(n) + (n+1)\,\Omega_{\Pi_{d-1}}(n)\,.$$

(c) The **Stirling numbers of the second kind** $S(n,k)$ count the number of partitions of n objects into k nonempty, unordered parts. Show that the following well-known recurrence holds:

$$S(n+1,k+1) \;=\; S(n,k) + (k+1)\,S(n,k+1)\,.$$

(d) Conclude that $\Omega_{\Pi_d}(n) = S(n+d,n)$.

(e) The **Stirling numbers of the first kind** $c(n,k)$ count the number of permutations of n objects having k cycles and they satisfy the recursion

$$c(n+1,k) \;=\; n\,c(n,k) + c(n,k-1)\,.$$

Show that $\Omega^{\circ}_{\Pi_d}(n) = c(n,n-d)$.

2.9 For fixed $k,n \in \mathbb{Z}_{>0}$ consider the map $g : B_k \to \mathbb{Z}_{>0}$ given by

$$g(T) \;=\; |T|^n.$$

Show that

$$k!\,S(n,k) \;=\; (g\,\mu_{B_k})([k])\,,$$

where $S(n,k)$ is the Stirling number of the second kind. (*Hint:* $k!\,S(n,k)$ counts surjective maps $[n] \to [k]$.)

2.10 ⬠ Compute the zeta polynomial of the poset D_{10} in Figure 1.12, appended by a maximal element.

2.11 Show that the zeta polynomial of the Boolean lattice B_d is $Z_{B_d}(n) = n^d$.

2.12 ⬠ Given a poset Π with $\hat{0}$ and $\hat{1}$, show that $Z_\Pi(n+1) - Z_\Pi(n)$ equals the number of multichains

$$\hat{0} \;=\; x_0 \;\preceq\; x_1 \;\preceq\; \cdots \;\preceq\; x_n \;\prec\; \hat{1}\,.$$

(*Hint:* Any multichain of length n yields a multichain of length $n+1$ by appending $\hat{1}$.)

2.13 ⬠ Prove Theorem 2.3.1: Every finite distributive lattice is isomorphic to a poset of order ideals of some poset. (*Hint:* Given a distributive lattice Π, consider the subposet Π' consisting of all **join irreducible** elements, i.e., those elements $a \neq \hat{0}$ that are not of the form $a = b \vee c$ for some $b, c \prec_\Pi a$. Show that Π is isomorphic to $\mathcal{J}(\Pi')$.)

2.14 State and prove a result analogous to Corollary 2.2.3 for distributive lattices.

2.15 Let Π be a finite graded poset that has a minimum $\hat{0}$ and a maximum $\hat{1}$, and define the **rank** $\mathrm{rk}_\Pi(x)$ of $x \in \Pi$ to be the length of $[\hat{0}, x]$, that is, the length of a saturated chain from $\hat{0}$ to x.

(a) Convince yourself that if y covers x, then $\mathrm{rk}_\Pi(y) = \mathrm{rk}_\Pi(x) + 1$.

(b) Prove that Π is Eulerian if and only if for all $x \prec y$ the interval $[x,y]$ has as many elements of even rank as of odd rank.

2.16 ☖ Prove (2.4.1): If $A_1, A_2, \ldots, A_n$ are finite sets, then

$$|A_1 \cup A_2 \cup \cdots \cup A_n| \;=\; \sum_{\varnothing \neq J \subseteq [n]} (-1)^{|I|-1} |A_J| \,,$$

where $A_J := \bigcap_{j \in J} A_j$.

2.17 Show that (2.4.4) defines a right action of $I(\Pi)$ on $\mathbb{C}^\Pi$. That is, $I(\Pi)$ gives rise to a vector space of linear transformations on $\mathbb{C}^\Pi$ and $(\alpha * \beta)f = \beta(\alpha f)$, for every $\alpha, \beta \in I(\Pi)$.

2.18 In this example you will encounter a prime application of the principle of inclusion–exclusion. Let $\mathfrak{S}_d$ be the set of bijections $\tau : [d] \to [d]$. An element $i \in [n]$ is a **fixed point** of τ if $\tau(i) = i$ and let

$$\mathrm{Fix}(\tau) \;:=\; \{i \,:\, \tau(i) = i\}$$

be the set of fixed points of τ. We wish to determine $d(n)$, the number of $\tau \in \mathfrak{S}_d$ such that are fixed-point free. This is called the **derangement number**. If $A_i = \{\tau : i \in \mathrm{Fix}(\tau)\}$, then

$$d(n) \;:=\; |\mathfrak{S}_d \setminus (A_1 \cup \cdots \cup A_d)| \,.$$

Determine $|A_I|$ for $I \subseteq [d]$ and use (2.4.2) to find a compact formula for $d(n)$.

2.19 ☖ Let S be a finite set. Show that

$$\sum_{T \subseteq S} (-1)^{|T|} \;=\; \begin{cases} 1 & \text{if } S = \varnothing, \\ 0 & \text{otherwise.} \end{cases}$$

Use this to complete the proof of Proposition 2.4.1.

2.20 ☖ Let $G = (V, E)$ be a simple graph and let $c : V \to [n]$ be a coloring. Prove that $F_G(c)$ defined in (2.4.5) is a flat and that, conversely, for each flat F there is a coloring c such that $F_G(c) = F$.

2.21 For a function $f : B_n \to \mathbb{C}$ define the multivariate polynomial

$$P_f(x_1, \ldots, x_n) \;=\; \sum_{I \subseteq [n]} f(I) \prod_{i \in I} x_i \,.$$

Show that

$$P_{f_\geq}(x_1, \ldots, x_n) \;=\; P_{f_=}(1 + x_1, 1 + x_2, \ldots, 1 + x_n) \,.$$

2.22 ☖ Prove Lemma 2.4.4: Let $\mathcal{L} = \mathcal{L}(A_1, \ldots, A_n; A)$ and $S \in \mathcal{L}$. If $A_I = S$ for $I \subseteq [n]$, then $I \subseteq J_S$. Moreover, for $T \in \mathcal{L}$

$$S \preceq T \quad\Longleftrightarrow\quad J_S \subseteq J_T \,.$$

Chapter 3

Polyhedral Geometry

One geometry cannot be more true than another; it can only be more convenient.
Jules Henri Poincaré

In this chapter we define the most convenient geometry for the combinatorial objects from Chapter 1. To give a first impression of how geometry naturally enters our combinatorial picture, we return to the problem of counting multisubsets of size d of $[n+1]$. Every such multiset corresponds to a d-tuple $(m_1+1, m_2+1, \dots, m_d+1) \in \mathbb{Z}^d$ such that

$$0 \le m_1 \le m_2 \le \cdots \le m_d \le n .$$

Forgetting about the integrality of the m_j gives a genuine geometric object containing the solutions to this system of $d+1$ linear inequalities:

$$n\,\triangle = \left\{ \mathbf{x} \in \mathbb{R}^d : 0 \le x_1 \le x_2 \le \cdots \le x_d \le n \right\}.$$

The d-multisubsets correspond exactly to the integer lattice points $n\,\triangle \cap \mathbb{Z}^d$. The set $n\,\triangle$ is a *polyhedron*: it is defined by finitely many linear inequalities. Polyhedra constitute a rich class of geometric objects—rich enough to capture much of the enumerative combinatorics that we pursue in this book.

Besides introducing machinery to handle polyhedra, our main emphasis in this chapter is on the *faces* of a given polyhedron. They form a poset that is naturally graded by dimension, and counting the faces in each dimension gives rise to the famous *Euler–Poincaré formula*. This identity is at play (often behind the scenes) in practically every combinatorial reciprocity theorem that we will encounter in later chapters.

3.1. Inequalities and Polyhedra

To help you ease into a geometric (rather than an algebraic) way of thinking, let's start over. A linear equation is of the form

$$a_1 x_1 + \cdots + a_d x_d \; = \; b \tag{3.1.1}$$

for some $a_1, \ldots, a_d, b \in \mathbb{R}$ and, as you will know, an integral part of Linear Algebra is to determine the set of solutions to systems of linear equations

$$\begin{aligned} a_{1,1} x_1 + \cdots + a_{1,d} x_d \; &= \; b_1 \\ a_{2,1} x_1 + \cdots + a_{2,d} x_d \; &= \; b_2 \\ &\vdots \\ a_{k,1} x_1 + \cdots + a_{k,d} x_d \; &= \; b_k \, . \end{aligned} \tag{3.1.2}$$

More compactly, we may write $\mathbf{A}\,\mathbf{x} = \mathbf{b}$, where $\mathbf{A} \in \mathbb{R}^{k \times d}$ is the matrix of coefficients and $\mathbf{b} \in \mathbb{R}^k$ collects the right-hand sides. The objects of interest in this chapter are the sets of solutions to finitely many linear *inequalities*: a **polyhedron** $\mathsf{Q} \subseteq \mathbb{R}^d$ is the set of solutions to a system

$$\begin{aligned} a_{1,1} x_1 + \cdots + a_{1,d} x_d \; &\le \; b_1 \\ a_{2,1} x_1 + \cdots + a_{2,d} x_d \; &\le \; b_2 \\ &\vdots \\ a_{k,1} x_1 + \cdots + a_{k,d} x_d \; &\le \; b_k \end{aligned} \tag{3.1.3}$$

for some $a_{ij}, b_i \in \mathbb{R}$ for $1 \le i \le k$ and $1 \le j \le d$. In compact form we can write

$$\mathsf{Q} \; = \; \left\{ \mathbf{x} \in \mathbb{R}^d \, : \, \mathbf{A}\,\mathbf{x} \le \mathbf{b} \right\} .$$

Note that an inequality involving $\ge$ still fits into the above form by simply multiplying both sides with -1. In particular $\mathbf{A}\,\mathbf{x} = \mathbf{b}$ is equivalent to $\mathbf{A}\,\mathbf{x} \le \mathbf{b}$ and $-\mathbf{A}\,\mathbf{x} \le -\mathbf{b}$, and so solution sets to linear equations partake in this endeavor. For example, for $n = 1$ and $d = 3$, the polyhedron $\triangle$ from the chapter prelude is the set of solutions to

$$\begin{bmatrix} -1 & & \\ 1 & -1 & \\ & 1 & -1 \\ & & 1 \end{bmatrix} \mathbf{x} \; \le \; \begin{bmatrix} 0 \\ 0 \\ 0 \\ 1 \end{bmatrix} . \tag{3.1.4}$$

This is illustrated in Figure 3.1. There are many systems of linear inequalities that yield the same polyhedron Q. We call Q a **rational** polyhedron if $\mathbf{A}$ and $\mathbf{b}$ can be chosen over the rational numbers.

What you might not be as familiar with is a geometric perspective on linear systems of equations. Borrowing from our geometric intuition in three

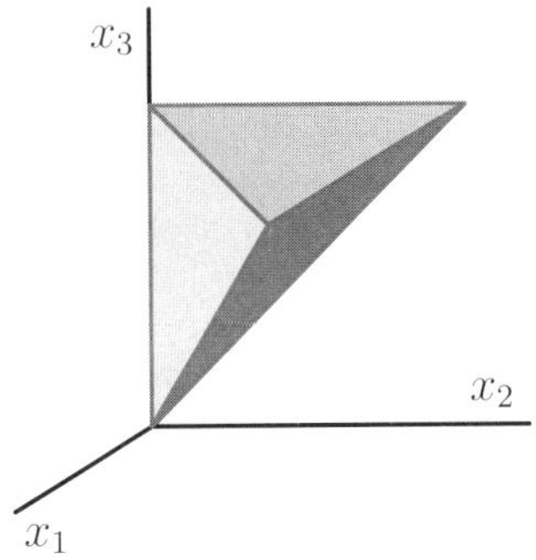

Figure 3.1. The polyhedron $\triangle$ given in (3.1.4).

dimensions, we call the set of solutions $\mathsf{H} \subset \mathbb{R}^d$ to a single linear equation (3.1.1) an **(affine) hyperplane**, provided $a_i \neq 0$ for some i. That is,

$$\mathsf{H} := \left\{ \mathbf{x} \in \mathbb{R}^d : \langle \mathbf{a}, \mathbf{x} \rangle = b \right\} \tag{3.1.5}$$

for some normal $\mathbf{a} \in \mathbb{R}^d \setminus \{\mathbf{0}\}$ and displacement $b \in \mathbb{R}$. Here $\langle\ ,\ \rangle$ denotes the standard inner product on $\mathbb{R}^d$, but any other one works just as well. We call H a **linear hyperplane** if $\mathbf{0} \in \mathsf{H}$ or, equivalently, $b = 0$. Hence, every affine hyperplane is of the form $\mathsf{H} = \mathbf{p} + \mathsf{H}_0$, where H_0 is a linear hyperplane and $\mathbf{p}$ is a base point. Thus, the set of solutions $\mathsf{L} \subseteq \mathbb{R}^d$ to (3.1.2), called an **affine subspace**, is of the form

$$\mathsf{L} = \mathsf{H}_1 \cap \mathsf{H}_2 \cap \cdots \cap \mathsf{H}_k ,$$

where H_i is the hyperplane defined by the i-th linear equation. By Exercise 3.1, either $\mathsf{L} = \varnothing$ or $\mathsf{L} = \mathbf{p} + \mathsf{L}_0$, where L_0 is a **linear subspace**, i.e., an intersection of linear hyperplanes, and $\mathbf{p}$ is a suitable translation.

The presentation of a hyperplane H given in (3.1.5) actually defines an **oriented** hyperplane, in the following sense. The two connected components of $\mathbb{R}^d \setminus \mathsf{H}$ are called **(open) halfspaces** and we can use the orientation to distinguish the **closed** halfspaces associated with H as

$$\begin{aligned} \mathsf{H}^{\geq} &:= \left\{ \mathbf{x} \in \mathbb{R}^d : \langle \mathbf{a}, \mathbf{x} \rangle \geq b \right\}, \\ \mathsf{H}^{\leq} &:= \left\{ \mathbf{x} \in \mathbb{R}^d : \langle \mathbf{a}, \mathbf{x} \rangle \leq b \right\}. \end{aligned} \tag{3.1.6}$$

Hence, a polyhedron $\mathsf{Q} \subseteq \mathbb{R}^d$ is the intersection of finitely many closed halfspaces

$$\mathsf{Q} = \mathsf{H}_1^{\leq} \cap \cdots \cap \mathsf{H}_k^{\leq} = \left\{ \mathbf{x} \in \mathbb{R}^d : \langle \mathbf{a}_i, \mathbf{x} \rangle \leq b_i \text{ for } 1 \leq i \leq k \right\}, \tag{3.1.7}$$

such as the one shown in Figure 3.2. We remark that, trivially, all affine

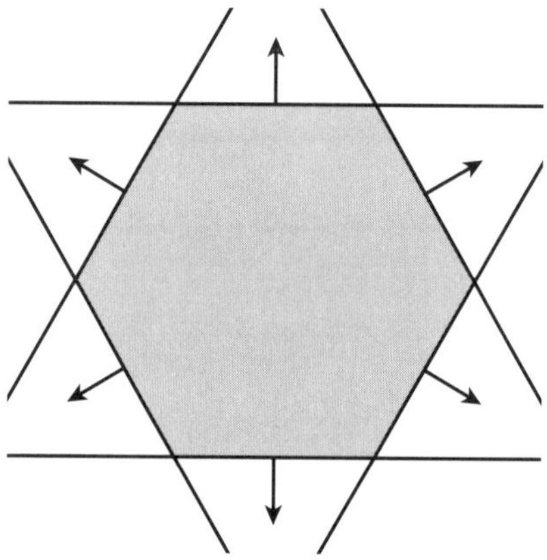

Figure 3.2. A bounded polyhedron in the plane. The arrows indicate the orientation of the hyperplanes.

subspaces, including $\mathbb{R}^d$ and $\varnothing$, are polyhedra. We call a polyhedron Q **proper** if it is not an affine space. For example,

$$\left\{ \mathbf{x} \in \mathbb{R}^3 : \begin{bmatrix} 1 & 1 & 0 \\ -1 & -1 & 0 \\ -1 & 0 & 0 \\ 0 & -1 & 0 \\ 0 & 0 & -1 \\ 0 & 0 & 1 \end{bmatrix} \begin{bmatrix} x_1 \\ x_2 \\ x_3 \end{bmatrix} \leq \begin{bmatrix} 1 \\ -1 \\ 0 \\ 0 \\ 0 \\ 1 \end{bmatrix} \right\} \tag{3.1.8}$$

is a square embedded in $\mathbb{R}^3$ which is pictured in Figure 3.3.

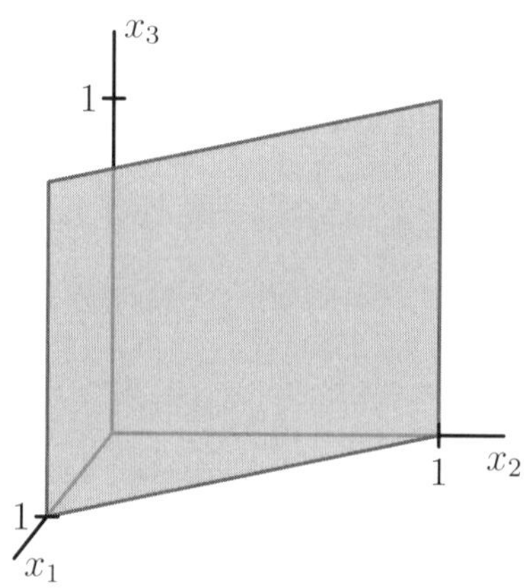

Figure 3.3. The square defined by (3.1.8).

A halfspace $\mathsf{H}_j^{\leq}$ (or linear inequality) is **irredundant** if

$$\bigcap_{i \neq j} \mathsf{H}_i^{\leq} \neq \mathsf{Q}\,.$$

You are welcome to verify that the four inequalities for $\triangle$ given in (3.1.4) are all irredundant. Of course, discarding redundant halfspaces one at a time leaves us with an irredundant presentation of a given polyhedron. Exercise 3.2 shows that, in general, the number of necessary halfspaces can depend on the order in which the given halfspaces are inspected. This is just one of the many situations that sets apart the study of linear inequalities systems from that of systems of linear equations.

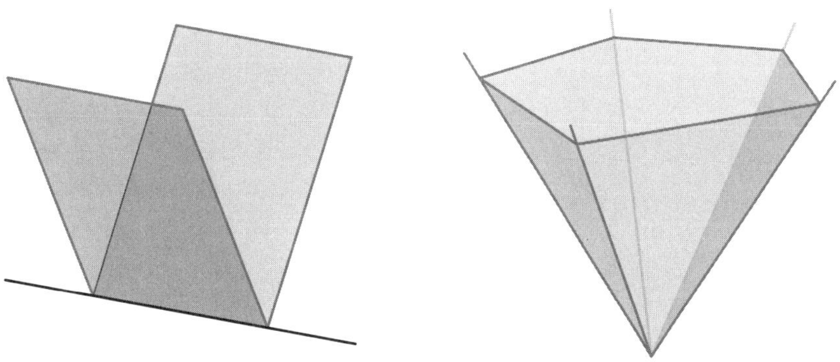

Figure 3.4. Two polyhedral cones, one of which is line free.

A polyhedron $\mathsf{C} \subseteq \mathbb{R}^d$ is a **polyhedral cone** if $\mu \mathbf{p} \in \mathsf{C}$ for any $\mathbf{p} \in \mathsf{C}$ and $\mu \geq 0$. See Figure 3.4 for two examples. In particular, every linear subspace is a polyhedral cone and Exercise 3.3 asks you to prove the following result.

Proposition 3.1.1. *A polyhedron $\mathsf{Q} \subseteq \mathbb{R}^d$ is a polyhedral cone if and only if it is of the form*

$$\mathsf{Q} \;=\; \{\mathbf{x} \in \mathbb{R}^d \;:\; \mathbf{A}\,\mathbf{x} \leq \mathbf{0}\}$$

for some matrix $\mathbf{A} \in \mathbb{R}^{k \times d}$, that is, Q is the intersection of finitely many linear halfspaces.

Except for $\mathsf{C} = \{\mathbf{0}\}$, polyhedral cones are examples of **unbounded** polyhedra.

The **recession cone** $\operatorname{rec}(\mathsf{Q})$ of a polyhedron $\mathsf{Q} \subseteq \mathbb{R}^d$ is the collection of directions in which to escape to infinity. More formally,

$$\operatorname{rec}(\mathsf{Q}) \;:=\; \left\{\mathbf{u} \in \mathbb{R}^d : \mathbf{p} + \mathbb{R}_{\geq 0}\,\mathbf{u} \subseteq \mathsf{Q} \text{ for some } \mathbf{p} \in \mathsf{Q}\right\}.$$

Here and in the future, we write $\mathbb{R}_{\geq 0}\mathbf{u}$ for the set $\{\lambda\mathbf{u} : \lambda \geq 0\}$. Figure 3.5 shows an unbounded polyhedron and its recession cone.

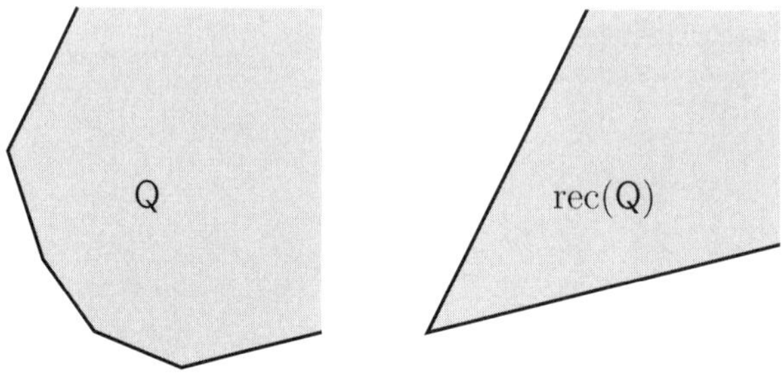

Figure 3.5. An unbounded 2-dimensional polyhedron Q and its recession cone.

That rec(Q) is a polyhedral cone and that, in fact, $\mathbf{p} + \text{rec}(Q) \subseteq Q$ for all $\mathbf{p} \in Q$ is the content of Exercise 3.4. As per Exercise 3.5, the following holds.

Proposition 3.1.2. *A nonempty polyhedron* $Q \subseteq \mathbb{R}^d$ *is bounded if and only if* $\text{rec}(Q) = \{\mathbf{0}\}$.

The relationship between general polyhedra and polyhedral cones is similar to that of affine and linear subspaces. For polyhedral cones as well as linear subspaces, the origin plays a distinguished role. Moreover, there is a natural construction that allows us to pass from polyhedra to polyhedral cones and back. For a closed set $S \subset \mathbb{R}^d$, we define its **homogenization** $\text{hom}(S)$ as the closure of the set

$$\left\{(\mathbf{x}, \lambda) \in \mathbb{R}^{d+1} \, : \, \lambda \geq 0, \; \mathbf{x} \in \lambda S\right\}. \tag{3.1.9}$$

In particular, the homogenization of a nonempty polyhedron $Q = \{\mathbf{x} \in \mathbb{R}^d : \mathbf{A}\,\mathbf{x} \leq \mathbf{b}\}$ is the polyhedral cone

$$\text{hom}(Q) \;=\; \left\{(\mathbf{x}, t) \in \mathbb{R}^{d+1} \, : \, t \geq 0, \; \mathbf{A}\,\mathbf{x} - t\mathbf{b} \leq \mathbf{0}\right\}.$$

For example, the homogenization of a pentagon is shown in Figure 3.6, and the homogenization of $\triangle$ in (3.1.4) is

$$\text{hom}(\triangle) \;=\; \left\{(\mathbf{x}, t) \in \mathbb{R}^4 \, : \, 0 \leq x_1 \leq x_2 \leq x_3 \leq t\right\}.$$

We can recover our polyhedron Q from its homogenization as the set of those points $\mathbf{y} \in \text{hom}(Q)$ for which $y_{d+1} = 1$ and rec(Q) is linearly isomorphic to $\text{hom}(Q) \cap \{y_{d+1} = 0\}$. (Two polyhedra Q and Q′ are **linearly isomorphic** if there is an invertible affine transformation mapping Q to Q′.)

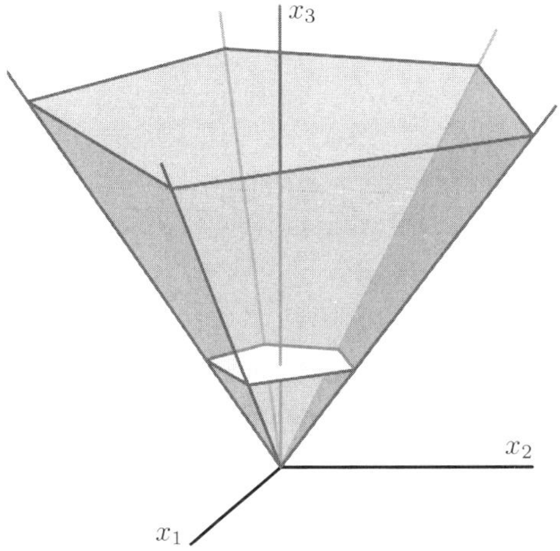

Figure 3.6. The homogenization of a pentagon.

Homogenization seems like a simple construction but it will come in quite handy in this and later chapters. For example, let

$$\mathsf{T} = \left\{ \mathbf{y} \in \mathbb{R}^3 : \mathbf{y} \geq 0,\ y_1 + y_2 + y_3 \leq 1 \right\}.$$

Then

$$\mathbb{R}^4_{\geq 0} \cong \operatorname{hom}(\mathsf{T}) \cong \operatorname{hom}(\triangle),$$

where we write $\mathbb{R}_{\geq 0} := \{a \in \mathbb{R} : a \geq 0\}$. Related to the recession cone is the **lineality space** $\operatorname{lineal}(\mathsf{Q})$ of a polyhedron Q. It is the inclusion-maximal linear subspace $\mathsf{L} \subseteq \mathbb{R}^d$ such that $\mathbf{p} + \mathsf{L} \subseteq \mathsf{Q}$ for some $\mathbf{p} \in \mathsf{Q}$. You are invited to prove the following proposition in Exercise 3.8.

Proposition 3.1.3. *Let* $\mathsf{Q} = \{\mathbf{x} \in \mathbb{R}^d : \mathbf{A}\,\mathbf{x} \leq \mathbf{b}\}$ *be a nonempty polyhedron. Then*

$$\operatorname{lineal}(\mathsf{Q}) = \operatorname{rec}(\mathsf{Q}) \cap (-\operatorname{rec}(\mathsf{Q})) = \left\{ \mathbf{x} \in \mathbb{R}^d : \mathbf{A}\,\mathbf{x} = \mathbf{0} \right\}.$$

In particular $\mathbf{p} + \operatorname{lineal}(\mathsf{Q}) \subseteq \mathsf{Q}$ *for all* $\mathbf{p} \in \mathsf{Q}$.

For example,

$$\left\{ \mathbf{x} \in \mathbb{R}^3 : \begin{bmatrix} 1 & -1 & -1 \\ -1 & -1 & 1 \end{bmatrix} \begin{bmatrix} x_1 \\ x_2 \\ x_3 \end{bmatrix} \leq \begin{bmatrix} 0 \\ 0 \end{bmatrix} \right\}$$

is a wedge with lineality space $\{\mathbf{x} \in \mathbb{R}^3 : x_1 = x_3,\ x_2 = 0\}$, which is a line. (A picture of this wedge, which we encourage you to draw, should look a bit like the left side of Figure 3.4.) We call a polyhedron **line free** if $\operatorname{lineal}(\mathsf{Q}) = \{\mathbf{0}\}$. Exercise 3.10 yields that a polyhedral cone C is line free if and only if $\mathbf{p}, -\mathbf{p} \in \mathsf{C}$ implies $\mathbf{p} = \mathbf{0}$. Hence, Q is line free if and only if $\hom(\mathsf{Q})$ is line free.

As in all geometric disciplines, a fundamental invariant of an object is its dimension. For a linear subspace L we know how to define $\dim \mathsf{L}$—courtesy of Linear Algebra—and, since the dimension should be independent of translation, this yields the dimension of any affine subspace $\mathsf{L} = \mathbf{p} + \mathsf{L}_0$. If $\mathsf{L} = \varnothing$, then we set $\dim \mathsf{L} := -1$. For a set $S \subseteq \mathbb{R}^d$, we define its **affine hull** $\operatorname{aff}(S)$ as the inclusion-minimal affine subspace of $\mathbb{R}^d$ that contains S

$$\operatorname{aff}(S) \;=\; \bigcap \left\{ \mathsf{H} \text{ hyperplane in } \mathbb{R}^d \;:\; S \subseteq \mathsf{H} \right\}. \tag{3.1.10}$$

We define the **dimension** of a polyhedron Q as $\dim \mathsf{Q} := \dim \operatorname{aff}(\mathsf{Q})$. When $\dim \mathsf{Q} = n$, we call Q an n-**polyhedron**. For example, the square defined by (3.1.8) has affine hull $\{\mathbf{x} \in \mathbb{R}^3 \,:\, x_1 + x_2 = 1\}$ and so (surprise!) its dimension is 2.

To justify this convention, we note in Exercise 3.11 that the **(topological) interior** of a polyhedron Q given in the form (3.1.7) is

$$\left\{ \mathbf{x} \in \mathbb{R}^d \;:\; \langle \mathbf{a}_i, \mathbf{x} \rangle < b_i \text{ for all } 1 \le i \le k \right\}. \tag{3.1.11}$$

However, this notion of *interior* is not intrinsic to Q but makes reference to the ambient space $\mathbb{R}^d$. For example, a triangle might or might not have an interior depending on whether we embed it in $\mathbb{R}^2$ or $\mathbb{R}^3$. Luckily, every polyhedron comes with a canonical embedding into its affine hull and we can define the **relative interior** of Q as the set of points of Q that are in the interior of Q relative to its embedding into $\operatorname{aff}(\mathsf{Q})$. Thus, $\operatorname{aff}(\mathsf{Q})$ is the affine subspace relative to which Q has a nonempty interior and this explains our definition of dimension. We will denote the relative interior of Q by Q°.[1] When Q is full dimensional, Q° is given by (3.1.11). In the case that Q is not full dimensional, we have to be a bit more careful (the details are the content of Exercise 3.12): assuming Q is given in the form (3.1.7), let $I := \{i \in [k] : \langle \mathbf{a}_i, \mathbf{x} \rangle = b_i \text{ for all } \mathbf{x} \in \mathsf{Q}\}$. Then

$$\begin{aligned} \mathsf{Q}^\circ \;&=\; \{\mathbf{x} \in \mathsf{Q} \;:\; \langle \mathbf{a}_i, \mathbf{x} \rangle < b_i \text{ for all } i \notin I\} \\ &=\; \operatorname{aff}(\mathsf{Q}) \cap \left\{ \mathbf{x} \in \mathbb{R}^d \;:\; \langle \mathbf{a}_i, \mathbf{x} \rangle < b_i \text{ for all } i \notin I \right\}. \end{aligned} \tag{3.1.12}$$

[1] A note on terminology: A polyhedron is, by definition, closed. However, we will sometimes talk about an *open polyhedron*, by which we mean the relative interior of a polyhedron. In a few instances we will simultaneously deal with polyhedra and open polyhedra, in which case we may use the superfluous term *closed polyhedron* to distinguish one from the other.

For instance, our running example, the square defined by (3.1.8), has relative interior

$$\left\{ \mathbf{x} \in \mathbb{R}^3 \, : \, x_1 + x_2 = 1, \, \begin{bmatrix} -1 & 0 & 0 \\ 0 & -1 & 0 \\ 0 & 0 & -1 \\ 0 & 0 & 1 \end{bmatrix} \begin{bmatrix} x_1 \\ x_2 \\ x_3 \end{bmatrix} < \begin{bmatrix} 0 \\ 0 \\ 0 \\ 1 \end{bmatrix} \right\}.$$

The **(relative) boundary** of Q is

$$\partial \mathsf{Q} \;:=\; \mathsf{Q} \setminus \mathsf{Q}^\circ \,.$$

With these definitions at hand, you will discover that many objects throughout mathematics turn out to be polyhedra. This is obvious for the **unit cube**

$$[0,1]^d \;=\; \left\{ \mathbf{x} \in \mathbb{R}^d : 0 \le x_i \le 1 \text{ for all } 1 \le i \le d \right\}$$

or its centrally-symmetric counterpart $[-1,1]^d = \mathbf{1} - 2[0,1]^d$. The latter is the unit ball in the ℓ_∞-norm and Exercise 3.13 concerns the ℓ_1-norm unit ball, the **cross polytope**

$$\Diamond_d \;:=\; \left\{ \mathbf{x} \in \mathbb{R}^d \, : \, \|\mathbf{x}\|_1 = |x_1| + \cdots + |x_d| \le 1 \right\}. \tag{3.1.13}$$

The 3-dimensional instance is pictured in Figure 3.7.

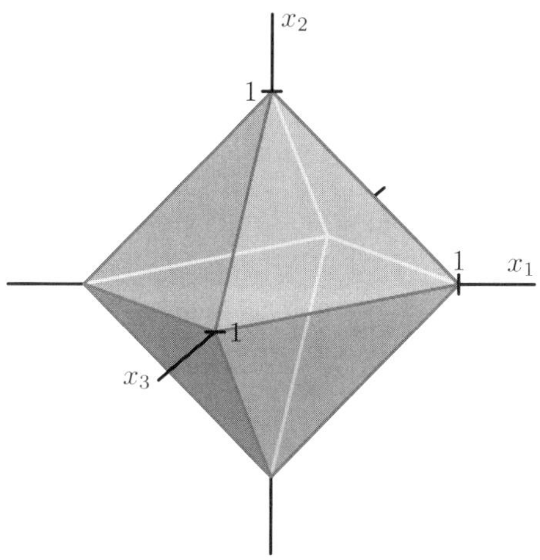

Figure 3.7. The 3-dimensional cross polytope.

You might have also noticed that the ideas from the prelude to this chapter can be generalized. For example, k-subsets S of $[n]$ not containing two consecutive numbers correspond precisely to the lattice points $\mathbf{x} \in \mathbb{Z}^k$ satisfying

$$0 < x_1 \quad \text{and} \quad x_i + 1 < x_{i+1} \text{ (for all } 1 \le i < k) \quad \text{and} \quad x_k < n+1 \,,$$

which are the lattice points in the interior of a bounded polyhedron in $\mathbb{R}^k$. This yields the geometric perspective that this book is set out to promote but, so far, it gives a description only in geometric terms. In the following section, we will see that polyhedra also yield a *generative* description, a way to intrinsically describe points in polyhedra, akin to presenting points in linear subspaces as linear combinations.

3.2. Polytopes, Cones, and Minkowski–Weyl

The term *polyhedron* appears in many parts of mathematics, unfortunately with different connotations. We should have been more careful in the previous section where we actually defined **convex polyhedra**. A set $S \subseteq \mathbb{R}^d$ is **convex** if for every $\mathbf{p}, \mathbf{q} \in S$, the line segment

$$[\mathbf{p}, \mathbf{q}] \;:=\; \{(1-\lambda)\,\mathbf{p} + \lambda\,\mathbf{q} \,:\, 0 \le \lambda \le 1\}$$

with endpoints $\mathbf{p}$ and $\mathbf{q}$ is contained in S. The intersection of any collection of convex sets is again convex and since halfspaces are convex, our polyhedra Q as defined via (3.1.7) are closed convex sets. As we won't be dealing with nonconvex polyhedra, we can safely drop the adjective *convex* and continue to refer to Q as a *polyhedron*.

For any set $S \subseteq \mathbb{R}^d$, there is a unique inclusion-minimal convex set $\operatorname{conv}(S)$ containing S, called the **convex hull** of S. The convex hull is simply the intersection of all convex sets containing S which, by Exercise 3.14, can be written as

$$\operatorname{conv}(S) \;=\; \left\{ \lambda_1\mathbf{v}_1 + \cdots + \lambda_k\mathbf{v}_k \,:\, \begin{array}{c} k \ge 0,\ \mathbf{v}_1, \dots, \mathbf{v}_k \in S \\ \lambda_1, \dots, \lambda_k \ge 0 \\ \lambda_1 + \cdots + \lambda_k = 1 \end{array} \right\}. \tag{3.2.1}$$

We will be mostly interested in the situation when S is finite: a convex set P is a **(convex) polytope** if $\mathsf{P} = \operatorname{conv}(S)$ for some finite set $S \subset \mathbb{R}^d$. Figure 3.8 illustrates the concept. We call P a **rational polytope** or **lattice polytope** whenever we can choose S in $\mathbb{Q}^d$ or $\mathbb{Z}^d$, respectively.

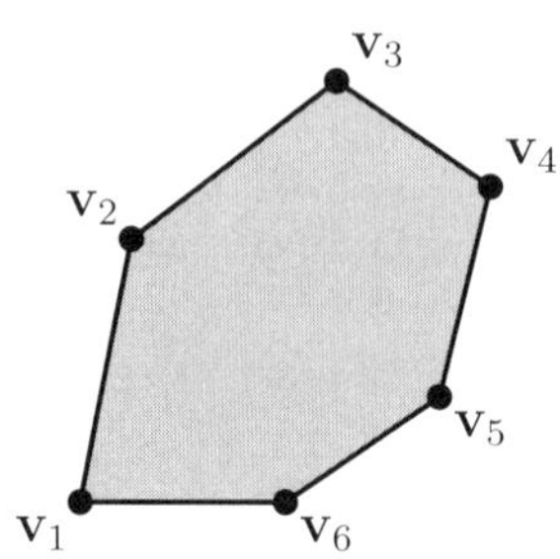

Figure 3.8. The convex hull of six points in the plane.

Analogously to the situation with polyhedra, we call a point $\mathbf{v} \in S$ a **vertex** of $\mathsf{P} = \text{conv}(S)$ if $\text{conv}(S \setminus \{\mathbf{v}\}) \neq \mathsf{P}$. Exercise 3.15 helps you to conclude that there is a unique inclusion-minimal set $V \subseteq S$ such that $\mathsf{P} = \text{conv}(V)$. We call V the **vertex set** of P and write $\text{vert}(\mathsf{P}) := V$.

A **convex cone** is a nonempty convex set $\mathsf{C} \subseteq \mathbb{R}^d$ such that $\mu\mathsf{C} \subseteq \mathsf{C}$ for all $\mu \geq 0$. Equivalently, a nonempty set C is a convex cone provided $\mu\mathbf{p} + \lambda\mathbf{q} \in \mathsf{C}$ for all $\mathbf{p}, \mathbf{q} \in \mathsf{C}$ and $\mu, \lambda \in \mathbb{R}_{\geq 0}$. Again, convex cones form a family of sets that is closed under intersection and we are therefore led to define the **conical hull** $\text{cone}(S)$ of a set S, which is the inclusion-minimal convex cone containing S. We call a convex cone C **finitely generated** provided $\mathsf{C} = \text{cone}(S)$ for some finite set S. If $S = \{\mathbf{s}_1, \mathbf{s}_2, \dots, \mathbf{s}_k\}$, we will also write

$$\text{cone}(S) = \mathbb{R}_{\geq 0}\,\mathbf{s}_1 + \mathbb{R}_{\geq 0}\,\mathbf{s}_2 + \cdots + \mathbb{R}_{\geq 0}\,\mathbf{s}_k$$

in accordance with Exercise 3.19. We call a finitely generated cone C **rational** if we can choose $S \subset \mathbb{Q}^d$ or, equivalently, $S \subset \mathbb{Z}^d$. An inclusion-minimal set $U \subseteq S$ such that $\mathsf{C} = \text{cone}(U)$ is called a set of **generators** for C. Similar to the case of polyhedra, U is typically not unique; see Exercise 3.17. However, there is an important class of cones for which we have uniqueness up to scaling.[2] A finitely generated convex cone C is **pointed** if there is some $\mathbf{w} \in \mathbb{R}^d$ such that

$$\langle \mathbf{w}, \mathbf{p} \rangle > 0 \quad \text{for all } \mathbf{p} \in \mathsf{C} \setminus \{\mathbf{0}\}\,. \tag{3.2.2}$$

Proposition 3.2.1. *Let $\mathsf{C} \subset \mathbb{R}^d$ be a finitely generated convex cone. If C is pointed, then C has a unique set of generators up to scaling.*

Proof. Set $\mathsf{H} := \{\mathbf{x} \in \mathbb{R}^d : \langle \mathbf{w}, \mathbf{x} \rangle = 1\}$. For every $\mathbf{p} \in \mathsf{C}$, there is a unique $\mu_{\mathbf{p}} > 0$ such that $\mu_{\mathbf{p}}\mathbf{p} \in \mathsf{H}$. Hence, every nonzero point in C has a representative in the convex set $\mathsf{P} := \mathsf{C} \cap \mathsf{H}$ or, said differently, $\mathsf{C} = \text{cone}(\mathsf{P})$.

Now let $\mathsf{C} = \mathbb{R}_{\geq 0}\,\mathbf{s}_1 + \cdots + \mathbb{R}_{\geq 0}\,\mathbf{s}_k$ for some $\mathbf{s}_1, \dots, \mathbf{s}_k \in \mathsf{C} \setminus \{\mathbf{0}\}$. We can assume that $\langle \mathbf{w}, \mathbf{s}_i \rangle = 1$ for all i and hence $\mathsf{P} = \text{conv}(\mathbf{s}_1, \dots, \mathbf{s}_k)$ is a polytope. It follows that every set of generators contains the vertices of P up to scaling and the claim follows. □

The link between finitely generated pointed cones and polytopes is reminiscent of polyhedra and polyhedral cones. Exercise 3.18 yields that the homogenization of a polytope $\mathsf{P} \subset \mathbb{R}^d$ is the pointed cone

$$\text{hom}(\mathsf{P}) = \text{cone}(\mathsf{P} \times \{1\}) = \text{cone}\{(\mathbf{v}, 1) : \mathbf{v} \in \text{vert}(\mathsf{P})\}\,. \tag{3.2.3}$$

In particular $\{(\mathbf{v}, 1) : \mathbf{v} \in \text{vert}(\mathsf{P})\}$ is a set of generators for $\text{hom}(\mathsf{P})$ and, as above, the polytope $\mathsf{P} \subset \mathbb{R}^d$ can be recovered by intersecting $\text{hom}(\mathsf{P}) \subset \mathbb{R}^{d+1}$ with the hyperplane $\{\mathbf{x} \in \mathbb{R}^{d+1} : x_{d+1} = 1\}$.

---

[2]Naturally, if $\mathsf{C} = \text{cone}(\mathbf{s}_1, \dots, \mathbf{s}_k)$, then $\mathsf{C} = \text{cone}(\mu_1\mathbf{s}_1, \dots, \mu_k\mathbf{s}_k)$ for any $\mu_1, \dots, \mu_k > 0$.

We can extend the notion of lineality space from polyhedra to general convex sets and to (finitely generated) convex cones. In particular, if $\mathrm{lineal}(\mathsf{C}) \neq \{\mathbf{0}\}$, then C cannot be pointed. Indeed, if $\mathbf{p} \in \mathrm{lineal}(\mathsf{C}) \setminus \{\mathbf{0}\}$, then $\pm\mathbf{p} \in \mathsf{C}$ and hence (3.2.2) cannot hold. In fact, the converse also holds.

Proposition 3.2.2. *Let $\mathsf{C} \subset \mathbb{R}^d$ be a convex cone. Then C is pointed if and only if C is line free.*

The key is the following central **separation theorem** from convex geometry.

Theorem 3.2.3. *Let $\mathsf{K} \subset \mathbb{R}^d$ be a closed convex set and $\mathbf{p} \in \mathbb{R}^d \setminus \mathsf{K}$. Then there is a hyperplane $\mathsf{H} = \{\mathbf{x} \in \mathbb{R}^d : \langle \mathbf{w}, \mathbf{x} \rangle = \delta\}$ such that $\mathsf{K} \subset \mathsf{H}^< := \mathsf{H}^\leq \backslash \mathsf{H}$ and $\mathbf{p} \in \mathsf{H}^>$.*

Such a hyperplane H is called a **separating hyperplane**. We defer the not-that-difficult proof to Exercise 3.21.

Proof of Proposition 3.2.2. Let $\mathsf{C} = \mathrm{cone}(\mathbf{s}_1, \dots, \mathbf{s}_k)$. By Exercise 3.20, C is line free if and only if the point $(\mathbf{0}, 1)$ is not contained in $\mathrm{hom}(\mathsf{C})$. By Theorem 3.2.3, this is equivalent to the existence of $(\mathbf{w}, w_{d+1}) \in \mathbb{R}^{d+1}$ and $\delta \in \mathbb{R}$ such that $\langle (\mathbf{w}, w_{d+1}), (\mathbf{0}, 1) \rangle = w_{d+1} < \delta$ and

$$\begin{aligned} \langle (\mathbf{w}, w_{d+1}), (\mathbf{s}_i, 1) \rangle &> \delta \\ \iff \quad \langle \mathbf{w}, \mathbf{s}_i \rangle &> \delta - w_{d+1} > 0 \end{aligned}$$

for all $i = 1, \dots, k$. This implies that C is pointed. □

We observe what we have actually done in the proof of Proposition 3.2.2. To a finitely generated cone $\mathsf{C} = \mathrm{cone}(\mathbf{s}_1, \dots, \mathbf{s}_k)$, we have associated a polyhedral cone

$$\mathsf{C}^\vee := \left\{ \mathbf{w} \in \mathbb{R}^d : \langle \mathbf{s}_i, \mathbf{w} \rangle \geq 0 \text{ for } i = 1, \dots, k \right\}.$$

The statement that we have shown is that C is line free if and only if $\mathsf{C}^\vee$ is full dimensional (and hence has a nonempty interior). The polyhedral cone $\mathsf{C}^\vee$ is called the cone **polar** to C and Exercise 3.22 explores more of this. We pause for a second to compare (bounded) polyhedra and polytopes to each other. Both are defined in terms of finite data; inequalities for the one, points for the other class. The fundamental difference is that (3.2.1) gives a direct mean to access all points in a polytope. This is quite different for polyhedra. On the other hand, the description of a polyhedron $\mathsf{Q} \subset \mathbb{R}^d$ in terms of inequalities gives a simple way to check if a given point $\mathbf{q} \in \mathbb{R}^d$ is contained in Q or not; for polytopes, this is by far not as straightforward (and, in fact, requires us to determine if a system of linear inequalities has a solution). We will soon see that these are the two sides of the same coin, but first we look at an example: considering the bounded polyhedron

$$\triangle = \left\{ \mathbf{x} \in \mathbb{R}^d : 0 \le x_1 \le x_2 \le \cdots \le x_d \le 1 \right\},$$

we define

$$\begin{aligned} \mathbf{u}_0 &= (1,1,1,\ldots,1), \\ \mathbf{u}_1 &= (0,1,1,\ldots,1), \\ \mathbf{u}_2 &= (0,0,1,\ldots,1), \\ &\vdots \\ \mathbf{u}_d &= (0,0,0,\ldots,0). \end{aligned}$$

It is evident that $\mathbf{u}_0, \ldots, \mathbf{u}_d \in \triangle$ and, by virtue of convexity, it follows that $\operatorname{conv}(\mathbf{u}_0, \ldots, \mathbf{u}_d) \subseteq \triangle$. We claim that this is actually an equality. Indeed, for a point $\mathbf{p} = (p_1, p_2, \ldots, p_d) \in \triangle$, we define $\lambda_i := p_{i+1} - p_i$ for $0 \le i \le d$, where we set $p_0 := 0$ and $p_{d+1} = 1$. With this, we observe that $\lambda_0, \ldots, \lambda_d \ge 0$ and

$$\mathbf{p} = \sum_{i=0}^{d} \lambda_i \mathbf{u}_i \quad \text{and} \quad \sum_{i=0}^{d} \lambda_i = p_{d+1} - p_0 = 1. \tag{3.2.4}$$

Thus, $\triangle$ is both a polyhedron and a (lattice) polytope. In fact, $\triangle$ belongs to a particular family of polytopes which we now introduce.

We recall that a collection of points $\mathbf{p}_0, \mathbf{p}_1, \ldots, \mathbf{p}_k$ is **affinely independent** if one of the following equivalent conditions holds:

i) if $\mu_0, \mu_1, \ldots, \mu_k \in \mathbb{R}$ satisfy

$$\sum_{i=0}^{k} \mu_i \, \mathbf{p}_i = 0 \quad \text{and} \quad \sum_{i=0}^{k} \mu_i = 0,$$

then $\mu_0 = \mu_1 = \cdots = \mu_k = 0$;

ii) the vectors

$$\begin{pmatrix} \mathbf{p}_0 \\ 1 \end{pmatrix} \begin{pmatrix} \mathbf{p}_1 \\ 1 \end{pmatrix} \cdots \begin{pmatrix} \mathbf{p}_k \\ 1 \end{pmatrix}$$

are linearly independent;

iii) the vectors $\mathbf{p}_1 - \mathbf{p}_0, \ldots, \mathbf{p}_k - \mathbf{p}_0$ are linearly independent.

Exercise 3.23 asks you to verify these equivalences and Exercise 3.24 gives a bit of context. If $\mathbf{p}_0, \ldots, \mathbf{p}_k \in \mathbb{R}^d$ are affinely independent points, then $\mathsf{P} = \operatorname{conv}(\mathbf{p}_0, \ldots, \mathbf{p}_k)$ is called a **simplex**. For $k = 0, 1, 2, 3$, this is a point, a segment, a triangle, and a tetrahedron, respectively. Using the affine hull, we can extend the notion of **dimension** to polytopes which helps us verify that $\operatorname{conv}(\mathbf{p}_0, \ldots, \mathbf{p}_k)$ is a polytope of dimension k. We can characterize simplices also in terms of their homogenizations: we call a cone $\mathsf{C} = \operatorname{cone}(\mathbf{s}_1, \ldots, \mathbf{s}_k)$

simplicial if its set of generators $\mathbf{s}_1, \dots, \mathbf{s}_k$ are linearly independent. In particular, it follows from (3.2.3) that P is a simplex if and only if $\hom(\mathsf{P})$ is simplicial. Exercise 3.16 together with the fact that $\triangle$ is a polytope as well as a polyhedron proves the next result.

Proposition 3.2.4. *Every simplex is a polyhedron. Likewise, every simplicial cone is a polyhedron.*

Polyhedra can be unbounded and thus it cannot be true that every polyhedron is also a polytope. To fix this, we need the following notion: the **Minkowski sum** of two convex sets $\mathsf{K}_1, \mathsf{K}_2 \subseteq \mathbb{R}^d$ is

$$\mathsf{K}_1 + \mathsf{K}_2 \ := \ \{\mathbf{p} + \mathbf{q} \, : \, \mathbf{p} \in \mathsf{K}_1, \ \mathbf{q} \in \mathsf{K}_2\} \, .$$

An example is depicted in Figure 3.9.

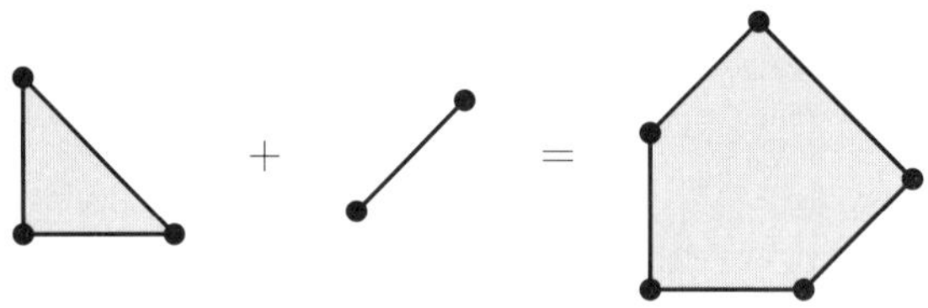

Figure 3.9. A Minkowski sum.

That $\mathsf{K}_1 + \mathsf{K}_2$ is again convex is the content of Exercise 3.25. Minkowski sums are key to the following fundamental theorem of polyhedral geometry, often called the **Minkowski–Weyl theorem**.

Theorem 3.2.5. *A set $\mathsf{Q} \subseteq \mathbb{R}^d$ is a polyhedron if and only if there exist a polytope P and a finitely generated cone C such that*

$$\mathsf{Q} \ = \ \mathsf{P} + \mathsf{C} \, .$$

In particular, C is the recession cone of Q and polytopes are precisely the bounded polyhedra.

Figure 3.10 illustrates Theorem 3.2.5 on the example from Figure 3.5. The Minkowski–Weyl theorem highlights the special role of polyhedra among all convex bodies. It states that polyhedra possess a discrete *intrinsic* description in terms of finitely many vertices and generators of P and C, respectively, as well as a discrete *extrinsic* description in the form of finitely many linear inequalities.

We already did much of the leg work towards a proof of Theorem 3.2.5. Exercise 3.26 reduces the claim to the statement that C is a finitely generated cone if and only if C is a polyhedral cone. Using Exercise 3.22, we see that it suffices to show that if C is a polyhedral cone, then C is a finitely generated convex cone.

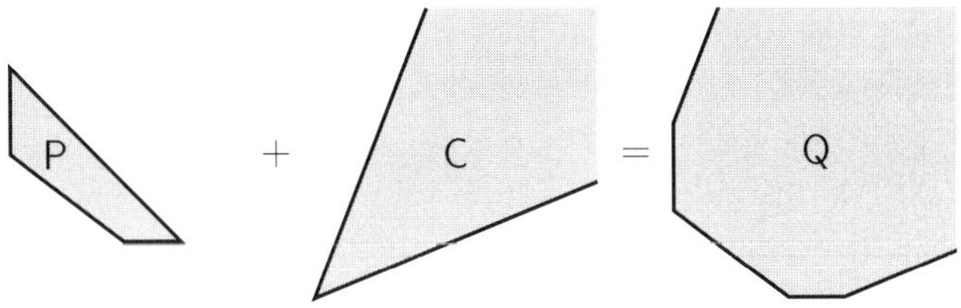

Figure 3.10. A decomposition of a polyhedron into a polytope and a cone.

Proposition 3.2.6. *If* $\mathsf{C} = \{\mathbf{x} \in \mathbb{R}^d : \langle \mathbf{a}_i, \mathbf{x}\rangle \leq 0 \text{ for } i = 1, \ldots, k\}$ *is a nonempty polyhedral cone, then* C *is a finitely generated cone.*

Proof. We prove the claim by induction on $\dim \mathsf{C}$ with the base cases $d \leq 2$ left to you (Exercise 3.29). We may assume that $\mathsf{C} \subset \mathbb{R}^d$ is full dimensional. By induction, the cones

$$\mathsf{C}_i \ := \ \{\mathbf{x} \in \mathsf{C} : \langle \mathbf{a}_i, \mathbf{x}\rangle = 0\} \ \subseteq \ \mathsf{C}$$

are finitely generated cones of dimension $< d$ for all $i = 1, \ldots, k$, and we let S_i be the set of generators of C_i. We claim that $\mathsf{C} = \mathrm{cone}(\bigcup_i S_i)$. To see this, let $\mathbf{p} \in \mathsf{C}$. If $\mathbf{p} \in \mathsf{C}_i$, we are done. Otherwise let $\mathbf{s} \in \bigcup_i S_i$ be arbitrary. Both $\mathbf{p}$ and $\mathbf{s}$ are in C; let $\lambda > 1$ be the smallest number such that $\mathbf{r} := (1-\lambda)\mathbf{s}+\lambda\mathbf{p}$ satisfies one linear inequality defining C with equality. We note that $\mathbf{p} \in \mathrm{cone}(\mathbf{s}, \mathbf{r})$ and $\mathbf{r} \in \mathsf{C}_i$ for some i. Thus, $\mathbf{p} \in \mathrm{cone}(\{\mathbf{s}\} \cup S_i)$, which proves the claim. □

The converse statement, that every finitely generated cone is a polyhedral cone, will be a byproduct of our considerations in Section 5.3, which even yield a practical algorithm. We close this section by reaping some of the nice consequences that the Minkowski–Weyl Theorem 3.2.5 entails.

It is clear from the definition that the image of a polytope or a finitely generated cone under a linear map $T : \mathbb{R}^d \to \mathbb{R}^e$ is a polytope or finitely generate cone, respectively. This is not so clear for polyhedra. Likewise, it is not easy to prove that the intersection of a polytope with an affine subspace is again a polytope. However, these become almost trivial (and left to Exercise 3.28).

Corollary 3.2.7. *Let* $\mathsf{Q} \subseteq \mathbb{R}^d$ *be a polyhedron and* $\phi(\mathbf{x}) = \mathbf{A}\,\mathbf{x}+\mathbf{b}$ *an affine projection* $\mathbb{R}^d \to \mathbb{R}^e$. *Then* $\phi(\mathsf{Q})$ *is a polyhedron. If* $\mathsf{P} \subset \mathbb{R}^d$ *is a polytope, then* $\mathsf{P} \cap \mathsf{Q}$ *is a polytope.*

3.3. Faces, Partially Ordered by Inclusion

In Proposition 3.2.6, we used the idea to focus on those points of a polyhedron Q that satisfy some linear inequality with equality. This leads to the notion of *faces* of a polyhedron, which we will study in depth in this section.

We call a hyperplane $\mathsf{H} = \{\mathbf{x} \in \mathbb{R}^d : \langle \mathbf{w}, \mathbf{x} \rangle = \delta\}$ **admissible** for a polyhedron $\mathsf{Q} \subseteq \mathbb{R}^d$ if $\mathsf{Q} \subseteq \mathsf{H}^{\leq}$. A **face** of Q is a subset of the form $\mathsf{F} = \mathsf{Q} \cap \mathsf{H}$, where H is an admissible hyperplane. We also decree that $\mathsf{F} = \varnothing$ and $\mathsf{F} = \mathsf{Q}$ are faces of Q. The reason for the former is that, unless $\mathsf{Q} = \mathbb{R}^d$, there is at least one admissible hyperplane H with $\mathsf{Q} \cap \mathsf{H} = \varnothing$. The reason for the latter is that the notion of face is then independent of the embedding: if we embed $\mathsf{Q} \times \{1\} \subset \mathbb{R}^{d+1}$, then Q is contained in a hyperplane and hence a face of itself. We call those faces that are neither empty nor Q itself the **proper** faces of Q. Admissible hyperplanes that yield nonempty faces are called **supporting** hyperplanes; Figure 3.11 shows two examples.

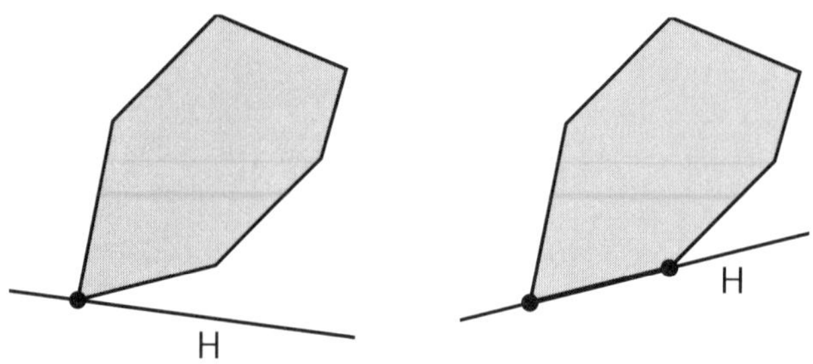

Figure 3.11. Two supporting lines, defining a vertex and an edge.

Every face F of Q is a polyhedron in its own right (Exercise 3.30) and comes with a dimension. Hence, we call F a **k-face** if F is a face of Q of dimension k. Some faces have special names: 0-faces are called **vertices**, bounded 1-faces are called **edges**, and unbounded 1-faces isomorphic to $\mathbb{R}_{\geq 0}$ are called **rays**. Exercise 3.31 shows that calling 0-faces vertices is consistent with our earlier definition and in light of Theorem 3.2.5, we call a polyhedron Q **pointed** if it has a vertex. If Q is a d-polyhedron, then faces of dimensions $d-2$ and $d-1$ are called **ridges** and **facets**, respectively. For $\mathsf{F} = \varnothing$, we have $\operatorname{aff}(\mathsf{F}) = \varnothing$ and hence $\varnothing$ is the unique face of Q of dimension -1.

As a polyhedron on its own, a face F itself has faces. The following result shows that *being a face of* is a transitive relation.

Proposition 3.3.1. *Let Q be a polyhedron and $\mathsf{F} \subseteq \mathsf{Q}$ a face. Then every face of F is also a face of Q.*

Proof. We only prove the statement in the case that Q is a polytope and leave the general case to Exercise 3.33. Let $\mathsf{F} \subset \mathsf{Q}$ be a proper face. There is a supporting hyperplane $\mathsf{H} = \{\mathbf{x} : \langle \mathbf{a}, \mathbf{x}\rangle = b\}$ such that $\mathsf{F} = \mathsf{Q} \cap \mathsf{H}$. Now let $\mathsf{H}' = \{\mathbf{x} : \langle \mathbf{a}', \mathbf{x}\rangle = b'\}$ be an admissible hyperplane for F such that $\mathsf{G} = \mathsf{H}' \cap \mathsf{F}$ is a face of F. Note that H' is only admissible for F and may as well meet Q in its interior. For $\varepsilon > 0$, we define $\mathbf{w} := \mathbf{a} + \varepsilon\mathbf{a}'$ and $\delta := b + \varepsilon\, b'$. We now verify that $\langle \mathbf{w}, \mathbf{x}\rangle \le \delta$ is satisfied for all points in Q and with equality precisely for the points in G. This proves the proposition. In fact, we only need to verify this for the vertices $\mathbf{v} \in \mathrm{vert}(\mathsf{Q})$.

If $\mathbf{v} \in \mathsf{F}$, then $\langle \mathbf{a}, \mathbf{v}\rangle = b$ and, since H' is supporting for F, both claims are true. If $\mathbf{v} \notin \mathsf{F}$, then $b - \langle \mathbf{a}, \mathbf{v}\rangle \ge \eta$ for some $\eta > 0$. Now, if we choose $\varepsilon > 0$ sufficiently small (see Exercise 3.34), then $\langle \mathbf{w}, \mathbf{x}\rangle < \delta$ will hold for all $\mathbf{v} \in \mathrm{vert}(\mathsf{Q}) \setminus \mathsf{F}$. Hence $\{\mathbf{x} : \langle \mathbf{w}, \mathbf{x}\rangle = \delta\}$ is supporting for Q and meets Q precisely in G. □

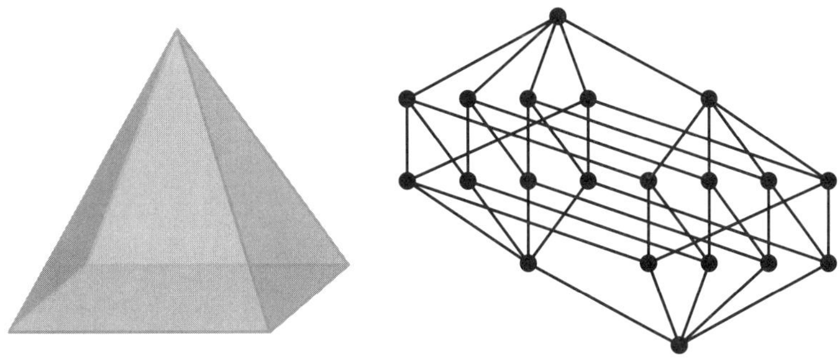

Figure 3.12. The face lattice of a square pyramid.

The collection $\Phi(\mathsf{Q})$ of faces of a polyhedron Q (including $\varnothing$ and Q) is partially ordered by inclusion and we call the poset $(\Phi(\mathsf{Q}), \subseteq)$ the **face lattice** of Q. Figure 3.12 gives an example of the face lattice of a square pyramid. To distinguish faces from arbitrary subsets of Q, we will often write $\mathsf{F} \preceq \mathsf{Q}$. Before we justify the further qualification of being a *lattice*, we note that Proposition 3.3.1 yields the following.

Corollary 3.3.2. *Let* Q *be a polyhedron and* $\mathsf{F} \subseteq \mathsf{Q}$ *a face. Then* $\Phi(\mathsf{F})$ *corresponds to the interval* $[\varnothing, \mathsf{F}] \subseteq \Phi(\mathsf{Q})$.

What this corollary implies, in turn, is that the face lattice of a polyhedron is a *graded* poset; see Exercise 3.35.

Corollary 3.3.3. *The face lattice of a polyhedron is a graded poset. If* Q *is a pointed polyhedron, then the rank of a face* F *is* $\dim \mathsf{F} + 1$.

In particular, a pointed d-polyhedron has a face of dimension k for every $k \le d$. This is not true for general polyhedra, however, it is not hard to see that a lineality space does not complicate things too much. For a polyhedron $\mathsf{Q} \subseteq \mathbb{R}^d$ with lineality space $\mathsf{L} = \operatorname{lineal}(\mathsf{Q})$, we write F/L for the projection of a face $\mathsf{F} \subseteq \mathsf{Q}$ in $\mathbb{R}^d/\mathsf{L}$. If you are not keen on quotient spaces, we can also identify F/L with $\mathsf{F} \cap \mathsf{L}^\perp$ or, equivalently, with the orthogonal projection of F onto $\mathsf{L}^\perp$. In either case Q/L is a pointed polyhedron of dimension $\dim \mathsf{Q} - \dim \mathsf{L}$ and the following result shows that the face lattice is retained; see Exercise 3.36.

Lemma 3.3.4. *Let Q be a polyhedron with lineality space L. The map $\Phi(\mathsf{Q}) \to \Phi(\mathsf{Q}/\mathsf{L})$ given by $\mathsf{F} \mapsto \mathsf{F}/\mathsf{L}$ is an isomorphism of face lattices.*

If Q is an affine subspace, then $\mathsf{Q}/\operatorname{lineal}(\mathsf{Q})$ is a point. We call a polyhedron **proper** if Q is not an affine subspace. A natural combinatorial statistic associated to a polytope (or any graded poset) is the number of elements of each rank. For a d-polyhedron Q, we define the **face numbers**

$$f_k \;=\; f_k(\mathsf{Q}) \;:=\; \text{number of faces of } \mathsf{Q} \text{ of dimension } k$$

for $-1 \le k \le d$. The face numbers are often recorded in the **f-vector**

$$f(\mathsf{Q}) \;:=\; (f_{-1}, f_0, f_1, \ldots, f_d)\ .$$

We notice that for every nonempty d-polyhedron Q, we always have $f_{-1} = f_d = 1$ and thus we take the liberty of omitting these entries from the f-vector whenever convenient.

The following proposition shows that $\Phi(\mathsf{Q})$ is a **meet semilattice**, that is, the meet of any two elements exists. Since every face is a subset of Q, Exercise 2.3 yields that $\Phi(\mathsf{Q})$ is indeed a lattice.

Proposition 3.3.5. *Let Q be a polyhedron and $\mathsf{F}, \mathsf{F}' \preceq \mathsf{Q}$ two faces. Then $\mathsf{F} \cap \mathsf{F}'$ is a face of both F and F'.*

Proof. Let $\mathsf{H} = \{\mathbf{x} : \langle \mathbf{a}, \mathbf{x}\rangle = b\}$ and $\mathsf{H}' = \{\mathbf{x} : \langle \mathbf{a}', \mathbf{x}\rangle = b'\}$ be admissible hyperplanes for F and F', respectively. Then

$$\mathsf{H}'' \;:=\; \{\mathbf{x} : \langle \mathbf{a} + \mathbf{a}', \mathbf{x}\rangle = b + b'\}$$

is admissible for Q, and $\mathsf{Q} \cap \mathsf{H}'' = \mathsf{F} \cap \mathsf{F}'$. □

For a polytope, every face is a polytope as well and is uniquely determined by its set of vertices. In poset-speak, we may also say that every face of a polytope is the join of vertices. Conversely, we want to show that every face is the intersection (or meet) of facets. The following result makes the connection between irredundant halfspaces and facets.

Proposition 3.3.6. *Let $\mathsf{Q} = \mathsf{H}_1^{\le} \cap \cdots \cap \mathsf{H}_m^{\le} \subset \mathbb{R}^d$ be a full-dimensional polyhedron given by irredundant halfspaces. Then $\mathsf{Q} \cap \mathsf{H}_i$ is a facet of Q for every $i = 1, \ldots, m$. Conversely, if F is a facet of Q, then $\mathsf{F} = \mathsf{Q} \cap \mathsf{H}_i$ for some i.*

We call a supporting hyperplane H **facet defining** for Q if $\mathsf{Q} \cap \mathsf{H}$ is a facet.

Proof. It is clear that $\mathsf{F} = \mathsf{Q} \cap \mathsf{H}_i$ is a face and we only have to show that $\dim \mathsf{F} = d - 1$. Consider $\mathsf{Q}' := \bigcap_{j \neq i} \mathsf{H}_j^{\leq}$. Since H_i is irredundant, $\mathsf{Q} \subsetneq \mathsf{Q}'$. In particular H_i meets Q' in its interior and Exercise 3.37 shows that $\dim \mathsf{F} = d - 1$.

For the second statement, assume that F is a facet. Pick any point $\mathbf{p}$ in the relative interior of F. Then $\mathbf{p} \in \partial \mathsf{Q}$ and hence $\mathbf{p} \in \mathsf{H}_i$ for some i. Since H_i is supporting for Q, it follows that $\mathsf{F} \subseteq \mathsf{H}_i$. Since $\operatorname{aff}(\mathsf{F})$ is a hyperplane, it follows that $\mathsf{F} = \mathsf{Q} \cap \mathsf{H}_i$. □

Proposition 3.3.7. *Let* Q *be a polyhedron and* $\mathsf{F} \subset \mathsf{Q}$ *a face. Then* F *is the intersection of all facets containing it.*

Proof. We prove the claim by induction on $\dim \mathsf{Q} - \dim \mathsf{F}$. If $\dim \mathsf{Q} - \dim \mathsf{F} = 1$, then F is a facet of Q and the statement is true. Now if $\dim \mathsf{F} < \dim \mathsf{Q} - 1$, then there is a facet $\mathsf{G} \subset \mathsf{Q}$ containing F. In particular $\dim \mathsf{G} - \dim \mathsf{F} < \dim \mathsf{Q} - \dim \mathsf{F}$ and hence F is an intersection of facets of G. However, an irredundant halfspace description of G can be obtained from that of Q and since G is an intersection of facets of Q, it follows from Proposition 3.3.6 that F is an intersection of facets of Q. □

With the ideas used in the proof of Proposition 3.3.6, we obtain a natural decomposition of Q into relatively open faces, illustrated in Figure 3.13.

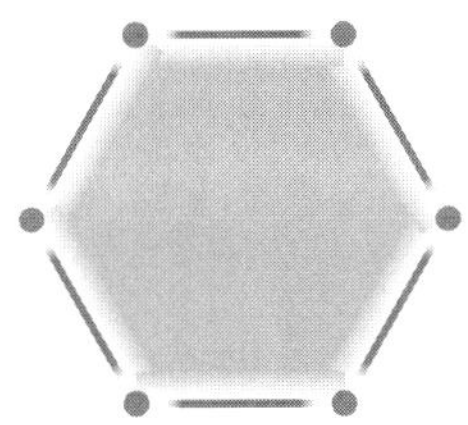

Figure 3.13. The face decomposition of Lemma 3.3.8.

Lemma 3.3.8. *Let* Q *be a polyhedron. For every point* $\mathbf{p} \in \mathsf{Q}$ *there is a unique face* F *of* Q *such that* $\mathbf{p} \in \mathsf{F}^\circ$. *Equivalently, we have the disjoint union*[3]

$$\mathsf{Q} = \biguplus_{\mathsf{F} \preceq \mathsf{Q}} \mathsf{F}^\circ .$$

[3]We use the symbol $\uplus$ to denote *disjoint* unions.

Proof. The inclusion $\supseteq$ is clear, so we have to argue $\subseteq$ and that the union is disjoint. Suppose Q is given as the intersection of irredundant halfspaces

$$\mathsf{Q} \; = \; \bigcap_{j=1}^{m} \mathsf{H}_j^{\le},$$

and $\mathbf{p} \in \mathsf{Q}$. After possibly renumbering the halfspaces, we may assume

$$\mathbf{p} \in \mathsf{H}_1^{=}, \dots, \mathsf{H}_k^{=} \qquad \text{and} \qquad \mathbf{p} \in \mathsf{H}_{k+1}^{<}, \dots, \mathsf{H}_m^{<},$$

where $0 \le k < m$. Thus

$$\mathsf{F} \; := \; \bigcap_{j=1}^{k} \mathsf{H}_j^{=} \;\cap\; \bigcap_{j=k+1}^{m} \mathsf{H}_j^{\le}$$

is a face of Q whose interior (see (3.1.12))

$$\mathsf{F}^{\circ} \; = \; \bigcap_{j=1}^{k} \mathsf{H}_j^{=} \;\cap\; \bigcap_{j=k+1}^{m} \mathsf{H}_j^{<}$$

contains $\mathbf{p}$. The uniqueness of F follows from Exercise 3.38 and the fact that F is, by construction, inclusion minimal. □

There is another reason for including the empty face in $\Phi(\mathsf{P})$. We recall that a homogenization of a polytope $\mathsf{P} \subset \mathbb{R}^d$ is the finitely generated cone $\hom(\mathsf{P}) = \operatorname{cone}(\mathsf{P} \times \{1\})$. The hyperplane $\mathsf{H} = \{\mathbf{y} : y_{d+1} = 1\}$ meets $\hom(\mathsf{P})$ in the interior and recovers P. In fact, H meets every face $\mathsf{F} \neq \{\mathbf{0}\}$ of $\hom(\mathsf{P})$ in the relative interior and Exercise 3.37 yields the following.

Proposition 3.3.9. *Let* P *be a nonempty polytope. Then as posets*

$$\Phi(\mathsf{P}) \;\cong\; \Phi(\hom(\mathsf{P})) \setminus \{\varnothing\}\,.$$

It is high time for an example. Let T be a $(d-1)$-dimensional simplex. The face lattice is clearly invariant under affine transformations and, by Exercise 3.16, we may assume that

$$\mathsf{T} \; = \; \operatorname{conv}(\mathbf{e}_1, \dots, \mathbf{e}_d) \; = \; \left\{ \mathbf{x} \in \mathbb{R}^d_{\ge 0} \; : \; x_1 + \cdots + x_d = 1 \right\}.$$

Using Proposition 3.3.9, we may as well determine the nonempty faces of $\hom(\mathsf{T}) = \mathbb{R}^d_{\ge 0}$, which shows that the face lattice of a $(d-1)$-dimensional simplex is isomorphic to the Boolean lattice on d elements. Hence $f_{i-1}(\mathsf{T}) = \binom{d}{i}$ for $i \ge 0$. An attractive class of polytopes—as we will see shortly—is given by the following definition: a polytope P is **simplicial** if every proper face of P is a simplex. In Exercise 3.39 you are asked to verify that the cross polytopes are simplicial polytopes.

There is a general construction technique that can be distilled from simplices. Let $\mathsf{P} \subset \mathbb{R}^d$ be a (possibly empty) polytope of dimension $< d$. For a point $\mathbf{v} \in \mathbb{R}^d \setminus \operatorname{aff}(\mathsf{P})$, we define

$$\mathbf{v} * \mathsf{P} \ := \ \operatorname{conv}(\mathsf{P} \cup \{\mathbf{v}\})$$

and call it the **pyramid** with apex $\mathbf{v}$ and base P. It is not hard to show that $\mathbf{v} * \mathsf{P}$ is linearly isomorphic to $\mathbf{v}' * \mathsf{P}$ for any other $\mathbf{v}' \in \mathbb{R}^d \setminus \operatorname{aff}(\mathsf{P})$. The following result shows that *taking pyramids* is a combinatorial construction in the sense that it is independent of the choice of $\mathbf{v}$ or the geometry of P, and you are asked to show this in Exercise 3.40.

Proposition 3.3.10. *Let $\mathsf{P}' = \mathbf{v} * \mathsf{P}$ be a pyramid. For each face $\mathsf{F} \preceq \mathsf{P}$, the polytope $\mathbf{v} * \mathsf{F}$ is a face of P'. Conversely, every face $\mathsf{F}' \preceq \mathsf{P}'$ is either a face of P or is of the form $\mathsf{F}' = \mathbf{v} * \mathsf{F}$ for some face $\mathsf{F} \preceq \mathsf{P}$.*

In particular, every simplex is obtained by taking iterated pyramids starting with a point: if $\mathsf{T} = \operatorname{conv}(\mathbf{v}_1, \dots, \mathbf{v}_k)$ is a simplex, then $\mathsf{F} = \operatorname{conv}(\mathbf{v}_1, \dots, \mathbf{v}_{k-1})$ is a simplex and $\mathsf{T} = \mathbf{v}_k * \mathsf{F}$. A more general construction, the **join** of polyhedra, is discussed in Exercise 3.44.

As a final example, we consider the face lattice of the cube

$$[-1,1]^d \ = \ \left\{ \mathbf{x} \in \mathbb{R}^d \ : \ -1 \le x_i \le 1 \text{ for } i = 1, \dots, d \right\}.$$

We already know that every face is the intersection of facets and, since the above presentation is irredundant, the facets are of the form

$$\mathsf{F}_i^- \ := \ \left\{ \mathbf{x} \in [-1,1]^d \ : \ x_i = -1 \right\},$$
$$\mathsf{F}_i^+ \ := \ \left\{ \mathbf{x} \in [-1,1]^d \ : \ x_i = 1 \right\}$$

for $i = 1, \dots, d$. In particular, $\mathsf{F}_i^- \cap \mathsf{F}_i^+ = \varnothing$. In fact, we can encode intersections of facets by setting, for $\sigma \in \{-, 0, +\}^d$,

$$\mathsf{F}_\sigma \ := \ \bigcap_{i \,:\, \sigma_i \neq 0} \mathsf{F}_i^{\sigma_i}.$$

Then F_σ is linearly isomorphic to $[-1,1]^k$, where $k = |\{i : \sigma_i = 0\}|$, and $\mathsf{F}_\sigma = \mathsf{F}_{\sigma'}$ if and only if $\sigma = \sigma'$. This suggests a combinatorial model for the face lattice of a d-dimensional cube. We can turn $\{-, 0, +\}$ into a partially ordered set by setting $- \prec 0$ and $+ \prec 0$. Then $\{-, 0, +\}^d$ is a direct product of posets (see Exercise 2.6) and we obtain the following (Exercise 3.45).

Proposition 3.3.11. *Let $d \ge 1$. Then, as posets,*

$$(\Phi([-1,1]^d) \setminus \{\varnothing\}, \subseteq) \ \cong \ (\{-, 0, +\}^d, \preceq).$$

We can view the cube as the d-fold Cartesian product

$$[-1,1]^d = [-1,1] \times [-1,1] \times \cdots \times [-1,1]$$

and generalize: Exercise 3.46 shows that taking the Cartesian product $\mathsf{Q} \times \mathsf{Q}'$ of two polyhedra is again a combinatorial construction.

There are plenty of similar combinatorial constructions, and some are discussed in Exercises 3.47 and 3.48 but most constructions depend on the actual geometry. For example, the Minkowski sum $\mathsf{Q} + \mathsf{Q}'$ of two polyhedra depends on how Q and Q' lie with respect to each other. However, if Q and Q' lie in complementary affine subspaces, then $\mathsf{Q} + \mathsf{Q}'$ is linearly isomorphic to $\mathsf{Q} \times \mathsf{Q}'$ and hence is combinatorial. In particular, if $\mathbf{u}_1, \dots, \mathbf{u}_k \in \mathbb{R}^d$ are linearly independent, then the **parallelepiped**

$$[0,1]\mathbf{u}_1 + \cdots + [0,1]\mathbf{u}_k := \{\mu_1 \mathbf{u}_1 + \cdots + \mu_1 \mathbf{u}_k : 0 \le \mu_1, \dots, \mu_k \le 1\}$$

is linearly isomorphic to a k-cube.

We are now well equipped to model counting problems in terms of polyhedra and manipulate them geometrically. In order to obtain results this way, we will next investigate the combinatorial structure of polyhedra, that is, face lattices of polyhedra.

3.4. The Euler Characteristic

We now come to an important notion that will allow us to relate geometry to combinatorics, the *Euler characteristic*. Our approach to the Euler characteristics of convex polyhedra is by way of sets built up from polyhedra. A set $S \subseteq \mathbb{R}^d$ is **polyconvex** if it is the union of finitely many relatively open polyhedra:

$$S = \mathsf{P}_1^\circ \cup \mathsf{P}_2^\circ \cup \cdots \cup \mathsf{P}_k^\circ,$$

where $\mathsf{P}_1, \dots, \mathsf{P}_k \subseteq \mathbb{R}^d$ are polyhedra. For example, a polyhedron is polyconvex: according to Lemma 3.3.8, we can write it as the (disjoint) union of its relatively open faces. Note, however, that our definition entails that polyconvex sets are not necessarily convex, not necessarily connected, and not necessarily closed. As we will see, they form a nice bag of sets to draw from, but not every reasonable set—e.g., the unit disc in the plane (Exercise 3.49)—is a polyconvex set.

We denote by PC_d the collection of polyconvex sets in $\mathbb{R}^d$. This is an infinite(!) poset under inclusion with minimal and maximal elements $\varnothing$ and $\mathbb{R}^d$, respectively. The intersection and the union of finitely many polyconvex sets are polyconvex, which renders PC_d a distributive lattice.

A map ϕ from PC_d to some Abelian group is a **valuation** if $\phi(\varnothing) = 0$ and

$$\phi(S \cup T) = \phi(S) + \phi(T) - \phi(S \cap T) \tag{3.4.1}$$

for all $S, T \in \mathsf{PC}_d$. Here's what we're after.

Theorem 3.4.1. *There exists a valuation $\chi : \mathsf{PC}_d \to \mathbb{Z}$ such that $\chi(\mathsf{P}) = 1$ for every nonempty closed polytope $\mathsf{P} \subset \mathbb{R}^d$.*

This is a nontrivial statement, as we cannot simply define $\chi(S) = 1$ whenever $S \neq \varnothing$. Indeed, if $\mathsf{P} \subset \mathbb{R}^d$ is a d-polytope and $\mathsf{H} = \{\mathbf{x} \in \mathbb{R}^d : \langle \mathbf{a}, \mathbf{x} \rangle = b\}$ is a hyperplane such that $\mathsf{H} \cap \mathsf{P}^\circ \neq \varnothing$, then

$$\mathsf{P}_1 := \{\mathbf{x} \in \mathsf{P} : \langle \mathbf{a}, \mathbf{x} \rangle < b\} \quad \text{and} \quad \mathsf{P}_2 := \{\mathbf{x} \in \mathsf{P} : \langle \mathbf{a}, \mathbf{x} \rangle \geq b\}$$

are nonempty polyconvex sets such that $\mathsf{P}_1 \cap \mathsf{P}_2 = \varnothing$ and thus

$$\chi(\mathsf{P}) = \chi(\mathsf{P}_1) + \chi(\mathsf{P}_2) .$$

Therefore, if χ is a valuation satisfying the conditions of Theorem 3.4.1, then necessarily $\chi(\mathsf{P}_1) = 0$.

The goal of this section is to construct a valuation χ satisfying the properties dictated by Theorem 3.4.1. The valuation property (3.4.1) will be the key to simplifying the computation of $\chi(S)$ for arbitrary polyconvex sets: if $S = \mathsf{P}_1 \cup \mathsf{P}_2 \cup \cdots \cup \mathsf{P}_k$, where each $\mathsf{P}_i \subseteq \mathbb{R}^d$ is a relatively open polyhedron, then by iterating (3.4.1) we obtain the *inclusion–exclusion formula* (for which we recall (2.4.1))

$$\begin{aligned} \chi(S) &= \sum_i \chi(\mathsf{P}_i) - \sum_{i<j} \chi(\mathsf{P}_i \cap \mathsf{P}_j) + \cdots \\ &= \sum_{\varnothing \neq I \subseteq [k]} (-1)^{|I|-1} \chi(\mathsf{P}_I) , \end{aligned} \tag{3.4.2}$$

where $\mathsf{P}_I := \bigcap_{i \in I} \mathsf{P}_i$. In particular, the value of $\chi(S)$ does not depend on the presentation of S as a union of relatively open polyhedra.

Here is a way to construct polyconvex sets. Let $\mathcal{H} = \{\mathsf{H}_1, \mathsf{H}_2, \dots, \mathsf{H}_n\}$ be an **arrangement** (i.e., a finite set) of (oriented) hyperplanes[4]

$$\mathsf{H}_i = \left\{\mathbf{x} \in \mathbb{R}^d : \langle \mathbf{a}_i, \mathbf{x} \rangle = b_i\right\}$$

in $\mathbb{R}^d$. An example of an arrangement of six hyperplanes (here: lines) in the plane is shown in Figure 3.14.

Continuing our definitions in (3.1.6), for a hyperplane H_i we denote by

$$\mathsf{H}_i^{>} := \left\{\mathbf{x} \in \mathbb{R}^d : \langle \mathbf{a}_i, \mathbf{x} \rangle > b_i\right\}$$

[4]As you will see, reorienting a hyperplane will leave all results unchanged.

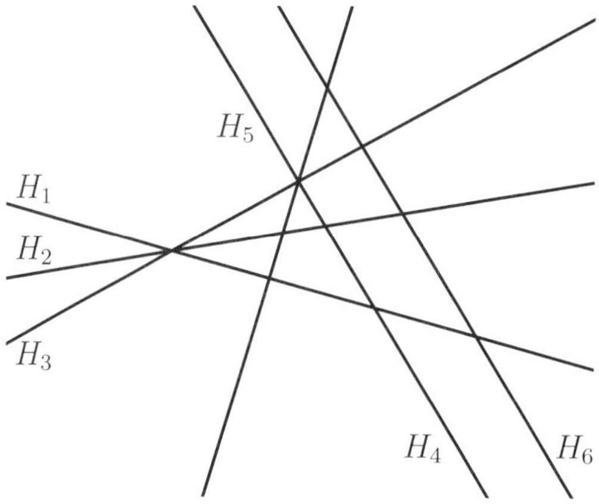

Figure 3.14. An arrangement of six lines in the plane.

the **open positive halfspace** bounded by H_i. We analogously define $\mathsf{H}_i^<$ and $\mathsf{H}_i^= := \mathsf{H}_i$. For each $\sigma \in \{<,=,>\}^n$, we obtain a (possibly empty) relatively open polyhedron

$$\mathsf{H}_\sigma \ := \ \mathsf{H}_1^{\sigma_1} \cap \mathsf{H}_2^{\sigma_2} \cap \cdots \cap \mathsf{H}_n^{\sigma_n}, \tag{3.4.3}$$

and these relatively open polyhedra partition $\mathbb{R}^d$. For example, the line arrangement in Figure 3.14 decomposes $\mathbb{R}^2$ into 57 relatively open polyhedra: 19 of dimension two, 28 of dimension one, and 10 of dimension zero. For a point $\mathbf{p} \in \mathbb{R}^d$, let $\sigma(\mathbf{p}) \in \{<,=,>\}^n$ record the position of $\mathbf{p}$ relative to the n hyperplanes; that is, $\mathsf{H}_{\sigma(\mathbf{p})}$ is the unique relatively open polyhedron among the H_σ's containing $\mathbf{p}$.

For a fixed hyperplane arrangement $\mathcal{H}$ in $\mathbb{R}^d$, we define the class of **$\mathcal{H}$-polyconvex sets** $\mathsf{PC}(\mathcal{H}) \subset \mathsf{PC}_d$ to consist of those sets that are finite unions of relatively open polyhedra of the form H_σ given by (3.4.3). That is, every $S \in \mathsf{PC}(\mathcal{H}) \setminus \{\varnothing\}$ has a representation

$$S \ = \ \mathsf{H}_{\sigma^1} \uplus \mathsf{H}_{\sigma^2} \uplus \cdots \uplus \mathsf{H}_{\sigma^k} \tag{3.4.4}$$

for some $\sigma^1, \sigma^2, \ldots, \sigma^k \in \{<,=,>\}^n$ such that $\mathsf{H}_{\sigma^j} \neq \varnothing$ for all $1 \le j \le k$. Note that the relatively open polyhedra H_{σ^j} are disjoint and thus the representation of S given in (3.4.4) is unique. For H_σ in the form (3.4.3), we define

$$\chi(\mathcal{H}, \mathsf{H}_\sigma) \ := \ (-1)^{\dim(\mathsf{H}_\sigma)} \tag{3.4.5}$$

and so, consequently, for $S \in \mathsf{PC}(\mathcal{H})$ in the form (3.4.4),

$$\chi(\mathcal{H}, S) \ := \ \sum_{j=1}^{k} (-1)^{\dim(\mathsf{H}_{\sigma^j})} .$$

The next result, whose proof we leave as Exercise 3.52, states that this function, together with $\chi(\mathcal{H}, \varnothing) := 0$, is a valuation.

Proposition 3.4.2. *The function $\chi(\mathcal{H}, \cdot) : \mathsf{PC}(\mathcal{H}) \to \mathbb{Z}$ is a valuation.*

We can consider $\chi(\mathcal{H}, S)$ as a function in two arguments, the arrangement $\mathcal{H}$ and the set $S \subseteq \mathbb{R}^d$; note that a set S is typically polyconvex with respect to various arrangements. It is a priori not clear how the value of $\chi(\mathcal{H}, S)$ changes when we change the arrangement. The power of our above definition is that it doesn't.

Lemma 3.4.3. *Let $\mathcal{H}_1, \mathcal{H}_2$ be two hyperplane arrangements in $\mathbb{R}^d$ and let $S \in \mathsf{PC}(\mathcal{H}_1) \cap \mathsf{PC}(\mathcal{H}_2)$. Then*

$$\chi(\mathcal{H}_1, S) \;=\; \chi(\mathcal{H}_2, S)\,.$$

Proof. It is sufficient to show that

$$\chi(\mathcal{H}_1, S) \;=\; \chi(\mathcal{H}_1 \cup \{\mathsf{H}\}, S)\,, \qquad \text{where} \qquad \mathsf{H} \in \mathcal{H}_2 \setminus \mathcal{H}_1\,. \tag{3.4.6}$$

Iterating this yields $\chi(\mathcal{H}_1, S) = \chi(\mathcal{H}_1 \cup \mathcal{H}_2, S)$ and, similarly, $\chi(\mathcal{H}_2, S) = \chi(\mathcal{H}_1 \cup \mathcal{H}_2, S)$, which proves the claim.

As a next simplifying measure, we observe that it suffices to show (3.4.6) for $S = \mathsf{H}_\sigma$ for some σ. Indeed, from the representation in (3.4.4) and the inclusion–exclusion formula (3.4.2), we then obtain

$$\chi\,(\mathcal{H}_1, S) \;=\; \chi\,(\mathcal{H}_1, \mathsf{H}_{\sigma^1}) + \chi\,(\mathcal{H}_1, \mathsf{H}_{\sigma^2}) + \cdots + \chi\,(\mathcal{H}_1, \mathsf{H}_{\sigma^k})\,.$$

Thus suppose that $S = \mathsf{H}_\sigma \in \mathsf{PC}(\mathcal{H}_1)$ and $\mathsf{H} \in \mathcal{H}_2 \setminus \mathcal{H}_1$. There are three possibilities how S can lie relative to H. The easy cases are $S \cap \mathsf{H} = S$ and $S \cap \mathsf{H} = \varnothing$. In both cases S is a relative open polyhedron H_σ with respect to $\mathcal{H}_1 \cup \{\mathsf{H}\}$ and

$$\chi\,(\mathcal{H}_1 \cup \{\mathsf{H}\}, S) \;=\; (-1)^{\dim S} \;=\; \chi(\mathcal{H}_1, S)\,.$$

The only interesting case is $\varnothing \neq S \cap \mathsf{H} \neq S$. Since S is relatively open, $S^< := S \cap \mathsf{H}^<$ and $S^> := S \cap \mathsf{H}^>$ are both nonempty, relatively open polyhedra of dimension $\dim S$, and $S^= := S \cap \mathsf{H}^=$ is relatively open of dimension $\dim S - 1$ (Exercise 3.37). Therefore,

$$S \;=\; S^< \uplus S^= \uplus S^>$$

is a presentation of S as an element of $\mathsf{PC}(\mathcal{H}_1 \cup \{\mathsf{H}\})$, and so

$$\begin{aligned}
&\chi(\mathcal{H}_1 \cup \{\mathsf{H}\}, S) \\
&\quad= \chi(\mathcal{H}_1 \cup \{\mathsf{H}\}, S^<) + \chi(\mathcal{H}_1 \cup \{\mathsf{H}\}, S^=) + \chi(\mathcal{H}_1 \cup \{\mathsf{H}\}, S^>) \\
&\quad= (-1)^{\dim S} + (-1)^{\dim S - 1} + (-1)^{\dim S} \\
&\quad= (-1)^{\dim S} \;=\; \chi(\mathcal{H}_1, S)\,.
\end{aligned}$$

□

The argument used in the above proof is typical when working with valuations. The valuation property (3.4.1) allows us to refine polyconvex sets by cutting them with hyperplanes and halfspaces. Clearly, there is no finite set of hyperplanes $\mathcal{H}$ such that $\mathsf{PC}_d = \mathsf{PC}(\mathcal{H})$, but as long as we only worry about finitely many polyconvex sets at a time, we can restrict ourselves to $\mathsf{PC}(\mathcal{H})$ for some $\mathcal{H}$.

Proposition 3.4.4. *Let $S \in \mathsf{PC}_d$ be a polyconvex set. Then there is a hyperplane arrangement $\mathcal{H}$ such that $S \in \mathsf{PC}(\mathcal{H})$.*

Proof. This should be intuitively clear. We can write $S = \mathsf{P}_1 \cup \mathsf{P}_2 \cup \cdots \cup \mathsf{P}_k$ for some relatively open polyhedra P_i. Now for each P_i there is a finite set of hyperplanes $\mathcal{H}_i := \{\mathsf{H}_1, \mathsf{H}_2, \ldots, \mathsf{H}_m\}$ such that $\mathsf{P}_i = \bigcap_j \mathsf{H}_j^{\sigma_j}$ for some $\sigma \in \{<,=,>\}^m$. Thus $\mathsf{P}_i \in \mathsf{PC}(\mathcal{H}_i)$, and by refining we conclude $S \in \mathsf{PC}(\mathcal{H}_1 \cup \mathcal{H}_2 \cup \cdots \cup \mathcal{H}_k)$. □

We can express the content of Proposition 3.4.4 more conceptually. For two hyperplane arrangements $\mathcal{H}_1$ and $\mathcal{H}_2$,

$$\mathcal{H}_1 \subseteq \mathcal{H}_2 \quad \Longrightarrow \quad \mathsf{PC}(\mathcal{H}_1) \subseteq \mathsf{PC}(\mathcal{H}_2)\,.$$

In more abstract terms then, PC_d is the union of $\mathsf{PC}(\mathcal{H})$ over all arrangements $\mathcal{H}$.

Lemma 3.4.3 and Proposition 3.4.4 get us one step closer to Theorem 3.4.1.

Proposition 3.4.5. *There is a unique valuation $\chi : \mathsf{PC}_d \to \mathbb{Z}$ such that for all $S \in \mathsf{PC}_d$,*

$$\chi(S) \;=\; \chi(\mathcal{H}, S)$$

for all hyperplane arrangements $\mathcal{H}$ for which $S \in \mathsf{PC}(\mathcal{H})$.

Proof. For a given $S \in \mathsf{PC}_d$, Lemma 3.4.3 implies that the definition

$$\chi(S) \;:=\; \chi(\mathcal{H}, S) \tag{3.4.7}$$

for every $\mathcal{H}$ such that $S \in \mathsf{PC}(\mathcal{H})$ is sound, and Proposition 3.4.4 ensures that there is such an $\mathcal{H}$. For uniqueness, we look back at (3.4.5) and conclude immediately that $\chi(\mathsf{P}) = (-1)^{\dim \mathsf{P}}$ for every relatively open polyhedron P. But then uniqueness follows for every polyconvex set, as we can write it as a disjoint union of finitely many relatively open polyhedra. □

We emphasize one part of the above proof for future reference.

Corollary 3.4.6. *If P is a nonempty relatively open polyhedron, then*

$$\chi(\mathsf{P}) \;=\; (-1)^{\dim \mathsf{P}}.$$

The valuation χ in Proposition 3.4.5 is the **Euler characteristic**, and for the rest of this book we mean the Euler characteristic of P when we write $\chi(\mathsf{P})$.

What is left to show to finish our proof of Theorem 3.4.1 is that $\chi(\mathsf{P}) = 1$ whenever P is a (nonempty) polytope. We first note that (3.4.5) gives us an effective way to compute the Euler characteristic of a polyhedron via face numbers: if Q is a polyhedron, then Lemma 3.3.8 states

$$\mathsf{Q} \;=\; \biguplus_{\mathsf{F} \preceq \mathsf{Q}} \mathsf{F}^\circ ,$$

where we recall that our notation $\mathsf{F} \preceq \mathsf{Q}$ means F is a face of Q. The inclusion–exclusion formula (3.4.2) and Corollary 3.4.6 give

$$\chi(\mathsf{Q}) \;=\; \sum_{\varnothing \prec \mathsf{F} \preceq \mathsf{Q}} (-1)^{\dim \mathsf{F}} \;=\; \sum_{i=0}^{\dim \mathsf{Q}} (-1)^i f_i(\mathsf{Q}) \,. \tag{3.4.8}$$

This is the **Euler–Poincaré formula**.

Now let $\mathsf{P} \subset \mathbb{R}^d$ be a nonempty polytope. To compute $\chi(\mathsf{P})$, we may assume that the origin is contained in the relative interior of P. (Any other point in the relative interior would work, but the origin is just too convenient.) For a nonempty face $\mathsf{F} \preceq \mathsf{P}$, let

$$\mathsf{C}_{\mathbf{0}}(\mathsf{F}) \;:=\; \bigcup_{t>0} t\,\mathsf{F}^\circ \;=\; \left\{ \mathbf{p} \in \mathbb{R}^d \, : \, \tfrac{1}{t}\mathbf{p} \in \mathsf{F}^\circ \text{ for some } t > 0 \right\}$$

and $\mathsf{C}_{\mathbf{0}}(\varnothing) := \{\mathbf{0}\}$. Figures 3.15 and 3.16 illustrate the following straightforward facts whose proofs we leave as Exercise 3.53.

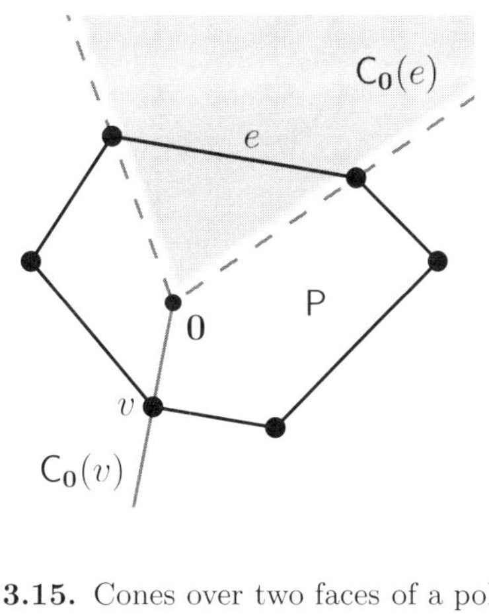

Figure 3.15. Cones over two faces of a polytope.

Proposition 3.4.7. *Let* P *be a nonempty polytope with* $\mathbf{0} \in \mathsf{P}^\circ$. *For each proper face* $\mathsf{F} \prec \mathsf{P}$, *the set* $\mathsf{C}_{\mathbf{0}}(\mathsf{F})$ *is a relatively open polyhedral cone of*

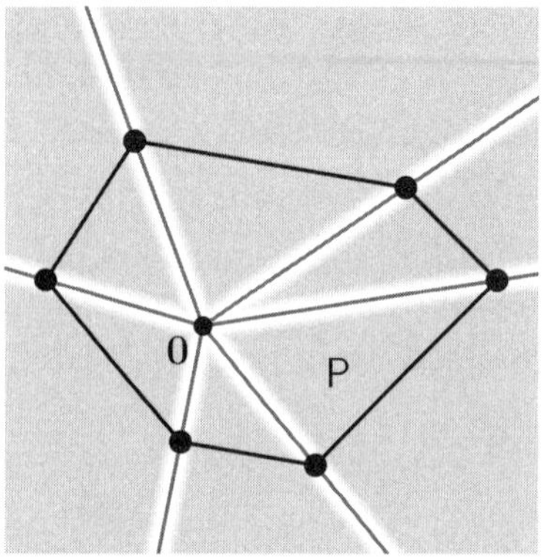

Figure 3.16. The decomposition given in Proposition 3.4.7.

dimension $\dim \mathsf{F} + 1$. *Furthermore,*

$$C_0(P) = \operatorname{aff}(P) = \biguplus_{F \prec P} C_0(F).$$

We can now finally complete the proof of Theorem 3.4.1.

Proof of Theorem 3.4.1. We will show that the Euler characteristic χ satisfies the properties stated in Theorem 3.4.1. Let P be a nonempty closed polytope of dimension d. By the Euler–Poincaré formula (3.4.8), the Euler characteristic is invariant under translation and thus we may assume that $\mathbf{0}$ is in the relative interior of P.

Proposition 3.4.7 yields two representations of the affine subspace aff(P), and computing the Euler characteristic using the two different representations gives

$$\chi(C_0(P)) = \chi(\operatorname{aff}(P)) = \sum_{F \prec P} \chi(C_0(F)) = 1 + \sum_{\varnothing \prec F \prec P} (-1)^{\dim F + 1}.$$

Both P° and $C_0(P) = \operatorname{aff}(P)$ are relatively open polyhedra of the same dimension and hence $\chi(C_0(P)) = \chi(P^\circ) = (-1)^d$, by Corollary 3.4.6. Thus

$$1 = \sum_{\varnothing \prec F \preceq P} (-1)^{\dim F} = \chi(P). \qquad \square$$

We do not know yet the Euler characteristic of an unbounded polyhedron Q. It turns out that this depends on whether Q is pointed or not. We start with the easier case.

Corollary 3.4.8. *If* Q *is a polyhedron with lineality space* $L = \operatorname{lineal}(Q)$, *then*

$$\chi(Q) = (-1)^{\dim L} \chi(Q/L).$$

This is pretty straightforward considering the relationship between faces and their dimensions of Q and Q/L given by Lemma 3.3.4; we leave the details to Exercise 3.54. In preparation for the case of a general pointed unbounded polyhedron, we first treat pointed *cones*.

Proposition 3.4.9. *If* $\mathsf{C} \subset \mathbb{R}^d$ *is a pointed cone, then* $\chi(\mathsf{C}) = 0$.

Proof. Let $\mathsf{C} = \operatorname{cone}(\mathbf{u}_1, \ldots, \mathbf{u}_m)$ for some $\mathbf{u}_1, \ldots, \mathbf{u}_m \in \mathbb{R}^d \setminus \{\mathbf{0}\}$. Since C is pointed, there is a supporting hyperplane

$$\mathsf{H}_0 := \left\{ \mathbf{x} \in \mathbb{R}^d : \langle \mathbf{a}, \mathbf{x} \rangle = 0 \right\}$$

such that $\mathsf{C} \subseteq \mathsf{H}_0^{\geq}$ and $\mathsf{C} \cap \mathsf{H}_0 = \{\mathbf{0}\}$. In particular, this means that $\langle \mathbf{a}, \mathbf{u}_i \rangle > 0$ for all i and by rescaling, if necessary, we may assume that $\langle \mathbf{a}, \mathbf{u}_i \rangle = \delta$ for some $\delta > 0$. Let $\mathsf{H}_\delta := \{\mathbf{x} \in \mathbb{R}^d : \langle \mathbf{a}, \mathbf{x} \rangle = \delta\}$ and consider

$$\begin{aligned} \overline{\mathsf{C}} &:= \mathsf{C} \cap \mathsf{H}_\delta^{\leq} \quad \text{and} \\ \mathsf{C}_\infty &:= \mathsf{C} \cap \mathsf{H}_\delta \,. \end{aligned}$$

(See Figure 3.17 for an illustration.) By construction, $\overline{\mathsf{C}}$ is a polytope and

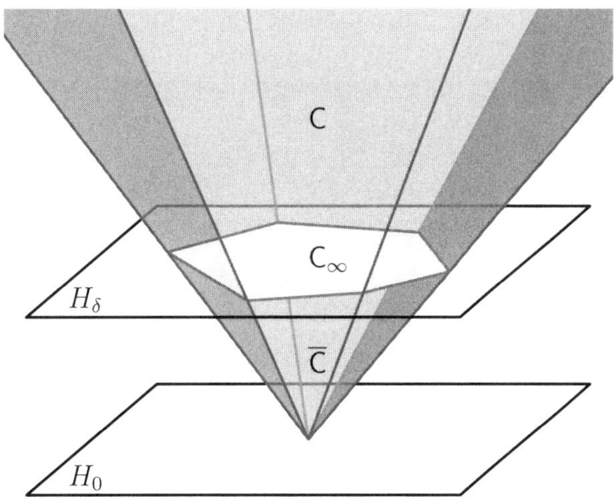

Figure 3.17. The polytopes $\overline{\mathsf{C}}$ and C_∞.

C_∞ is a face of $\overline{\mathsf{C}}$ (and thus also a polytope). Moreover, each unbounded face of C (here this means every nonempty face $\mathsf{F} \neq \{\mathbf{0}\}$) meets H_δ. Thus, each k-face F of C gives rise to a k-face of $\overline{\mathsf{C}}$ and, if F is unbounded, a $(k-1)$-face of C_∞. These are all the faces of $\overline{\mathsf{C}}$ and C_∞ (Exercise 3.55). We thus compute

$$\begin{aligned}\chi(\mathsf{C}) &= \sum_{\mathsf{F}\preceq\mathsf{C}}(-1)^{\dim\mathsf{F}} = \sum_{\mathsf{F}\preceq\overline{\mathsf{C}}}(-1)^{\dim\mathsf{F}} + \sum_{\mathsf{F}\preceq\mathsf{C}_\infty}(-1)^{\dim\mathsf{F}+1} \\ &= \chi(\overline{\mathsf{C}}) - \chi(\mathsf{C}_\infty) = 0\,,\end{aligned} \tag{3.4.9}$$

where the last equality follows from Theorem 3.4.1. □

The general case of a pointed unbounded polyhedron is not much different from that of pointed cones. We recall from Theorem 3.2.5 that every polyhedron Q is of the form $\mathsf{Q} = \mathsf{P} + \mathsf{C}$, where P is a polytope and C is a polyhedral cone. Exercise 3.32 says that for each nonempty face $\mathsf{F} \preceq \mathsf{Q}$ there are unique faces $\mathsf{F}' \preceq \mathsf{P}$ and $\mathsf{F}'' \preceq \mathsf{C}$ such that $\mathsf{F} = \mathsf{F}' + \mathsf{F}''$. In particular, F is an unbounded face of Q if and only if F'' is an unbounded face of C.

Corollary 3.4.10. *If Q is a pointed unbounded polyhedron, then $\chi(\mathsf{Q}) = 0$.*

Proof. We extend the idea of the proof of Proposition 3.4.9: we will find a hyperplane H such that H is disjoint from any bounded face but the intersection of H with an unbounded face $\mathsf{F} \preceq \mathsf{Q}$ yields a polytope of dimension $\dim \mathsf{F} - 1$.

Let $\mathsf{Q} = \mathsf{P} + \mathsf{C}$, where P is a polytope and C is a pointed cone. Let $\mathsf{H}_\delta = \{\mathbf{x} : \langle \mathbf{a}, \mathbf{x}\rangle = \delta\}$ be the hyperplane constructed for the cone C in the proof of Proposition 3.4.9. This time we choose $\delta > 0$ sufficiently large so that $\mathsf{P} \cap \mathsf{H}_\delta = \varnothing$. Thus, if $F \prec \mathsf{Q}$ is a bounded face, then $\mathsf{F} = \mathsf{F}' + \{\mathbf{0}\}$ for some face F' of P, and $\mathsf{F} \cap \mathsf{H}_\delta = \varnothing$. See Figure 3.18 for an illustration.

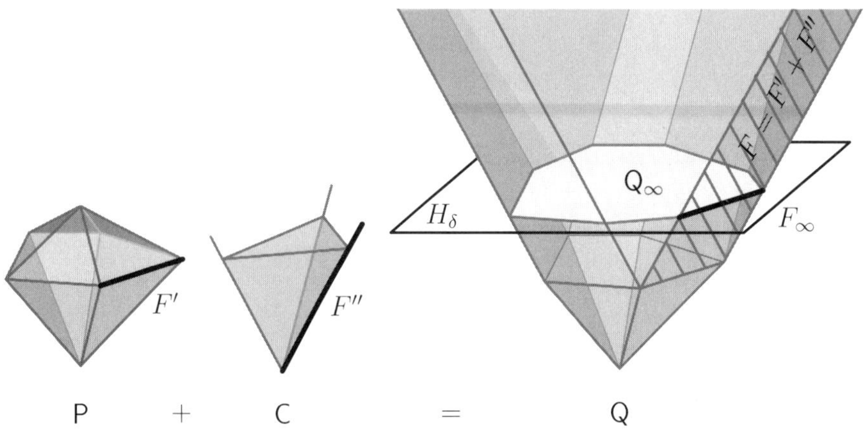

Figure 3.18. "Compactifying" a polyhedron.

Let $\mathsf{F} = \mathsf{F}' + \mathsf{F}''$ be an unbounded face of Q for some $\mathsf{F}' \preceq \mathsf{P}$ and $\mathsf{F}'' \preceq \mathsf{C}$, and let $\mathsf{F}_\infty := \mathsf{F} \cap \mathsf{H}_\delta$. Note that F_∞ is bounded, since otherwise there exist

$\mathbf{p} \in \mathsf{F}_\infty$ and $\mathbf{u} \neq 0$ such that $\mathbf{p} + t\mathbf{u} \in \mathsf{F}_\infty \subseteq \mathsf{F}$ for all $t \geq 0$; this implies that $\mathbf{u} \in \mathsf{F}''$, but by construction $\langle \mathbf{a}, \mathbf{u} \rangle > 0$ and thus there is only one t for which $\mathbf{p} + t\mathbf{u} \in \mathsf{H}_\delta$.

As for the dimension of F_∞, we only have to show the impossibility of $\dim \mathsf{F}_\infty < \dim \mathsf{F} - 1$. This can only happen if H_δ meets F in the boundary. But for each $\mathbf{p} \in \mathsf{F}'$ and $\mathbf{u} \in \mathsf{F}'' \setminus \{\mathbf{0}\}$ we have that $\mathbf{p} + \mathbf{0}$ and $\mathbf{p} + t\mathbf{u}$ are points in F, and for $t > 0$ sufficiently large they lie on different sides of H_δ.

Hence, $\overline{\mathsf{Q}} := \mathsf{Q} \cap \mathsf{H}_\delta^{\leq}$ and $\mathsf{Q}_\infty := \mathsf{Q} \cap \mathsf{H}_\delta$ are both polytopes such that each k-face of Q yields a k-face of $\overline{\mathsf{Q}}$ and a $(k-1)$-face of Q_∞, provided it was unbounded. A computation exactly analogous to (3.4.9) then gives

$$\begin{aligned} \chi(\mathsf{Q}) &= \sum_{\mathsf{F} \preceq \mathsf{Q}} (-1)^{\dim \mathsf{F}} = \sum_{\mathsf{F} \preceq \overline{\mathsf{Q}}} (-1)^{\dim \mathsf{F}} + \sum_{\mathsf{F} \preceq \mathsf{Q}_\infty} (-1)^{\dim \mathsf{F} + 1} \\ &= \chi(\overline{\mathsf{Q}}) - \chi(\mathsf{Q}_\infty) = 0 . \end{aligned}$$

□

We summarize the contents of Theorem 3.4.1 and Corollaries 3.4.8 and 3.4.10 as follows.

Theorem 3.4.11. *Let* $\mathsf{Q} = \mathsf{P} + \mathsf{C} + \mathsf{L}$ *be a polyhedron, where* P *is a polytope,* C *is a pointed cone, and* $\mathsf{L} = \operatorname{lineal}(\mathsf{Q})$. *Then*

$$\chi(\mathsf{Q}) = \begin{cases} (-1)^{\dim \mathsf{L}} & \text{if } \mathsf{C} = \{\mathbf{0}\} , \\ 0 & \text{otherwise.} \end{cases}$$

3.5. Möbius Functions of Face Lattices

The Euler characteristic is a fundamental concept throughout mathematics. In the context of geometric combinatorics it ties together the combinatorics and the geometry of polyhedra in an elegant way. First evidence of this is provided by the central result of this section: the Möbius function of the face lattice of a polyhedron can be computed in terms of the Euler characteristic.

Theorem 3.5.1. *Let* Q *be a polyhedron with face lattice* $\Phi = \Phi(\mathsf{Q})$. *For faces* $\mathsf{F}, \mathsf{G} \in \Phi$ *with* $\varnothing \prec \mathsf{F} \preceq \mathsf{G}$,

$$\mu_\Phi(\mathsf{F}, \mathsf{G}) = (-1)^{\dim \mathsf{G} - \dim \mathsf{F}} ,$$

and $\mu_\Phi(\varnothing, \mathsf{G}) = (-1)^{\dim \mathsf{G} + 1} \chi(\mathsf{G})$ *for* $\mathsf{G} \neq \varnothing$.

Towards a proof of this result, let $\psi : \Phi \times \Phi \to \mathbb{Z}$ be the map stipulated in Theorem 3.5.1, i.e.,

$$\psi(\mathsf{F}, \mathsf{G}) := \begin{cases} (-1)^{\dim \mathsf{G} - \dim \mathsf{F}} & \text{if } \varnothing \prec \mathsf{F} \preceq \mathsf{G} , \\ (-1)^{\dim \mathsf{G} + 1} \chi(\mathsf{G}) & \text{if } \varnothing = \mathsf{F} \prec \mathsf{G} . \end{cases}$$

We recall from Section 2.2 that the Möbius function μ_Φ is the inverse of the zeta function ζ_Φ and hence is unique. Thus, to prove the claim in Theorem 3.5.1, namely, that $\mu_\Phi(\mathsf{F},\mathsf{G}) = \psi(\mathsf{F},\mathsf{G})$, it is sufficient to show that ψ satisfies the defining relations (2.2.1) for the Möbius function. That is, we have to show that $\psi(\mathsf{F},\mathsf{F}) = 1$ and, for $\mathsf{F} \prec \mathsf{G}$,

$$\sum_{\mathsf{F}\preceq\mathsf{K}\preceq\mathsf{G}} \psi(\mathsf{K},\mathsf{G}) \;=\; 0\,. \tag{3.5.1}$$

That $\psi(\mathsf{F},\mathsf{F}) = 1$ is evident from the definition, so the meat lies in (3.5.1). Here is a first calculation which shows that we are on the right track by thinking of Euler characteristics: for $\mathsf{F} = \varnothing$, we compute

$$\begin{aligned}\sum_{\varnothing\preceq\mathsf{K}\preceq\mathsf{G}} \psi(\mathsf{K},\mathsf{G}) &= \psi(\varnothing,\mathsf{G}) + \sum_{\varnothing\prec\mathsf{K}\preceq\mathsf{G}} (-1)^{\dim\mathsf{G}-\dim\mathsf{K}} \\ &= \psi(\varnothing,\mathsf{G}) + (-1)^{\dim\mathsf{G}}\chi(\mathsf{G}) \\ &= 0\,.\end{aligned}$$

For the general case $\varnothing \prec \mathsf{F} \preceq \mathsf{G} \preceq \mathsf{Q}$, we would like to make the same argument but unfortunately we do not know if the interval $[\mathsf{F},\mathsf{G}] \subseteq \Phi$ is isomorphic to the face lattice of a polyhedron (see, however, Exercise 3.57). We will do something else instead and take a route that emphasizes the general geometric idea of modelling geometric objects locally by simpler ones. Namely, we will associate to each face F a polyhedral cone that captures the structure around F. We already used this idea in the proof of Theorem 3.4.1.

Let $\mathsf{Q} \subseteq \mathbb{R}^d$ be a polyhedron and $\mathbf{q} \in \mathbb{R}^d$. The **tangent cone** of Q at $\mathbf{q}$ is defined by

$$\mathrm{T}_{\mathbf{q}}(\mathsf{Q}) \;:=\; \{\mathbf{q}+\mathbf{u} \,:\, \mathbf{q}+\varepsilon\mathbf{u} \in \mathsf{Q} \text{ for all } \varepsilon > 0 \text{ sufficiently small}\}\,.$$

See Figure 3.19 for examples.

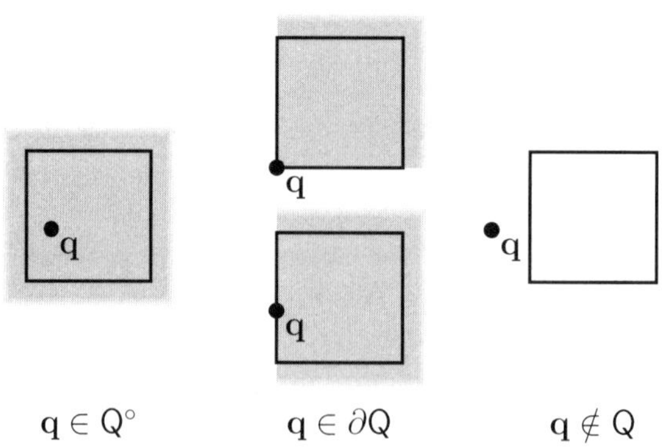

Figure 3.19. Sample tangent cones of a quadrilateral.

By definition $\mathrm{T_q(Q)} = \varnothing$ if $\mathbf{q} \notin \mathsf{Q}$ and $\mathrm{T_q(Q)} = \mathrm{aff}(\mathsf{Q})$ if $\mathbf{q} \in \mathsf{Q}^\circ$. More generally, the following result says that $\mathrm{T_q(Q)}$ is the translate of a polyhedral cone, which justifies the name *tangent cone*.

Proposition 3.5.2. *Let* $\mathsf{Q} = \{\mathbf{x} \in \mathbb{R}^d : \langle \mathbf{a}_i, \mathbf{x} \rangle \le b_i,\ i \in [n]\}$ *be a polyhedron and let* $\mathbf{q} \in \partial \mathsf{Q}$. *Then*

$$\mathrm{T_q(Q)} \;=\; \left\{\mathbf{x} \in \mathbb{R}^d \;:\; \langle \mathbf{a}_i, \mathbf{x} \rangle \le b_i \text{ for all } i \text{ with } \langle \mathbf{a}_i, \mathbf{q} \rangle = b_i \right\}. \tag{3.5.2}$$

In particular, if $\mathbf{p}$ *and* $\mathbf{q}$ *are both contained in the relative interior of a face* $\mathsf{F} \preceq \mathsf{Q}$, *then* $\mathrm{T_q(Q)} = \mathrm{T_p(Q)}$.

Proof. We observe that $\mathrm{T_q(Q)} = \mathrm{T_{q-r}(Q - r)}$ for all $\mathbf{r} \in \mathbb{R}^d$, and so we may assume that $\mathbf{q} = \mathbf{0}$. Let $I := \{i \in [n] : b_i = 0\}$; note that, since $\mathbf{0} \in \mathsf{Q}$, we have $b_i > 0$ for $i \notin I$.

By definition, $\mathbf{u} \in \mathrm{T_0(Q)}$ if and only if

$$\begin{aligned} \varepsilon \langle \mathbf{a}_i, \mathbf{u} \rangle \;=\; \langle \mathbf{a}_i, \varepsilon \mathbf{u} \rangle \;&\le\; b_i \qquad \text{for } i \notin I \quad \text{and} \\ \varepsilon \langle \mathbf{a}_i, \mathbf{u} \rangle \;&\le\; 0 \qquad \text{for } i \in I\,, \end{aligned}$$

for sufficiently small $\varepsilon > 0$. The latter condition just says $\langle \mathbf{a}_i, \mathbf{u} \rangle \le 0$ for $i \in I$, and the former condition can always be satisfied for a given $\mathbf{u}$, since $b_i > 0$ for $i \notin I$. This proves the first claim.

For the second claim, we note from Lemma 3.3.8 that $\mathbf{p} \in \mathsf{Q}$ is contained in the relative interior of the same face as $\mathbf{q} = \mathbf{0}$ if and only if $I = \{i \in [n] : \langle \mathbf{a}_i, \mathbf{p} \rangle = b_i\}$. Hence (3.5.2) gives $\mathrm{T_p(Q)} = \mathrm{T_q(Q)}$. □

Proposition 3.5.2 prompts the definition of tangent cones of faces: for a nonempty face $\mathsf{F} \preceq \mathsf{Q}$ we define the **tangent cone** of Q at F as

$$\mathrm{T_F(Q)} \;:=\; \mathrm{T_q(Q)}$$

for any point $\mathbf{q} \in \mathsf{F}^\circ$. We set $\mathrm{T}_\varnothing(\mathsf{Q}) = \mathsf{Q}$ and refer to Exercise 3.58 for a justification.

The next result solidifies our claim that the tangent cone of Q at F models the facial structure of Q around F.

Lemma 3.5.3. *Let* Q *be a polyhedron and* $\mathsf{F} \preceq \mathsf{Q}$ *a nonempty face. The tangent cone* $\mathrm{T_F(Q)}$ *is the translate of a polyhedral cone of dimension* $\dim \mathsf{Q}$ *with lineality space parallel to* $\mathrm{T_F(F)} = \mathrm{aff}(\mathsf{F})$. *The faces of* Q *that contain* F *are in bijection with the nonempty faces of* $\mathrm{T_F(Q)}$ *via*

$$\mathsf{G} \;\mapsto\; \mathrm{T_F(G)}\,.$$

Proof. The claims follow from the representation (3.5.2) and what we learned about polyhedra in Section 3.1. We may assume that

$$\mathsf{Q} \;=\; \left\{\mathbf{x} \in \mathbb{R}^d \;:\; \langle \mathbf{a}_i, \mathbf{x} \rangle \le b_i \text{ for all } i \in [n]\right\}$$

is a full-dimensional polyhedron. Let $\mathbf{p} \in \mathsf{F}^\circ$ and $I = \{i \in [n] : \langle \mathbf{a}_i, \mathbf{p} \rangle = b_i\}$. Then from (3.5.2) it follows that $\mathrm{T}_\mathsf{F}(\mathsf{Q})$ is given by a subset of the inequalities defining Q and hence $\mathsf{Q} \subseteq \mathrm{T}_\mathsf{F}(\mathsf{Q})$; in particular, $\mathrm{T}_\mathsf{F}(\mathsf{Q})$ is full dimensional. When we translate $\mathrm{T}_\mathsf{F}(\mathsf{Q})$ by $-\mathbf{p}$, we obtain

$$\mathrm{T}_{\mathsf{F}-\mathbf{p}}(\mathsf{Q} - \mathbf{p}) \;=\; \left\{ \mathbf{x} \in \mathbb{R}^d \,:\, \langle \mathbf{a}_i, \mathbf{x} \rangle \le 0 \text{ for all } i \in I \right\},$$

a polyhedral cone with lineality space

$$\mathsf{L} \;=\; \left\{ \mathbf{x} \in \mathbb{R}^d \,:\, \langle \mathbf{a}_i, \mathbf{x} \rangle = 0 \text{ for all } i \in I \right\} \;=\; \operatorname{aff}(\mathsf{F}) - \mathbf{p}\,.$$

For the last claim, we use Exercise 3.50, which says that for each face $\mathsf{G} \preceq \mathsf{Q}$ that contains F, there is an inclusion-maximal subset $J \subseteq I$ such that

$$\mathsf{G} \;=\; \{\mathbf{x} \in \mathsf{Q} : \langle \mathbf{a}_i, \mathbf{x} \rangle = b_i \text{ for all } i \in J\}\,.$$

In particular, J defines a face of $\mathrm{T}_\mathsf{F}(\mathsf{Q})$, namely,

$$\{\mathbf{x} \in \mathrm{T}_\mathsf{F}(\mathsf{Q}) \,:\, \langle \mathbf{a}_i, \mathbf{x} \rangle = b_i \text{ for all } i \in J\}\,,$$

which, by Proposition 3.5.2, is exactly $\mathrm{T}_\mathsf{F}(\mathsf{G})$. To see that the map $\mathsf{G} \mapsto \mathrm{T}_\mathsf{F}(\mathsf{G})$ is a bijection, we observe that the map that sends a face $\hat{\mathsf{G}} \preceq \mathrm{T}_\mathsf{F}(\mathsf{Q})$ to $\hat{\mathsf{G}} \cap \mathsf{Q}$ is an inverse. □

The gist of Lemma 3.5.3 is that for $\varnothing \prec \mathsf{F} \preceq \mathsf{G} \preceq \mathsf{Q}$, the posets $[\mathsf{F}, \mathsf{G}]$ (considered as an interval in $\Phi(\mathsf{Q})$) and $\Phi(\mathrm{T}_\mathsf{F}(\mathsf{G})) \setminus \{\varnothing\}$ are isomorphic. With these preparations at hand, we can prove Theorem 3.5.1.

Proof of Theorem 3.5.1. Let $\mathsf{F} \prec \mathsf{G}$ be two faces of Q. We already treated the case $F = \varnothing$, so we may assume that F is nonempty. To show that ψ satisfies (3.5.1), we use Lemma 3.5.3 to replace the sum in (3.5.1) over faces in the interval $[\mathsf{F}, \mathsf{G}]$ in $\Phi(\mathsf{Q})$ by the nonempty faces of $\mathrm{T}_\mathsf{F}(\mathsf{G})$:

$$\sum_{\mathsf{F} \preceq \mathsf{K} \preceq \mathsf{G}} \psi(\mathsf{K}, \mathsf{G}) \;=\; \sum_{\mathrm{T}_\mathsf{F}(\mathsf{F}) \preceq \mathrm{T}_\mathsf{F}(\mathsf{K}) \preceq \mathrm{T}_\mathsf{F}(\mathsf{G})} (-1)^{\dim \mathsf{G} - \dim \mathsf{K}}. \tag{3.5.3}$$

By Lemma 3.5.3, the tangent cone $\mathrm{T}_\mathsf{F}(\mathsf{G})$ is the translate of a polyhedral cone with lineality space $\mathsf{L} = \operatorname{aff}(\mathsf{F}) - \mathbf{p}$ for $\mathbf{p} \in \mathsf{F}^\circ$. Furthermore, we claim that $\mathrm{T}_\mathsf{F}(\mathsf{G})/\mathsf{L}$ is unbounded. Indeed, otherwise $\mathrm{T}_\mathsf{F}(\mathsf{G}) = \operatorname{aff}(\mathsf{F}) = \mathrm{T}_\mathsf{F}(\mathsf{F})$, which would imply, using again Lemma 3.5.3, that $\mathsf{F} = \mathsf{G}$, contradicting our assumption $\mathsf{F} \prec \mathsf{G}$.

Hence, we can continue our computation by noting that the right-hand side of (3.5.3) equals the Euler characteristic of the line-free polyhedral cone $\mathrm{T}_\mathsf{F}(\mathsf{G})/\mathsf{L}$, and so, by Corollary 3.4.8,

$$\sum_{\mathrm{T}_\mathsf{F}(\mathsf{F}) \preceq \mathrm{T}_\mathsf{F}(\mathsf{K}) \preceq \mathrm{T}_\mathsf{F}(\mathsf{G})} (-1)^{\dim \mathsf{G} - \dim \mathsf{K}} \;=\; (-1)^{\dim \mathsf{G}} \chi(\mathrm{T}_\mathsf{F}(\mathsf{G})/\mathsf{L}) \;=\; 0\,. \qquad \square$$

We will use tangent cones again in Chapter 5 to compute the Möbius function of (seemingly) more complicated objects. Yet another application of tangent cones is given in Section 3.7.

We finish this section by returning to a theme of Section 2.3, from which we recall the notion of an Eulerian poset Π, i.e., one that comes with Möbius function

$$\mu_\Pi(x,y) \;=\; (-1)^{l_\Pi(x,y)},$$

where $l_\Pi(x,y)$ is the length of the interval $[x,y]$. The length of an interval $[\mathsf{F},\mathsf{G}]$ in a face lattice is $\dim \mathsf{G} - \dim \mathsf{F}$, and so with Theorem 3.5.1 we conclude:

Corollary 3.5.4. *The face lattice $\Phi(\mathsf{P})$ of a polytope P is Eulerian.*

This allows us to derive an important corollary to Theorem 2.3.3, the reciprocity theorem for zeta polynomials of Eulerian posets, which we will apply to face lattices of a special class of polytopes. We will compute the zeta polynomial $Z_{\Phi(\mathsf{P})}(n)$ of the face lattice of a d-polytope P via Proposition 2.1.3, for which we have to count multichains

$$\varnothing = \mathsf{F}_0 \preceq \mathsf{F}_1 \preceq \cdots \preceq \mathsf{F}_n = \mathsf{P} \tag{3.5.4}$$

formed by faces of P. It is useful to consider the difference[5]

$$\Delta Z_{\Phi(\mathsf{P})}(n) \;:=\; Z_{\Phi(\mathsf{P})}(n+1) - Z_{\Phi(\mathsf{P})}(n)\,,$$

because Exercise 2.12 implies that $\Delta Z_{\Phi(\mathsf{P})}(n)$ equals the number of multichains

$$\varnothing = \mathsf{F}_0 \preceq \mathsf{F}_1 \preceq \cdots \preceq \mathsf{F}_n \prec \mathsf{P}\,. \tag{3.5.5}$$

We recall that a polytope is simplicial if all of its proper faces are simplices (equivalently, if all of its *facets* are simplices). If P is simplicial, the multichains in (3.5.5) are of a special form, namely, $\mathsf{F}_1, \mathsf{F}_2, \dots, \mathsf{F}_n$ are all simplices, and consequently (Exercise 3.59)

$$\Delta Z_{\Phi(\mathsf{P})}(n) \;=\; 1 + \sum_{\varnothing \prec \mathsf{F} \prec \mathsf{P}} n^{\dim(\mathsf{F})+1} \;=\; \sum_{k=0}^{d} f_{k-1}(\mathsf{P})\, n^k, \tag{3.5.6}$$

where we set $f_{-1}(\mathsf{P}) := 1$, accounting for the empty face $\varnothing$. In conjunction with Corollary 3.5.4, Theorem 2.3.3 now implies

$$\begin{aligned}
(-1)^d \Delta Z_{\Phi(\mathsf{P})}(-n) \;&=\; (-1)^d \left(Z_{\Phi(\mathsf{P})}(-n+1) - Z_{\Phi(\mathsf{P})}(-n) \right) \\
&=\; -Z_{\Phi(\mathsf{P})}(n-1) + Z_{\Phi(\mathsf{P})}(n) \\
&=\; \Delta Z_{\Phi(\mathsf{P})}(n-1)\,. \qquad (3.5.7)
\end{aligned}$$

[5]The difference operator Δ will play a prominent role in Chapter 4.

Via (3.5.6) this yields relations among the face numbers. Namely, since

$$\begin{aligned} \Delta Z_{\Phi(\mathsf{P})}(n-1) &= \sum_{k=0}^{d} f_{k-1}\,(n-1)^k = \sum_{k=0}^{d} f_{k-1} \sum_{j=0}^{k} (-1)^{k-j} \binom{k}{j} n^j \\ &= \sum_{j=0}^{d} n^j \sum_{k=j}^{d} (-1)^{k-j} \binom{k}{j} f_{k-1}\,, \end{aligned}$$

we can rephrase (3.5.7) as follows.

Theorem 3.5.5. *For a simplicial d-polytope* P *and* $0 \le j \le d$,

$$f_{j-1}(\mathsf{P}) = \sum_{k=j}^{d} (-1)^{d-k} \binom{k}{j} f_{k-1}(\mathsf{P})\,.$$

These face-number identities for simplicial polytopes are the **Dehn–Sommerville relations**. The case $j = 0$ recovers the Euler–Poincaré formula (3.4.8).

3.6. Uniqueness of the Euler Characteristics and Zaslavsky's Theorem

It is a valid question if the Euler characteristic is the *unique* valuation on PC_d with the properties of Theorem 3.4.1, i.e.,

$$\chi(\mathsf{P}) = 1 \tag{3.6.1}$$

for every nonempty closed polytope $\mathsf{P} \subset \mathbb{R}^d$. The answer is *no*: as we will see shortly, (3.6.1) does not determine $\chi(\mathsf{Q})$ for an unbounded polyhedron Q.

Here is the situation on the real line: the building blocks for polyconvex sets are points $\{p\}$ with $p \in \mathbb{R}$ and open intervals of the form (a,b) with $a, b \in \mathbb{R} \cup \{\pm\infty\}$. Let $\overline{\chi}$ be a valuation on polyconvex sets in $\mathbb{R}$ that satisfies the properties of Theorem 3.4.1. We need to define $\overline{\chi}(\{p\}) = 1$ and from

$$1 = \overline{\chi}([a,b]) = \overline{\chi}((a,b)) + \overline{\chi}(\{a\}) + \overline{\chi}(\{b\})$$

for finite a and b, we infer that $\overline{\chi}((a,b)) = -1$. The interesting part now comes from unbounded intervals. For example, we can set $\overline{\chi}((a,\infty)) = \overline{\chi}((-\infty,a)) = 0$ and check that this indeed defines a valuation: the value of $\overline{\chi}$ on a closed unbounded interval $[a,\infty)$ is

$$\overline{\chi}([a,\infty)) = \overline{\chi}(\{a\}) + \overline{\chi}((a,\infty)) = 1\,,$$

in contrast to $\chi([a,\infty)) = 0$ (by Proposition 3.4.9). What about the value on $\mathbb{R} = (-\infty,\infty)$? It is easy to see (Exercise 3.60) that $\overline{\chi}(\mathbb{R}) = 1$.

This was a proof (crawling on hands and knees) for the case $d = 1$ of the following important result.

Theorem 3.6.1. *There is a unique valuation* $\overline{\chi} : \mathsf{PC}_d \to \mathbb{Z}$ *of polyconvex sets in* $\mathbb{R}^d$ *such that* $\overline{\chi}(\mathsf{Q}) = 1$ *for every closed polyhedron* $\mathsf{Q} \neq \varnothing$.

We can prove this theorem using the same arguments as in Section 3.4 by way of $\mathcal{H}$-polyconvex sets and, in particular, Lemma 3.4.3. However, as in Section 3.4, we will need to make a choice for the value of $\overline{\chi}$ on relatively open polyhedra. On the other hand, if $\overline{\chi}$ is unique, then there isn't really a choice for $\overline{\chi}(\mathsf{Q}^\circ)$.

Proposition 3.6.2. *Suppose that* $\overline{\chi}$ *is a valuation such that* $\overline{\chi}(\mathsf{Q}) = 1$ *for all nonempty closed polyhedra* Q. *If* Q *is a closed polyhedron with lineality space* $\mathsf{L} = \operatorname{lineal}(\mathsf{Q})$, *then*

$$\overline{\chi}(\mathsf{Q}^\circ) \;=\; \begin{cases} (-1)^{\dim(\mathsf{Q}/\mathsf{L})} & \text{if } \mathsf{Q}/\mathsf{L} \text{ is bounded,} \\ 0 & \text{otherwise.} \end{cases}$$

Proof. Let Q be a nonempty polyhedron. We can think of $\overline{\chi}$ as a function on the face lattice $\Phi = \Phi(\mathsf{Q})$ given by

$$\overline{\chi}(\mathsf{G}) \;=\; \sum_{\mathsf{F} \preceq \mathsf{G}} \overline{\chi}(\mathsf{F}^\circ) \tag{3.6.2}$$

and $\overline{\chi}(\varnothing) = 0$. We apply Möbius inversion (Theorem 2.4.2) to (3.6.2) to obtain

$$\overline{\chi}(\mathsf{G}^\circ) \;=\; \sum_{\mathsf{F} \preceq \mathsf{G}} \overline{\chi}(\mathsf{F})\, \mu_\Phi(\mathsf{F},\mathsf{G}) \;=\; \sum_{\varnothing \prec \mathsf{F} \preceq \mathsf{G}} \mu_\Phi(\mathsf{F},\mathsf{G}) \;=\; -\mu_\Phi(\varnothing,\mathsf{G})\,.$$

The result now follows from Theorem 3.5.1. □

Proof of Theorem 3.6.1. To show that $\overline{\chi}$ is a valuation on PC_d, we revisit the argumentation of Section 3.4 by way of $\mathcal{H}$-polyconvex sets. The value of $\overline{\chi}$ on relatively open polyhedra is given by Proposition 3.6.2. We explicitly give the crucial step (analogous to Lemma 3.4.3), namely: let Q be a relatively open polyhedron and H a hyperplane such that $\mathsf{Q} \cap \mathsf{H} \neq \varnothing$. Define the three nonempty polyhedra $\mathsf{Q}^<, \mathsf{Q}^=, \mathsf{Q}^>$ as the intersection of Q with $\mathsf{H}^<, \mathsf{H}^=, \mathsf{H}^>$, respectively. If Q is bounded, then we are exactly in the situation of Lemma 3.4.3. Now let Q be unbounded. The closure of $\mathsf{Q}^=$ is a face of the closure of both $\mathsf{Q}^<$ and $\mathsf{Q}^>$. Hence, if one of them is bounded, then so is $\mathsf{Q}^=$. Thus

$$\overline{\chi}(\mathsf{Q}) \;=\; \overline{\chi}(\mathsf{Q}^<) + \overline{\chi}(\mathsf{Q}^=) + \overline{\chi}(\mathsf{Q}^>)\,.$$

That $\overline{\chi}(\mathsf{Q}) = 1$ for all closed polyhedra Q follows from (3.6.2). This also shows uniqueness: by Proposition 3.6.2, the value on relatively open polyhedra is uniquely determined. By Proposition 3.4.4, for every polyconvex set S there is a hyperplane arrangement $\mathcal{H}$ such that S is $\mathcal{H}$-polyconvex and hence can be represented as a union of disjoint relatively open polyhedra. □

Table 3.1. Evaluations of χ and $\overline{\chi}$ at a polyhedron $\mathsf{Q} = \mathsf{P} + \mathsf{C} + \mathsf{L}$, where P is a polytope, C is a pointed cone, and $\mathsf{L} = \operatorname{lineal}(\mathsf{Q})$.

	χ	$\overline{\chi}$
$\mathsf{Q} = \mathsf{P} + \mathsf{C} + \mathsf{L}$ with $\mathsf{C} \neq \{\mathbf{0}\}$	0	1
$\mathsf{Q} = \mathsf{P} + \mathsf{L}$	$(-1)^{\dim \mathsf{L}}$	1
$\mathsf{Q}^\circ = (\mathsf{P} + \mathsf{C} + \mathsf{L})^\circ$ with $\mathsf{C} \neq \{\mathbf{0}\}$	$(-1)^{\dim \mathsf{Q}}$	0
$\mathsf{Q}^\circ = (\mathsf{P} + \mathsf{L})^\circ$	$(-1)^{\dim \mathsf{Q}}$	$(-1)^{\dim \mathsf{P}}$

Table 3.1 compares the two "Euler characteristics" we have established in this chapter. There is merit in having several "Euler characteristics". We will illustrate this in the remainder of this section with one of the gems of geometric combinatorics—*Zaslavsky's theorem*, Theorem 3.6.4 below. To state it, we need the notion of characteristic polynomials.

Let Π be a graded poset with minimum $\hat{0}$. The **characteristic polynomial** of Π is

$$\chi_\Pi(n) \ := \ \sum_{x \in \Pi} \mu_\Pi\left(\hat{0}, x\right) n^{\operatorname{rk}(\Pi) - \operatorname{rk}(x)},$$

where $\operatorname{rk}(x) = l(\hat{0}, x)$, the rank of $x \in \Pi$. In many situations, the characteristic polynomial captures interesting combinatorial information about the poset. Here is a simple example.

Proposition 3.6.3. *Let* P *be a polytope and* $\chi_\Phi(n)$ *the characteristic polynomial of the face lattice* $\Phi = \Phi(\mathsf{P})$. *Then* $(-1)^{\dim(\mathsf{P})+1}\chi_\Phi(-1)$ *equals the number of faces of* P *and* $\chi_\Phi(1) = 1 - \chi(\mathsf{P}) = 0$.

Proof. We recall from Theorem 3.5.1 that the Möbius function of Φ is given by $\mu_\Phi(\mathsf{F}, \mathsf{G}) = (-1)^{\dim \mathsf{G} - \dim \mathsf{F}}$ whenever $\mathsf{F} \preceq \mathsf{G}$ are faces of P. The rank of a face G is given by $\operatorname{rk}(\mathsf{G}) = \dim(\mathsf{G}) + 1$. Hence

$$\chi_\Phi(n) \ = \ \sum_{\varnothing \preceq \mathsf{F} \preceq \mathsf{P}} (-1)^{\dim \mathsf{F}+1} \, n^{\dim \mathsf{P} - \dim \mathsf{F}}. \tag{3.6.3}$$

Thus, the evaluation of $(-1)^{\dim \mathsf{P}+1}\chi_\Phi(n)$ at $n = -1$ simply counts the number of faces and for $n = 1$, equation (3.6.3) reduces to $1 - \chi(\mathsf{P})$ via the Euler–Poincaré formula (3.4.8). □

Let $\mathcal{H} = \{\mathsf{H}_1, \mathsf{H}_2, \dots, \mathsf{H}_k\}$ be an arrangement of hyperplanes in $\mathbb{R}^d$. A **flat** of $\mathcal{H}$ is a nonempty affine subspace $\mathsf{F} \subset \mathbb{R}^d$ of the form

$$\mathsf{F} \ = \ \mathsf{H}_{i_1} \cap \mathsf{H}_{i_2} \cap \cdots \cap \mathsf{H}_{i_k}$$

for some $1 \le i_1, \dots, i_k \le n$. The **intersection poset** $\mathcal{L}(\mathcal{H})$ of $\mathcal{H}$ is the collection of flats of $\mathcal{H}$ ordered by *reverse* inclusion. That is, for two flats

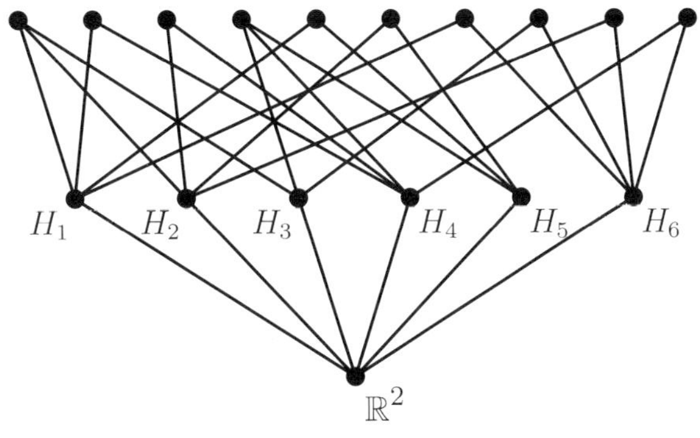

Figure 3.20. The intersection poset for the line arrangement in Figure 3.14.

$\mathsf{F}, \mathsf{G} \in \mathcal{L}(\mathcal{H})$, we have $\mathsf{F} \preceq \mathsf{G}$ if $\mathsf{G} \subseteq \mathsf{F}$. (Figure 3.20 shows the intersection poset of the arrangement of six lines in Figure 3.14.) The minimal element is the empty intersection given by $\hat{0} = \mathbb{R}^d$. There is a maximal element precisely if all hyperplanes have a point in common, in which case we may assume that all hyperplanes pass through the origin and call $\mathcal{H}$ **central**. An example of a central arrangement (consisting of the three coordinate

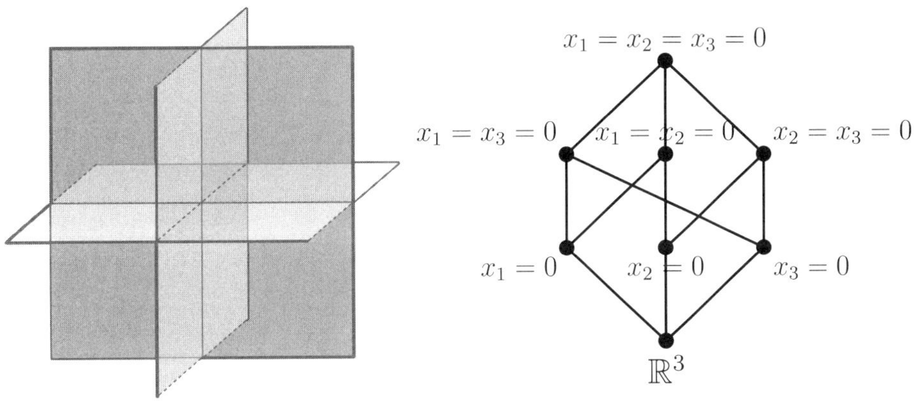

Figure 3.21. An arrangement of three coordinate hyperplanes and its intersection poset.

hyperplanes $x_1 = 0$, $x_2 = 0$, and $x_3 = 0$) is given in Figure 3.21. Let $\mathsf{L}_{\mathcal{H}} \subseteq \mathbb{R}^d$ be the inclusion-maximal linear subspace that is parallel to all hyperplanes, that is, $\mathsf{H}_i + \mathsf{L}_{\mathcal{H}} \subseteq \mathsf{H}_i$ for all $1 \leq i \leq k$. We call $\mathsf{L}_{\mathcal{H}}$ the **lineality space** of $\mathcal{H}$. The arrangement $\mathcal{H}$ is **essential** if $\mathsf{L}_{\mathcal{H}} = \{0\}$. If $\mathsf{L}_{\mathcal{H}} \neq \{0\}$, then $\mathcal{H}' := \{\mathsf{H}'_i := \mathsf{H}_i \cap \mathsf{L}_{\mathcal{H}}^{\perp} : i = 1, \ldots, k\}$ is essential and $\mathcal{L}(\mathcal{H}') \cong \mathcal{L}(\mathcal{H})$.

The **characteristic polynomial** of $\mathcal{H}$ is

$$\chi_{\mathcal{H}}(n) \;:=\; \sum_{\mathsf{F}\in\mathcal{L}(\mathcal{H})} \mu_{\mathcal{L}(\mathcal{H})}\big(\mathbb{R}^d, \mathsf{F}\big)\, n^{\dim \mathsf{F}} \;=\; n^{\dim \mathsf{L}_{\mathcal{H}}} \chi_{\mathcal{L}(\mathcal{H})}(n)\,,$$

where the last equality follows from $\mathrm{rk}_{\mathcal{L}(\mathcal{H})}(\mathsf{F}) = d - \dim \mathsf{F}$. For example, the characteristic polynomial of the arrangement of six lines in Figure 3.14 is $\chi_{\mathcal{H}}(n) = n^2 - 6n + 12$.

As mentioned in Section 3.4, the hyperplane arrangement $\mathcal{H}$ decomposes $\mathbb{R}^d$ into relatively open polyhedra: for each point $\mathbf{p} \in \mathbb{R}^d$ there is a unique $\sigma \in \{<,=,>\}^k$ such that

$$\mathbf{p} \;\in\; \mathsf{H}_\sigma \;=\; \mathsf{H}_1^{\sigma_1} \cap \mathsf{H}_2^{\sigma_2} \cap \cdots \cap \mathsf{H}_k^{\sigma_k}.$$

A (closed) **region** of $\mathcal{H}$ is (the closure of) a full-dimensional open polyhedron in this decomposition. A region can be unbounded or **relatively bounded**, that is, H_σ has lineality space $\mathsf{L}_{\mathcal{H}}$ and $\mathsf{H}_\sigma \cap \mathsf{L}_{\mathcal{H}}^\perp$ is bounded. In particular, if $\mathcal{H}$ is essential, then the regions are bounded or unbounded.

We define $r(\mathcal{H})$ to be the number of all regions and $b(\mathcal{H})$ to be the number of relatively bounded regions of $\mathcal{H}$. For example, the arrangement in Figure 3.14 has 19 regions, seven of which are bounded. Exercises 3.65–3.68 give a few more examples. Our proof of the following famous theorem makes use of the fact that we have two "Euler characteristics" at our disposal.

Theorem 3.6.4. *Let $\mathcal{H}$ be a hyperplane arrangement in $\mathbb{R}^d$. Then*

$$r(\mathcal{H}) \;=\; (-1)^d\, \chi_{\mathcal{H}}(-1) \qquad \text{and} \qquad b(\mathcal{H}) \;=\; (-1)^{d-\dim \mathsf{L}_{\mathcal{H}}}\, \chi_{\mathcal{H}}(1)\,.$$

Proof. We denote by $\bigcup \mathcal{H} = \mathsf{H}_1 \cup \cdots \cup \mathsf{H}_k$ the union of all hyperplanes. Then $\mathbb{R}^d \setminus \bigcup \mathcal{H} = R_1 \uplus R_2 \uplus \cdots \uplus R_m$, where $R_1, \dots, R_m$ are the regions of $\mathcal{H}$. This is a polyconvex set, and since every region is a d-dimensional open polyhedron with the same lineality space $\mathsf{L}_{\mathcal{H}}$,

$$r(B) \;=\; (-1)^d \chi\left(\mathbb{R}^d \setminus \bigcup \mathcal{H}\right) \quad\text{and}\quad b(B) \;=\; (-1)^{d-\dim \mathsf{L}_{\mathcal{H}}}\, \overline{\chi}\left(\mathbb{R}^d \setminus \bigcup \mathcal{H}\right),$$

by Corollary 3.4.6 and Proposition 3.6.2. Let $\phi : \mathsf{PC}_d \to \mathbb{R}$ be a valuation. Then

$$\phi\left(\mathbb{R}^d \setminus \bigcup \mathcal{H}\right) \;=\; \phi(\mathbb{R}^d) - \phi\left(\mathsf{H}_1 \cup \mathsf{H}_2 \cup \cdots \cup \mathsf{H}_k\right) \;=\; \sum_{I \subseteq [k]} (-1)^{|I|}\, \phi\left(\mathsf{H}_I\right)$$

$$(3.6.4) \qquad\qquad = \sum_{\mathsf{F}\in\mathcal{L}(\mathcal{H})} \mu_{\mathcal{L}(\mathcal{H})}\left(\hat{0}, \mathsf{F}\right) \phi(\mathsf{F})\,,$$

where the penultimate equation is the inclusion–exclusion formula (3.4.2) with $\mathsf{H}_I := \bigcap_{i \in I} \mathsf{H}_i$, and the last equation follows from Theorem 2.4.5. Since each $\mathsf{F} \in \mathcal{L}(\mathcal{H})$ is an affine subspace and hence a closed polyhedron, we have $\overline{\chi}(\mathsf{F}) = 1$, and so for $\phi = \overline{\chi}$ in (3.6.4) we obtain the evaluation $\chi_{\mathcal{H}}(1)$.

Similarly, for $\phi = \chi$, we have $\chi(\mathsf{F}) = (-1)^{\dim \mathsf{F}}$ and hence (3.6.4) for $\phi = \chi$ yields $\chi_{\mathcal{H}}(-1)$. □

3.7. The Brianchon–Gram Relation

In this final section we want to once again allure to Euler characteristics to prove an elegant result in geometric combinatorics, the *Brianchon–Gram relation*—Theorem 3.7.1 below. This is a simple geometric equation that can be thought of as a close relative to the Euler–Poincaré formula (3.4.8). It is best stated in terms of indicator functions: for a subset $S \subseteq \mathbb{R}^d$, the **indicator function** of S is the function $[S] : \mathbb{R}^d \to \mathbb{Z}$ defined by

$$[S](\mathbf{p}) \ := \ \begin{cases} 1 & \text{if } \mathbf{p} \in S, \\ 0 & \text{otherwise.} \end{cases}$$

Theorem 3.7.1. *Let* $\mathsf{P} \subset \mathbb{R}^d$ *be a polytope. Then*

$$[\mathsf{P}] \ = \sum_{\varnothing \prec \mathsf{F} \preceq \mathsf{P}} (-1)^{\dim \mathsf{F}} \left[\mathrm{T}_{\mathsf{F}}(\mathsf{P})\right]. \tag{3.7.1}$$

Here $\mathrm{T}_{\mathsf{F}}(\mathsf{P})$ is the tangent cone of P at a face F, a notion that we popularized in Section 3.5 in connection with the Möbius function of a face lattice. The Brianchon–Gram relation is a simple truth which we will see in action in Section 5.4. We devote the remainder of this section to its verification.

We first note that Lemma 3.5.3 implies $\mathsf{P} \subseteq \mathrm{T}_{\mathsf{F}}(\mathsf{P})$ for all $\mathsf{F} \neq \varnothing$. Hence, for a point $\mathbf{p} \in \mathsf{P}$, we compute

$$\sum_{\varnothing \prec \mathsf{F} \preceq \mathsf{P}} (-1)^{\dim \mathsf{F}} \left[\mathrm{T}_{\mathsf{F}}(\mathsf{P})\right](\mathbf{p}) \ = \sum_{\varnothing \prec \mathsf{F} \preceq \mathsf{P}} (-1)^{\dim \mathsf{F}} \ = \ \chi(\mathsf{P}) \ = \ 1\,,$$

by Theorem 3.4.1 and the Euler–Poincaré formula (3.4.8). Thus it remains to prove that the right-hand side of (3.7.1) evaluates to 0 for all $\mathbf{p} \notin \mathsf{P}$. This is clear when $\mathbf{p} \notin \operatorname{aff}(\mathsf{P})$, and so we may assume that P is full dimensional.

We call a point $\mathbf{p} \in \mathbb{R}^d$ **beneath** a face F of the polyhedron P if $\mathbf{p} \in \mathrm{T}_{\mathsf{F}}(\mathsf{P})$ and **beyond** F otherwise. From the definition of tangent cones, it follows that $\mathbf{p}$ is beyond F if and only if

$$[\mathbf{p}, \mathbf{q}] \cap \mathsf{P} \ = \ \{\mathbf{q}\} \qquad \text{for all } \mathbf{q} \in F$$

(Exercise 3.62). In this situation we also say that (the points in) F is (are) **visible** from $\mathbf{p}$. Figure 3.22 shows two examples.

We collect the points in P belonging to faces that $\mathbf{p}$ is beyond—equivalently, those points visible from $\mathbf{p}$—in the set[6]

$$|\mathrm{Vis}_{\mathbf{p}}(\mathsf{P})| \ := \ \bigcup \left\{ \mathsf{F}^\circ \, : \, \mathbf{p} \text{ is beyond } \mathsf{F} \prec \mathsf{P} \right\}. \tag{3.7.2}$$

[6]The maybe-funny-looking notation $|\mathrm{Vis}_{\mathbf{p}}(\mathsf{P})|$ will be explained in Chapter 5.

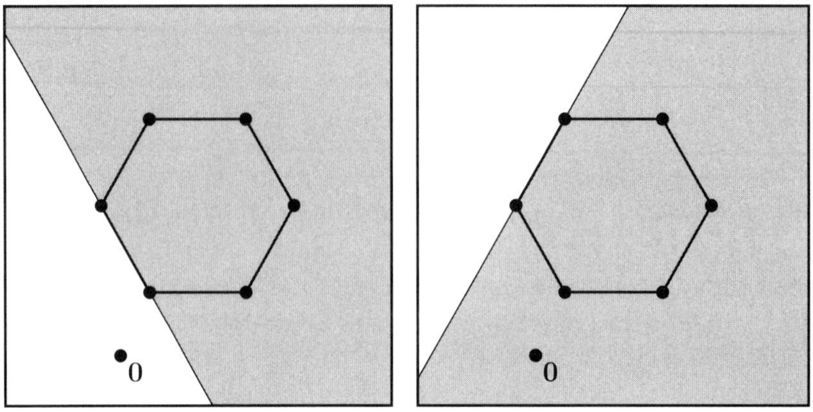

Figure 3.22. Two edges of a hexagon, one that is visible from $\mathbf{0}$ and one that is not.

Since P is full dimensional, there is no point beyond P and thus $|\mathrm{Vis}_{\mathbf{p}}(\mathsf{P})|$ is a subset of the boundary.

We remark that $|\mathrm{Vis}_{\mathbf{p}}(\mathsf{P})|$ is a polyconvex set, and for $\mathbf{p} \in \mathbb{R}^d \setminus \mathsf{P}$ the right-hand side of (3.7.1) equals

$$\sum_{\substack{\mathsf{F} \preceq \mathsf{P} \\ \mathbf{p} \text{ beneath } \mathsf{F}}} (-1)^{\dim \mathsf{F}} = \sum_{\mathsf{F} \preceq \mathsf{P}} (-1)^{\dim \mathsf{F}} - \sum_{\substack{\mathsf{F} \prec \mathsf{P} \\ \mathbf{p} \text{ beyond } \mathsf{F}}} (-1)^{\dim \mathsf{F}}$$
$$= \chi(\mathsf{P}) - \chi(|\mathrm{Vis}_{\mathbf{p}}(\mathsf{P})|)\,.$$

We thus want to show that $|\mathrm{Vis}_{\mathbf{p}}(\mathsf{P})|$ has Euler characteristic 1 whenever $\mathbf{p} \notin \mathsf{P}$. By translating both the point $\mathbf{p}$ and the polytope P by $-\mathbf{p}$, we may assume that $\mathbf{p} = \mathbf{0}$. Let $\mathsf{C} := \mathrm{cone}(\mathsf{P})$; see Figure 3.23 for a sketch. Since $\mathbf{0} \notin \mathsf{P}$, the cone C is pointed. The following construction is reminiscent of our proof of Theorem 3.4.1 in Section 3.4; it is illustrated in Figure 3.24.

Proposition 3.7.2. *Let* P *be a full-dimensional polytope and* $\mathsf{C} := \mathrm{cone}(\mathsf{P})$. *Then*

$$\mathsf{C} \setminus \{\mathbf{0}\} = \biguplus \{\mathrm{cone}(\mathsf{F})^\circ : \mathbf{0} \text{ is beyond } \mathsf{F} \prec \mathsf{P}\}\,.$$

Proof. A point $\mathbf{u} \in \mathbb{R}^d$ is contained in $\mathsf{C} \setminus \{\mathbf{0}\}$ if and only if $\lambda\mathbf{u} \in \mathsf{P}$ for some $\lambda > 0$. Since P is compact, there are a minimal such λ and a unique face F such that $\lambda\mathbf{u} \in \mathsf{F}^\circ$, by Lemma 3.3.8. Checking the definition shows that $\mathbf{0}$ is beyond F. □

Now to finish the proof of Theorem 3.7.1, if $\mathsf{F} \prec \mathsf{P}$ is a proper face such that $\mathbf{0}$ is beyond F, then $\mathrm{cone}(\mathsf{F})^\circ$ is a relatively open polyhedral cone of

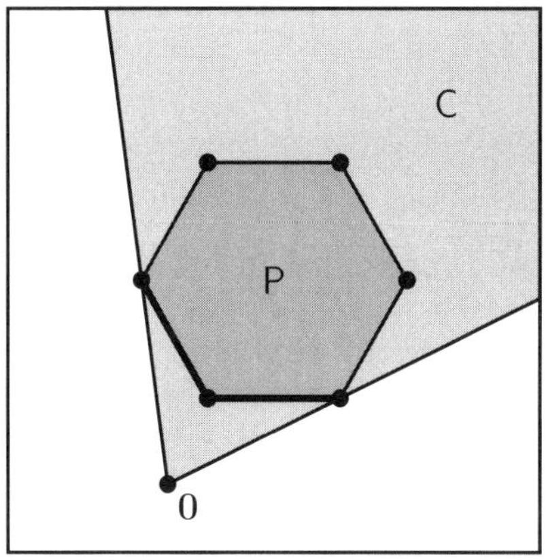

Figure 3.23. The cone $\mathsf{C} = \operatorname{cone}(\mathsf{P})$ and the faces of P beyond $\mathbf{0}$. Note that these contain (visibly!) precisely the points in P that are visible from $\mathbf{0}$.

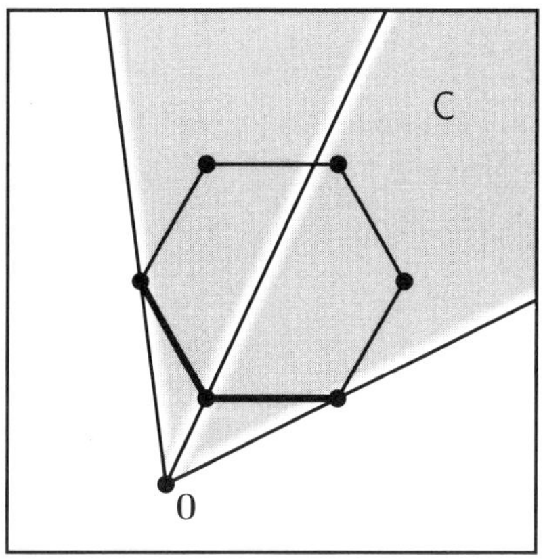

Figure 3.24. The decomposition of Proposition 3.7.2.

dimension $\dim \mathsf{F} + 1$. Hence, with Proposition 3.7.2,

$$\chi(\mathsf{C}) = \chi(\{\mathbf{0}\}) + \sum_{\substack{\mathsf{F} \prec \mathsf{P} \\ \mathbf{0} \notin \mathrm{T}_{\mathsf{F}}(\mathsf{P})}} (-1)^{\dim \mathsf{F}+1} = 1 - \chi(|\mathrm{Vis}_{\mathbf{0}}(\mathsf{P})|).$$

Since C is pointed, we conclude, with the help of Proposition 3.4.9, that $\chi(|\mathrm{Vis}_{\mathbf{0}}(\mathsf{P})|) = 1$, which was the missing piece to finish the proof of Theorem 3.7.1.

We will encounter the set $|\mathrm{Vis}_{\mathbf{p}}(\mathsf{P})|$ again in Section 5.3.

Notes

Three-dimensional polyhedra, i.e., three-dimensional convex geometric objects bounded by planes, have been admired since the dawn of time. They have been intensively studied by the ancient Greeks, including Plato and Archimedes, but they have been thought about even earlier. The algebraic perspective with linear inequalities allows for a generalization to arbitrary dimensions. This was initiated by Hermann Minkowski [**124**], who was also the first to systematically investigate separation theorems such as Theorem 3.2.3. The classical approach to separation is via *nearest-point maps*, which are implicit in Exercise 3.21. Linear inequalities are very versatile, and polyhedra constitute the geometric perspective on *linear programming*. A linear program is an optimization problem of the form

$$(3.7.3) \qquad \begin{array}{rrcll} \max & \langle \mathbf{c}, \mathbf{x} \rangle & & & \\ \text{subject to} & \langle \mathbf{a}_i, \mathbf{x} \rangle & \leq & b_i & \text{for } i = 1, \ldots, k. \end{array}$$

Linear programming and hence polyhedra are of utmost importance in the field of operations research. *Polyhedral combinatorics* is the field (or, we think, the art) of modelling a combinatorial optimization problem as a linear program (3.7.3). To give a trivial example, assume that the elements in $[d]$ have a weight or cost $c_i \in \mathbb{R}$ for $i = 1, \ldots, d$. For a given $k \geq 1$, how do we find a k-subset of $[d]$ that minimizes the total weight/cost? Well, we simply maximize the linear function $-\mathbf{c} = (-c_1, \ldots, -c_n)$ over the polytope

$$\triangle(d, k) \;:=\; \left\{ \mathbf{x} \in [0,1]^d \;:\; x_1 + \cdots + x_d = k \right\}.$$

It is not difficult to see that the maximum is attained at a vertex $\mathbf{v} \in \mathbb{Z}^d$ that decodes a k-subset of minimum total weight/cost. Since this is independent of $\mathbf{c}$, this shows that $\triangle(d, k)$ is a lattice polytope. We will have more to say about $\triangle(d, k)$ in Section 5.7. For more on the fascinating subject of polyhedral combinatorics, we refer to [**152**] and [**153**]. Theorem 3.2.5 is due to Minkowski [**124**]. It perfectly matches our intuition in three dimensions but, as you can see, the (algebraic) proof is quite nontrivial. A formal proof was given by Weyl [**182**]. The passage from linear inequalities to vertices and generators and back can be done effectively: in Section 5.3 we will outline an algorithm to compute the facet-defining inequalities of $\mathsf{P} = \operatorname{conv}(V)$ for given V. We have to be careful with what we mean by *effectively*. Complexity theoretically, this is bad and it is easy to write down simple inequalities that will bring any computer to its knees. Nevertheless, it is amazing how well the algorithm works in practice, and we invite you to try it out with the computer systems `polymake` [**68**] or SAGE [**55**].

Polarity for convex cones as in Exercise 3.22 saved us from proving both directions of Theorem 3.2.5 independently. Polarity for polyhedra and general convex sets is a powerful tool that (sadly) does not play much of a role for this book. For much more on polyhedra, we refer to [**78**, **190**] for the combinatorial side as well as [**15**, **76**] for a more metric touch.

The 3-dimensional case of the Euler–Poincaré formula (3.4.8) was proved by Leonard Euler in 1752 [**61**, **62**]. The full (i.e., higher-dimensional) version of (3.4.8) was discovered by Ludwig Schläfli in 1852 (though published only in 1902 [**150**]), but Schläfli's proof implicitly assumed that every polytope is shellable (as did numerous proofs of (3.4.8) that followed Schläfli's), a fact that was established only in 1971 by Heinz Bruggesser and Peter Mani [**41**]. The first airtight proof of (3.4.8) (in 1893, using tools from algebraic topology) is due to Henri Poincaré [**137**] (see also, e.g., [**81**, Theorem 2.44]). Despite its simplicity, the Euler characteristic is a very powerful valuation that brings together the geometry of polyconvex sets and their combinatorics; see [**99**]. Our approach to the Euler characteristic was inspired by [**112**].

There is a shorter path to the Möbius function of polyhedra hinted at in Exercises 3.56 and 3.57, but tangent cones will be important for us throughout. For example, there is no Brianchon–Gram relation without them.

It is interesting to note that the Euler–Poincaré formula (3.4.8) is the only linear relation, up to scaling, satisfied by all f-vectors of d-dimensional polytopes; see, e.g., [**78**, Section 8.1]. This is in stark contrast to f-vectors of simplicial polytopes and the Dehn–Sommerville relations (Theorem 3.5.5), named after Max Dehn, who proved their 5-dimensional instance in 1905 [**52**], and D. M. Y. Sommerville, who established the general case in 1927 [**158**]. That the Dehn–Sommerville relations follow from the reciprocity theorem for the zeta polynomial of the face lattice of a simplicial polytope was realized by Richard Stanley in [**162**] at the inception of zeta polynomials.

As already mentioned in Chapter 1, classifying face numbers is a major research problem. In dimension 3 this question is answered by Steinitz's theorem [**176**]: the f-vectors of 3-polytopes are the lattice points in a (translate of) a 2-dimensional polyhedral cone; see [**190**, Lecture 4]. The classification question in dimension 4 is still open. For special classes, such as the class of simplicial polytopes, there is a complete characterization of f-vectors. This is the *g-Theorem*, which was conjectured by Peter McMullen [**119**] and proved by Louis Billera and Carl Lee [**30**] and Stanley [**164**]; see also [**121**]. A main part in this characterization is a description of the conical hull of the set of f-vectors of simplicial d-polytopes. The Dehn–Sommerville relations in terms of h-vectors, defined in Section 5.6, turn out to be very important for that. Theorem 3.6.4 was part of Thomas Zaslavsky's Ph.D. thesis, which started the modern theory of hyperplane arrangements [**187**]. An approach to characteristic polynomials of hyperplane arrangements and, more generally, subspace arrangements by way of valuations is [**56**]. A nice survey article on the combinatorics of hyperplane arrangements is [**169**]. The study of *complex* hyperplane arrangement, that is, arrangements of codimension-1 subspaces in $\mathbb{C}^d$, gives rise to numerous interesting topological considerations [**130**]; for starters, a complex hyperplane does not separate $\mathbb{C}^d$ into two connected components.

The 3-dimensional case of the Brianchon–Gram relation (Theorem 3.7.1) was discovered by Charles Julien Brianchon in 1837 [**38**] and—as far as we know—independently reproved by Jørgen Gram in 1874 [**73**]. It is not clear who first proved the general d-dimensional case of the Brianchon–Gram relation; the oldest proofs we could find were from the 1960s [**78, 100, 134, 155**].

Exercises

3.1 ⬠ Show that every nonempty affine subspace $\mathsf{L} \subseteq \mathbb{R}^d$ is of the form $\mathsf{L} = \mathbf{p} + \mathsf{L}_0$ for some linear subspace $\mathsf{L}_0 \subseteq \mathbb{R}^d$ and $\mathbf{p} \in \mathbb{R}^d$.

3.2 ⬠ Let $\mathsf{Q} = \bigcap_{i=1}^{m} \mathsf{H}_i^{\leq}$ be a (possibly empty) polyhedron. An irredundant presentation is a set $I \subseteq [m]$ such that $\mathsf{Q} = \bigcap_{i \in I} \mathsf{H}_i^{\leq}$.

(a) Find a polyhedron Q that has two irredundant presentations I, I' with $|I| \neq |I'|$. (*Hint:* Even a point in the plane will do.)

(b) For $\mathsf{Q} \subseteq \mathbb{R}^d$ an affine subspace of dimension k, can you determine the possible sizes $|I|$?

3.3 ⬠ Prove that a polyhedron $\mathsf{Q} \subseteq \mathbb{R}^d$ is a polyhedral cone if and only if

$$\mathsf{Q} = \left\{ \mathbf{x} \in \mathbb{R}^d : \mathbf{A}\,\mathbf{x} \leq \mathbf{0} \right\}$$

for some matrix $\mathbf{A}$.

3.4 ⬠ Let $\mathsf{Q} = \{\mathbf{x} : A\,\mathbf{x} \leq b\}$ be a nonempty polyhedron.

(a) Show that

$$\operatorname{rec}(\mathsf{Q}) = \{\mathbf{x} : A\,\mathbf{x} \leq \mathbf{0}\} .$$

(b) Infer that $\mathbf{p} + \mathbb{R}\mathbf{u} \subseteq \mathsf{Q}$ for all $\mathbf{p} \in \mathsf{Q}$.

(c) What goes wrong if Q is empty?

3.5 ⬠ Use Exercises 3.4 to prove Proposition 3.1.2.

3.6 What is the homogenization of an affine subspace?

3.7 Let $\mathsf{Q}, \mathsf{Q}' \subseteq \mathbb{R}^d$.

(a) Show that if $\mathsf{Q} \cong \mathsf{Q}'$, then $\operatorname{hom}(\mathsf{Q}) \cong \operatorname{hom}(\mathsf{Q}')$.

(b) Show that the converse is not true.

(c) Let $\mathbf{A} \in \mathbb{R}^{d\times d}$ be invertible, $\mathbf{b}, \mathbf{c} \in \mathbb{R}^d$, and $\delta \in \mathbb{R}$ such that $(\mathbf{c}, \delta) \neq \mathbf{0}$. A **projective transformation** T is defined as

$$T(\mathbf{x}) \ := \ \frac{\mathbf{A}\,\mathbf{x} + \mathbf{b}}{\langle \mathbf{c}, \mathbf{x}\rangle + \delta}\,.$$

Thus, T is defined and invertible outside the hyperplane $\mathsf{H}^\infty := \{\mathbf{x} : \langle \mathbf{c}, \mathbf{x}\rangle + \delta = 0\}$. A projective transform T is **admissible** for a polyhedron Q if $\mathsf{Q}^\circ \cap \mathsf{H}^\infty = \varnothing$. A polyhedron Q' is **projectively isomorphic** to Q if there is an admissible projective transformation T such that $T(\mathsf{Q}) = \mathsf{Q}'$. Show that Q' is projectively isomorphic to Q if and only if $\hom(\mathsf{Q}')$ is linearly isomorphic to $\hom(\mathsf{Q})$.

3.8 ⬠ Let $\mathsf{Q} = \{\mathbf{x} \in \mathbb{R}^d : \mathbf{A}\,\mathbf{x} \leq \mathbf{b}\}$ be a nonempty polyhedron. Show that $\operatorname{lineal}(\mathsf{Q}) = \ker(\mathbf{A})$. Infer that $\mathbf{p} + \operatorname{lineal}(\mathsf{Q}) \subseteq \mathsf{Q}$ for all $\mathbf{p} \in \mathsf{Q}$.

3.9 The definition of lineality spaces makes sense for arbitrary convex sets K in $\mathbb{R}^d$. Show that in this more general situation, convexity implies that $\mathbf{p} + \operatorname{lineal}(\mathsf{K}) \subseteq \mathsf{K}$ for all $\mathbf{p} \in \mathsf{K}$.

3.10 ⬠ Show that a polyhedral cone C is pointed if and only if $\mathbf{p}, -\mathbf{p} \in \mathsf{C}$ implies $\mathbf{p} = \mathbf{0}$.

3.11 ⬠ Let $\mathsf{Q} = \{\mathbf{x} \in \mathbb{R}^d : \langle \mathbf{a}_i, \mathbf{x}\rangle \leq b_i \text{ for } i = 1, \dots, k\}$ be a polyhedron. For $\mathbf{p} \in \mathbb{R}^d$ and $\varepsilon > 0$, let $B(\mathbf{p}, \varepsilon)$ be the ball of radius ε centered at $\mathbf{p}$. A point $\mathbf{p} \in \mathsf{Q}$ is an **interior point** of Q if $B(\mathbf{p}, \varepsilon) \subseteq \mathsf{Q}$ for some $\varepsilon > 0$. Show that (3.1.11) equals the set of interior points of Q.

3.12 ⬠ Let $\mathsf{Q} = \{\mathbf{x} \in \mathbb{R}^d : \langle \mathbf{a}_i, \mathbf{x}\rangle \leq b_i \text{ for } i \in [k]\}$ be a polyhedron, and define $I := \{i \in [k] : \langle \mathbf{a}_i, \mathbf{p}\rangle = b_i \text{ for all } \mathbf{p} \in \mathsf{Q}\}$.

(a) Show that $\operatorname{aff}(\mathsf{Q}) = \{\mathbf{x} \in \mathbb{R}^d : \langle \mathbf{a}_i, \mathbf{x}\rangle = b_i \text{ for all } i \in I\}$.

(b) Let $\mathsf{L} = \operatorname{aff}(\mathsf{Q})$. Show that a point $\mathbf{p}$ is in the relative interior of Q if and only if

$$B(\mathbf{p}, \varepsilon) \cap \mathsf{L} \ \subseteq \ \mathsf{Q} \cap \mathsf{L}$$

for some $\varepsilon > 0$. (*Hint:* Exercise 3.11.)

(c) Show that $\mathsf{Q}^\circ = \{\mathbf{x} \in \mathsf{Q} : \langle \mathbf{a}_i, \mathbf{x}\rangle < b_i \text{ for all } i \notin I\}$.

3.13 ⬠ Show that the unit ball in the ℓ_1-norm is a bounded polyhedron.

3.14 ⬠ Let $S \subset \mathbb{R}^d$.

(a) Show that

$$\mathsf{K} \ = \ \left\{ \lambda_1\mathbf{v}_1 + \cdots + \lambda_k\mathbf{v}_k \ : \ \begin{array}{c} k \geq 0, \ \mathbf{v}_1, \dots, \mathbf{v}_k \in S \\ \lambda_1, \dots, \lambda_k \geq 0 \\ \lambda_1 + \cdots + \lambda_k = 1 \end{array} \right\}$$

is a convex set containing S.

(b) Show that if K' is a convex set containing S, then $\mathsf{K} \subseteq \mathsf{K}'$. Conclude the validity of (3.2.1).

3.15 ⬠ Let $\mathsf{P} = \operatorname{conv}(S) \subset \mathbb{R}^d$ be a polytope.

(a) Let $\mathbf{p}, \mathbf{q} \in S$. Show that if $\mathsf{P} = \operatorname{conv}(S \setminus \{\mathbf{p}\}) = \operatorname{conv}(S \setminus \{\mathbf{q}\})$, then $\mathsf{P} = \operatorname{conv}(S \setminus \{\mathbf{p}, \mathbf{q}\})$.

(b) Conclude there is a unique inclusion-minimal set $V \subseteq S$ such that $\mathsf{P} = \operatorname{conv}(V)$.

3.16 ⬠ Show that if $\triangle, \triangle'$ are two simplices of the same dimension, then there is an affine transformation T with $T(\triangle) = \triangle'$.

3.17 Show that every linear subspace is a finitely generated convex cone. What are the possible sizes of inclusion-minimal sets of generators?

3.18 ⬠ Let $S \subset \mathbb{R}^d$ be a compact and convex set. Show that

$$\operatorname{hom}(S) \;=\; \operatorname{cone}(S \times \{1\}) .$$

3.19 ⬠ Let $\mathbf{s}_1, \dots, \mathbf{s}_k \subset \mathbb{R}^d$. Show that

$$\operatorname{cone}(\mathbf{s}_1, \dots, \mathbf{s}_k) = \{\mu_1 \mathbf{s}_1 + \cdots + \mu_k \mathbf{s}_k \,:\, \mu_1, \dots, \mu_k \geq 0\} .$$

3.20 ⬠ Let $\mathsf{C} = \operatorname{cone}(\mathbf{u}_1, \dots, \mathbf{u}_k)$.

(a) Show that $\operatorname{lineal}(\mathsf{C}) \neq \{\mathbf{0}\}$ if and only if there is some $\mathbf{p} \in \mathsf{C} \setminus \{\mathbf{0}\}$ with $-\mathbf{p} \in \mathsf{C}$.

(b) Show that there is some $\mathbf{p} \neq \mathbf{0}$ with $\pm\mathbf{p} \in \mathsf{C}$ if and only if there are $\mu_1, \dots, \mu_k \geq 0$ not all zero such that

$$\mathbf{0} \;=\; \mu_1 \mathbf{u}_1 + \cdots + \mu_k \mathbf{u}_k .$$

3.21 ⬠ Let $\mathsf{K} \subset \mathbb{R}^d$ be a closed convex set and $\mathbf{p} \in \mathbb{R}^d \setminus \mathsf{K}$.

(a) Show that there is a unique point $\mathbf{q} \in \mathsf{K}$ such that $\|\mathbf{q} - \mathbf{p}\|_2 \leq \|\mathbf{q}' - \mathbf{p}\|_2$ for all $\mathbf{q}' \in \mathsf{K}$. (*Hint:* Reduce to the case that K is bounded (and hence compact) for existence and convexity for uniqueness.)

(b) Show that

$$\mathsf{H} \;:=\; \left\{\mathbf{x} \in \mathbb{R}^d \,:\, \|\mathbf{x} - \mathbf{q}\|_2 = \|\mathbf{x} - \mathbf{p}\|_2\right\}$$

is a hyperplane.

(c) Conclude that $\mathcal{H}$ is a separating hyperplane, which then proves the Separation Theorem 3.2.3.

3.22 ⬠ For $S \subseteq \mathbb{R}^d$, we define the **cone polar** of S as

$$S^\vee \;:=\; \{\mathbf{y} \in \mathbb{R}^d \,:\, \langle \mathbf{y}, \mathbf{x} \rangle \leq 0 \text{ for all } \mathbf{x} \in S\} .$$

(a) Show that $S^\vee$ is a closed convex cone.

(b) Show that if $\mathsf{C} = \operatorname{cone}(\mathbf{u}_1, \dots, \mathbf{u}_k)$, then

$$\mathsf{C}^\vee \;=\; \left\{\mathbf{y} \in \mathbb{R}^d \,:\, \langle \mathbf{u}_i, \mathbf{y} \rangle \leq 0 \text{ for } i = 1, \dots, k\right\} .$$

(c) Show that if C is a closed convex cone, then $(\mathsf{C}^\vee)^\vee = \mathsf{C}$. (*Hint:* Use the Separation Theorem 3.2.3 and modify the hyperplane so that it passes through the origin.)

(d) Assume the following statement: *If* C *is a polyhedral cone, then* C *is a finitely generated cone.* Show that the converse then also holds: *If* C *is a finitely generated cone, then* C *is a polyhedral cone.*

3.23 ⬠ Verify that the three conditions for affine independence on page 63 are equivalent.

3.24 Recall that an affine subspace is of the form $\mathsf{L} = \mathbf{p} + \mathsf{L}_0$, where L_0 is a linear subspace.

(a) Let $\mathbf{p}_1, \ldots, \mathbf{p}_k \in \mathsf{L}$ and $\alpha_1, \ldots, \alpha_k \in \mathbb{R}$ such that $\alpha_1 + \cdots + \alpha_k = 1$. Show that $\alpha_1\mathbf{p}_1 + \cdots + \alpha_k\mathbf{p}_k \in \mathsf{L}$. This is called an **affine linear combination**.

(b) For $\mathbf{p}_1, \ldots, \mathbf{p}_k \in \mathbb{R}^d$, show that $\operatorname{aff}(\mathbf{p}_1, \ldots, \mathbf{p}_k)$, the intersection of all affine subspaces containing $\mathbf{p}_1, \ldots, \mathbf{p}_k$, is exactly the set of affine linear combinations of $\mathbf{p}_1, \ldots, \mathbf{p}_k$.

(c) Show that if L is of dimension k, then there are points $\mathbf{p}_0, \ldots, \mathbf{p}_k$ such that every point in L can be expressed by a unique affine linear combination of $\mathbf{p}_0, \ldots, \mathbf{p}_k$.

3.25 ⬠ Show that if $\mathsf{K}_1, \mathsf{K}_2 \subseteq \mathbb{R}^d$ are convex, then so is the Minkowski sum $\mathsf{K}_1 + \mathsf{K}_2$. Show that if $\mathsf{K}_1, \mathsf{K}_2$ are polytopes, then $\mathsf{K}_1 + \mathsf{K}_2$ is a polytope.

3.26 ⬠ Let $\mathsf{Q} = \mathsf{P} + \mathsf{C} \subset \mathbb{R}^d$, where $\mathsf{P} = \operatorname{conv}(\mathbf{v}_1, \ldots, \mathbf{v}_r)$ is a polytope and $\mathsf{C} = \operatorname{cone}(\mathbf{u}_1, \ldots, \mathbf{u}_s)$ is a finitely generated cone. Show that the homogenization Q is given by

$$\operatorname{hom}(\mathsf{Q}) \;=\; \operatorname{cone}\left\{\begin{pmatrix}\mathbf{v}_1\\1\end{pmatrix}, \ldots, \begin{pmatrix}\mathbf{v}_r\\1\end{pmatrix}, \begin{pmatrix}\mathbf{u}_1\\0\end{pmatrix}, \ldots, \begin{pmatrix}\mathbf{u}_s\\0\end{pmatrix}\right\}.$$

3.27 Show that if $\mathsf{Q} = \{\mathbf{x} \in \mathbb{R}^d : \langle \mathbf{a}_i, \mathbf{x}\rangle \le b_i\}$ is a full-dimensional polyhedron, then it has a unique irredundant description, up to scaling.

3.28 ⬠ Prove Corollary 3.2.7: Let $\mathsf{Q} \subset \mathbb{R}^d$ be a polyhedron and $\phi(\mathbf{x}) = \mathbf{A}\,\mathbf{x} + \mathbf{b}$ an affine projection $\mathbb{R}^d \to \mathbb{R}^e$. Then $\phi(\mathsf{Q})$ is a polyhedron. If $\mathsf{P} \subset \mathbb{R}^d$ is a polytope, then $\mathsf{P} \cap \mathsf{Q}$ is a polytope.

3.29 ⬠ Show that every polyhedral cone of dimension at most 2 is finitely generated.

3.30 ⬠ Prove that a face of a (bounded) polyhedron is again a (bounded) polyhedron.

3.31 ⬠ Let $V \subset \mathbb{R}^d$ be a finite set and $\mathsf{P} = \operatorname{conv}(V)$ the associated polytope. Let $\mathbf{v} \in V$ be arbitrary.

(a) Show that if $\{\mathbf{v}\}$ is a face of P, then $\mathsf{P} \neq \operatorname{conv}(V \setminus \mathbf{v})$.

(b) Show that if $\mathsf{P} \neq \operatorname{conv}(V \setminus \mathbf{v})$, then $\{\mathbf{v}\}$ is a face of P. (*Hint:* Use Exercise 3.21 and modify the hyperplane to be supporting at $\mathbf{v}$.)

3.32 ⬠ Recall from Theorem 3.2.5 that every polyhedron Q is of the form $\mathsf{Q} = \mathsf{P} + \mathsf{C}$, where P is a polytope and C is a polyhedral cone. Prove that for each nonempty face $\mathsf{F} \preceq \mathsf{Q}$, there are unique faces $\mathsf{F}' \preceq \mathsf{P}$ and $\mathsf{F}'' \preceq \mathsf{C}$ such that $\mathsf{F} = \mathsf{F}' + \mathsf{F}''$.

3.33 ⬠ Prove Proposition 3.3.1 for the general case that Q is a polyhedron.

3.34 ⬠ In the proof of Proposition 3.3.1, let

$$\nu \ := \ \max\left(\{\langle \mathbf{a}', \mathbf{v}\rangle - b' \,:\, \mathbf{v} \in \operatorname{vert}(\mathsf{Q}) \setminus \mathsf{F}\} \cup \{0\}\right).$$

Show that it suffices to choose

$$0 \ < \ \varepsilon \ < \ \frac{\eta}{\nu}.$$

(In case $\nu = 0$, H' is already supporting for Q and there is no restriction on ε.)

3.35 ⬠ Prove Corollary 3.3.3: The face lattice of a polyhedron is a graded poset. If Q is a pointed polyhedron, then the rank of a face F is $\dim \mathsf{F} + 1$.

3.36 ⬠ Prove Lemma 3.3.4: Let Q be a polyhedron with lineality space L. The map $\Phi(\mathsf{Q}) \to \Phi(\mathsf{Q}/\mathsf{L})$ given by $\mathsf{F} \mapsto \mathsf{F}/\mathsf{L}$ is an isomorphism of face lattices.

3.37 ⬠ Let $\mathsf{Q} \subset \mathbb{R}^d$ be a nonempty full-dimensional polyhedron and H a hyperplane such that $\mathsf{Q}^\circ \cap \mathsf{H} \neq \varnothing$. Show that $\dim(\mathsf{Q}^\circ \cap \mathsf{H}) = \dim \mathsf{Q} - 1$. (*Hint:* Show that $\operatorname{aff}(\mathsf{Q}^\circ \cap \mathsf{H}) = \mathsf{H}$.)

3.38 ⬠ Prove that, if F and G are faces of a polyhedron Q that are unrelated in the face lattice (i.e., $\mathsf{F} \not\subseteq \mathsf{G}$ and $\mathsf{G} \not\subseteq \mathsf{F}$), then

$$\mathsf{F}^\circ \cap \mathsf{G}^\circ \ = \ \varnothing.$$

3.39 Show that cross-polytopes are simplicial polytopes.

3.40 ⬠ Prove Proposition 3.3.10: Let $\mathsf{P}' = \mathbf{v} * \mathsf{P}$ be a pyramid. For each face $\mathsf{F} \preceq \mathsf{P}$, the polytope $\mathbf{v} * \mathsf{F}$ is a face of P'. Conversely, every face $\mathsf{F}' \preceq \mathsf{P}'$ is either a face of P or is of the form $\mathsf{F}' = \mathbf{v} * \mathsf{F}$ for some face $\mathsf{F} \preceq \mathsf{P}$.

3.41 Let $\mathsf{P} = \operatorname{conv}(\mathbf{p}_1, \dots, \mathbf{p}_m)$ be a polytope. Show that a point $\mathbf{q}$ is in the relative interior of P if there are $\lambda_1, \dots, \lambda_m > 0$ such that

$$\mathbf{q} = \lambda_1 \mathbf{p}_1 + \cdots + \lambda_m \mathbf{p}_m \quad \text{and} \quad \lambda_1 + \cdots + \lambda_m = 1.$$

When is this an *if and only if* condition?

3.42 Prove that, if $\mathsf{C} = \operatorname{cone}(\mathbf{s}_1, \mathbf{s}_2, \dots, \mathbf{s}_m)$ (written in an irredundant form), then

$$\mathsf{C}^\circ \ = \ \{\lambda_1 \mathbf{s}_1 + \lambda_2 \mathbf{s}_2 + \cdots + \lambda_m \mathbf{s}_m \,:\, \lambda_1, \lambda_2, \dots, \lambda_m > 0\}.$$

3.43 Let $\operatorname{vert}(\mathsf{Q}_1) = \{\mathbf{u}_1, \mathbf{u}_2, \ldots, \mathbf{u}_m\}$ and $\operatorname{vert}(\mathsf{Q}_2) = \{\mathbf{v}_1, \mathbf{v}_2, \ldots, \mathbf{v}_n\}$, and consider a point $\mathbf{s}+\mathbf{t} \in \mathsf{Q}_1+\mathsf{Q}_2$. Then there are coefficients $\lambda_1, \ldots, \lambda_m \geq 0$ and $\mu_1, \ldots, \mu_n \geq 0$ such that $\sum_i \lambda_i = \sum_j \mu_j = 1$ and

$$\mathbf{s}+\mathbf{t} \;=\; \sum_{i=1}^{m} \lambda_i \mathbf{u}_i \;+\; \sum_{j=1}^{n} \mu_j \mathbf{v}_j \,.$$

Now set $\alpha_{ij} = \lambda_i \, \mu_j \geq 0$ for $(i,j) \in [m] \times [n]$. Prove that

$$\mathbf{s}+\mathbf{t} \;=\; \sum_{(i,j)\in[m]\times[n]} \alpha_{ij}(\mathbf{u}_i + \mathbf{v}_j)\,.$$

3.44 Recall that two affine subspaces $\mathsf{L}, \mathsf{L}' \subset \mathbb{R}^d$ are **skew** if L and L are not parallel (equivalently, $\dim \mathsf{L} + \mathsf{L}' = \dim \mathsf{L} + \dim \mathsf{L}'$) and $\mathsf{L} \cap \mathsf{L} = \varnothing$. Let $\mathsf{Q}, \mathsf{Q}' \subset \mathbb{R}^d$ such that $\operatorname{aff}(\mathsf{Q})$ and $\operatorname{aff}(\mathsf{Q}')$ are skew. We define the **join** as

$$\mathsf{Q} * \mathsf{Q}' \;:=\; \operatorname{conv}(\mathsf{Q} \cup \mathsf{Q}')\,.$$

Show that

$$\Phi(\mathsf{Q} * \mathsf{Q}') \;\cong\; \Phi(\mathsf{Q}) \times \Phi(\mathsf{Q}')\,,$$

where the latter is the direct product of face posets; see Exercise 2.6. Deduce that

$$f_i(\mathsf{Q} * \mathsf{Q}') \;=\; \sum_{\substack{k,l\geq -1\\ k+l=i-1}} f_k(\mathsf{Q})\, f_l(\mathsf{Q}')\,.$$

3.45 $\pentagon$ Verify Proposition 3.3.11: $(\Phi([-1,1]^d)\setminus\{\varnothing\},\subseteq) \;\cong\; (\{-,0,+\}^d,\preceq)$.

3.46 $\pentagon$ Let $\mathsf{Q} \subset \mathbb{R}^d$, $\mathsf{Q}' \subset \mathbb{R}^e$ be two polyhedra.

(a) Show that $\mathsf{Q} \times \mathsf{Q}'$ is a polyhedron.

(b) Show that

$$\Phi(\mathsf{Q} \times \mathsf{Q}')\setminus\{\varnothing\} \;\cong\; (\Phi(\mathsf{Q})\setminus\{\varnothing\}) \times (\Phi(\mathsf{Q}')\setminus\{\varnothing\})\,.$$

(c) Deduce that

$$f_i(\mathsf{Q} \times \mathsf{Q}') \;=\; \sum_{\substack{k,l\geq 0\\ k+l=i}} f_k(\mathsf{Q})\, f_l(\mathsf{Q}')\,.$$

3.47 Let $\mathsf{Q}, \mathsf{Q}' \subset \mathbb{R}^d$ be polyhedra such that $\mathsf{Q}^\circ \cap (\mathsf{Q}')^\circ = \{\mathbf{p}\}$ for some $\mathbf{p} \in \mathbb{R}^d$. We define the **direct sum** or **free sum**

$$\mathsf{Q} \oplus \mathsf{Q}' \;:=\; \operatorname{conv}(\mathsf{Q} \cup \mathsf{Q}')\,.$$

(a) Show that $\mathsf{Q} \oplus \mathsf{Q}'$ is a polyhedron.

(b) Show that

$$\Phi(\mathsf{Q} \oplus \mathsf{Q}')\setminus\{\mathsf{Q} \oplus \mathsf{Q}'\} \;\cong\; (\Phi(\mathsf{Q})\setminus\{\mathsf{Q}\}) \times (\Phi(\mathsf{Q}')\setminus\{\mathsf{Q}'\})\,.$$

(c) Show that the convex hull of

$$\mathsf{Q} \times \{\mathbf{0}\} \times \{-1\} \cup \{\mathbf{0}\} \times \mathsf{Q}' \times \{+1\}$$

is a join in the sense of Exercise 3.44. Can you find a relation between $\mathsf{Q} * \mathsf{Q}'$, $\mathsf{Q} \times \mathsf{Q}'$, and $\mathsf{Q} \oplus \mathsf{Q}'$?

3.48 Let $\mathsf{Q} = \{\mathbf{x} : \mathbf{A}\,\mathbf{x} \leq \mathbf{b}\}$ be a nonempty polyhedron and let $\mathsf{H} = \{\mathbf{x} : \langle \mathbf{w}, \mathbf{x} \rangle = \delta\}$ be a supporting hyperplane with face F. We define the **wedge** of Q at F as the polyhedron

$$\operatorname{wedge}(\mathsf{Q}, \mathsf{F}) \;:=\; \{(\mathbf{x}, t) \,:\, \mathbf{A}\,\mathbf{x} \leq \mathbf{b},\; 0 \leq t \leq \delta - \langle \mathbf{w}, \mathbf{x} \rangle\}\,.$$

Show that this is a combinatorial construction and determine the face lattice of $\operatorname{wedge}(\mathsf{Q}, \mathsf{F})$.

3.49 ⬠ Show that the unit disc $\{(x, y) \in \mathbb{R}^2 : x^2 + y^2 \leq 1\}$ is not a polyconvex set.

3.50 ⬠ Let F be a nonempty face of the polyhedron

$$\mathsf{Q} \;=\; \left\{\mathbf{x} \in \mathbb{R}^d \;:\; \langle \mathbf{a}_i, \mathbf{x} \rangle \leq b_i \text{ for } i \in [k]\right\}$$

and let G be a face of Q that contains F. There is a subset $I \subseteq [k]$ of indices such that

$$\mathsf{F} \;=\; \{\mathbf{x} \in \mathsf{Q} \,:\, \langle \mathbf{a}_i, \mathbf{x} \rangle = b_i \text{ for } i \in I\}\,.$$

Show that there exists an inclusion-maximal subset $J \subseteq I$ such that

$$\mathsf{G} \;=\; \{\mathbf{x} \in \mathsf{Q} \,:\, \langle \mathbf{a}_i, \mathbf{x} \rangle = b_i \text{ for } i \in J\}\,.$$

3.51 Suppose V is the set of vertices of the polytope P, and F is a face of P. Prove that F is the convex hull of $V \cap \mathsf{F}$.

3.52 ⬠ Let $\mathcal{H}$ be an arrangement of hyperplanes in $\mathbb{R}^d$ and $\mathsf{PC}(\mathcal{H})$ the collection of $\mathcal{H}$-polyconvex sets.

(a) Let $\mathbb{G}$ be some fixed Abelian group. Let ϕ be a map that assigns any nonempty relatively polyhedron H_σ a value $\phi(\mathsf{H}_\sigma) \in \mathbb{G}$. For $S \in \mathsf{PC}(\mathcal{H})$ define

$$\phi(S) \;:=\; \phi(\mathsf{H}_{\sigma^1}) + \phi(\mathsf{H}_{\sigma^2}) + \cdots + \phi(\mathsf{H}_{\sigma^k})$$

through (3.4.4). Show that $\phi : \mathsf{PC}(\mathcal{H}) \to \mathbb{G}$ is a valuation on $\mathsf{PC}(\mathcal{H})$.

(b) Conclude that the function $\chi(\mathcal{H}, \cdot) : \mathsf{PC}(\mathcal{H}) \to \mathbb{Z}$ is a valuation.

3.53 ⬠ Prove Proposition 3.4.7: Let P be a nonempty polytope with $\mathbf{0} \in \mathsf{P}^\circ$. For every proper face $\mathsf{F} \prec \mathsf{P}$, the set $\mathsf{C}_\mathbf{0}(\mathsf{F})$ is a relatively open polyhedral cone of dimension $\dim \mathsf{F} + 1$. Furthermore,

$$\mathsf{C}_\mathbf{0}(\mathsf{P}) \;=\; \operatorname{aff}(\mathsf{P}) \;=\; \biguplus_{\mathsf{F} \prec \mathsf{P}} \mathsf{C}_\mathbf{0}(\mathsf{F})\,.$$

3.54 ☖ Prove Corollary 3.4.8: If Q is a polyhedron with lineality space $\mathsf{L} = \mathrm{lineal}(\mathsf{Q})$, then

$$\chi(\mathsf{Q}) \;=\; (-1)^{\dim \mathsf{L}}\, \chi(\mathsf{Q}/\mathsf{L})\,.$$

3.55 ☖ Verify our assertions about faces in the proof of Proposition 3.4.9, namely, that each k-face F of C gives rise to a k-face of $\overline{\mathsf{C}}$ and, if F is unbounded, a $(k-1)$-face of C_∞, and that these are all the faces of $\overline{\mathsf{C}}$ and C_∞.

3.56 Let $\mathsf{Q} \subset \mathbb{R}^d$ be a pointed polyhedron of dimension d. Let $\mathbf{v}$ be a vertex of Q and $\mathsf{H} = \{\mathbf{x} \in \mathbb{R}^d : \langle \mathbf{a}, \mathbf{x}\rangle = \delta\}$ a hyperplane such that $\mathsf{Q} \subseteq \mathsf{H}^{\leq}$ and $\{\mathbf{v}\} = \mathsf{Q} \cap \mathsf{H}$. For $\varepsilon > 0$, define $\mathsf{H}_\varepsilon = \{\mathbf{x} \in \mathbb{R}^d : \langle \mathbf{a}, \mathbf{x}\rangle = \delta - \varepsilon\}$.

(a) For every sufficiently small $\varepsilon > 0$, show that the polytope

$$\mathsf{P}(\mathsf{H}, \mathbf{v}) \;:=\; \mathsf{Q} \cap \mathsf{H}_\varepsilon$$

has the following property: There is a bijection between the k-faces of $\mathsf{P}(\mathsf{H}, \mathbf{v})$ and the $(k+1)$-faces of Q containing $\mathbf{v}$, for all $k = -1, \dots, d-1$.

This shows that the face lattice $\Phi(\mathsf{P}(\mathsf{H}, \mathbf{v}))$ is isomorphic to the interval $[\mathbf{v}, \mathsf{Q}]$ in $\Phi(\mathsf{Q})$ and, in particular, independent of the choice of H. We call any such polytope $\mathsf{P}(\mathsf{H}, \mathbf{v})$ the **vertex figure** of Q at $\mathbf{v}$ and denote it by $\mathsf{Q}/\mathbf{v}$.

(b) Use Exercise 3.7 to prove that $\mathsf{P}(\mathsf{H}, \mathbf{v})$ and $\mathsf{P}(\mathsf{H}', \mathbf{v})$ are projectively isomorphic for any two supporting hyperplanes H, H' for $\mathbf{v}$. (This justifies calling $\mathsf{Q}/\mathbf{v}$ *the* vertex figure of Q at $\mathbf{v}$.)

(c) Show that $\Phi(\mathsf{Q}/\mathbf{v})$ is isomorphic to the interval $[\mathbf{v}, \mathsf{Q}]$ in $\Phi(\mathsf{Q})$.

3.57 Continuing Exercise 3.56, we call a polytope P a **face figure** of Q at the face G if $\Phi(\mathsf{P})$ is isomorphic to the interval $[\mathsf{G}, \mathsf{Q}]$ in $\Phi(\mathsf{Q})$. We denote any such polytope by Q/G.

(a) Verify that the following inductive construction yields a face figure: If $\dim \mathsf{G} = 0$, then Q/G is the vertex figure of Exercise 3.56. If $\dim \mathsf{G} > 0$, pick a face $\mathsf{F} \prec \mathsf{G}$ of dimension $\dim \mathsf{G} - 1$. Then Q/F exists and, in particular, there is a vertex $\mathbf{v}_\mathsf{G}$ of Q/F corresponding to G. We define the face figure Q/G as the vertex figure $(\mathsf{Q}/\mathsf{F})/\mathbf{v}_\mathsf{G}$.

(b) As an application of face figures, show that the Möbius function of a polyhedron is given by $\mu_\Phi(\mathsf{F}, \mathsf{G}) = \chi((\mathsf{G}/\mathsf{F})^\circ)$.

3.58 ☖ Let $\mathsf{Q} \subseteq \mathbb{R}^d$ be a polyhedron and $\mathsf{F} \preceq \mathsf{Q}$ a face. Show that

$$\mathrm{hom}(\mathrm{T}_\mathsf{F}(\mathsf{Q})) = \mathrm{T}_{\mathrm{hom}(\mathsf{F})}(\mathrm{hom}(\mathsf{Q}))\,.$$

3.59 ☖ Show (3.5.6): if P is simplicial, then

$$\Delta Z_{\Phi(\mathsf{P})}(n) \;=\; 1 + \sum_{\varnothing \prec \mathsf{F} \prec \mathsf{P}} n^{\dim(\mathsf{F})+1} \;=\; \sum_{k=0}^{d} f_{k-1}\, n^k.$$

3.60 ○ Show that $\overline{\chi}(\mathbb{R}) = 1$.

3.61 Real-valued valuations $\phi : \mathsf{PC}_d \to \mathbb{R}$ on $\mathbb{R}^d$ form a vector space Val and valuations that satisfy the properties of Theorem 3.4.1 constitute a vector subspace $U \subseteq \mathsf{Val}$. For the real line, what is the dimension of U?

3.62 ○ Prove that $\mathbf{p} \in \mathbb{R}^d$ is beyond the face F of a given polyhedron if and only if all points in F are visible from $\mathbf{p}$.

3.63 Let $\mathsf{P} \subset \mathbb{R}^d$ be a full-dimensional polytope. Show that for every $\mathbf{p} \in \mathbb{R}^d$, there is at least one facet F such that $\mathbf{p}$ is beneath F.

3.64 ○ Prove that $\mathcal{L}(\mathcal{H})$ is a meet semilattice, that is, any two elements in $\mathcal{L}(\mathcal{H})$ have a meet. Furthermore, show that $\mathcal{L}(\mathcal{H})$ is a lattice if and only if $\mathcal{H}$ is central.

3.65 Let $\mathcal{H} = \{x_j = 0 : 1 \le j \le d\}$, the d-dimensional **Boolean arrangement** consisting of the coordinate hyperplanes in $\mathbb{R}^d$. (Figure 3.21 shows an example.) Show that $r(\mathcal{H}) = 2^d$.

3.66 Let $\mathcal{H} = \{x_j = x_k : 1 \le j < k \le d\}$, the d-dimensional **real braid arrangement**. Show that $r(\mathcal{H}) = d!$.

3.67 Let $\mathcal{H}$ be an arrangement in $\mathbb{R}^d$ consisting of k hyperplanes in **general position**, i.e., each j-dimensional flat of $\mathcal{H}$ is the intersection of exactly $d-j$ hyperplanes, and any $d-j$ hyperplanes intersect in a j-dimensional flat. (One example is pictured in Figure 3.25.) Then

$$r(\mathcal{H}) = \binom{k}{d} + \binom{k}{d-1} + \cdots + \binom{k}{1} + \binom{k}{0}.$$

What can you say about $b(\mathcal{H})$?

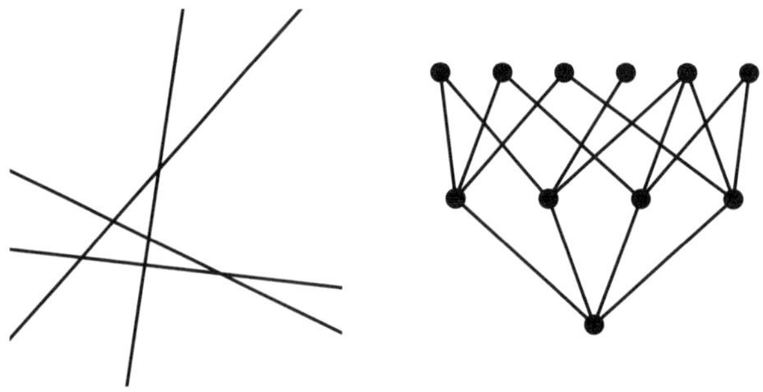

Figure 3.25. An arrangement of four lines in general position and its intersection poset.

3.68 Let $\mathcal{H}$ be an arrangement in $\mathbb{R}^d$ consisting of k hyperplanes. Show that

$$r(\mathcal{H}) \le \binom{k}{d} + \binom{k}{d-1} + \cdots + \binom{k}{1} + \binom{k}{0},$$

that is: the number of regions of $\mathcal{H}$ is bounded by the number of regions created by k hyperplanes in $\mathbb{R}^d$ in general position.

3.69 Let $\mathsf{P} \subset \mathbb{R}^d$ be a full-dimensional polytope, $\mathbf{v} \in \mathbb{R}^d \setminus \mathsf{P}$, and $\mathsf{F}_1, \dots, \mathsf{F}_m$ the facets of P.

(a) Assuming that $\mathsf{F}_i = \mathsf{P} \cap \{\mathbf{x} : \langle \mathbf{a}_i, \mathbf{x} \rangle = b_i\}$ and $\langle \mathbf{a}_i, \mathbf{p} \rangle \leq b_i$ for all $\mathbf{p} \in \mathsf{P}$, show that $\mathbf{v}$ is beyond F_i if and only if $\langle \mathbf{a}_i, \mathbf{v} \rangle > b_i$.

(b) Show that $\mathbf{v}$ is beyond a face F of P if and only if $\mathbf{v}$ is beyond some facet F_i containing F.

3.70 Give an elementary proof for the Brianchon–Gram relation (Theorem 3.7.1) for simplices.

Chapter 4

Rational Generating Functions

The mathematical phenomenon always develops out of simple arithmetic, so useful in everyday life, out of numbers, those weapons of the gods: the gods are there, behind the wall, at play with numbers.
Le Corbusier

We now return to a theme started in Chapter 1: counting functions that are polynomials. Before we can ask about possible interpretations of these counting functions at negative integers—the theme of this book—, we need structural results such as Proposition 1.1.1, which says that the chromatic polynomial is indeed a polynomial. We hope that we conveyed the message in Chapter 1 that such a structural result can be quite nontrivial for a given counting function. Another example is given by the zeta polynomials of Section 2.3—their definition $Z_\Pi(n) := \zeta_\Pi^n(\hat{0}, \hat{1})$ certainly does not hint at the fact that $Z_\Pi(n)$ is indeed a polynomial. One of our goals in this chapter is to develop machinery that allows us to detect and study polynomials. We will do so alongside introducing several families of counting functions and, naturally, we will discover a number of combinatorial reciprocity theorems along the way.

4.1. Matrix Powers and the Calculus of Polynomials

To warm up, we generalize in some sense the zeta polynomials from Section 2.3. A matrix $\mathbf{A} \in \mathbb{C}^{d \times d}$ is **unipotent** if $\mathbf{A} = \mathbf{I} + \mathbf{B}$, where $\mathbf{I}$ is the $d \times d$ identity matrix and there exists a positive integer k such that $\mathbf{B}^k = \mathbf{0}$

(that is, B is **nilpotent**). The zeta functions from Chapter 2 are our motivating examples of unipotent matrices. We recall that, thinking of the zeta function of a poset Π as a matrix, the entry $\zeta_\Pi(\hat{0},\hat{1})$ was crucial in Chapter 2—the analogous entry in powers of ζ_Π gave rise to the zeta polynomial (emphasis on *polynomial*!)

$$Z_\Pi(n) \;=\; \zeta_\Pi^n\left(\hat{0},\hat{1}\right).$$

We obtain a polynomial the same way from any unipotent matrix.

Proposition 4.1.1. *Let* $\mathbf{A} \in \mathbb{C}^{d\times d}$ *be a unipotent matrix, fix indices* $i,j \in [d]$, *and consider the sequence* $f(n) := (\mathbf{A}^n)_{ij}$ *formed by the* (i,j)*-entries of the* n*-th powers of* $\mathbf{A}$. *Then* $f(n)$ *agrees with a polynomial in* n.

Proof. We essentially repeat the argument behind (2.1.3) which gave rise to Proposition 2.1.5. Suppose $\mathbf{A} = \mathbf{I} + \mathbf{B}$, where $\mathbf{B}^k = \mathbf{0}$. Then

$$f(n) \;=\; ((\mathbf{I}+\mathbf{B})^n)_{ij} \;=\; \sum_{m=0}^{n} \binom{n}{m} (\mathbf{B}^m)_{ij} \;=\; \sum_{m=0}^{k-1} \binom{n}{m} (\mathbf{B}^m)_{ij}\,,$$

which is a polynomial in n. Here the second equality is the binomial theorem; see Exercise 2.1. □

At the heart of the above proof is the fact that $\binom{n}{m}$ is a polynomial in n, and in fact, viewing this binomial coefficient as a polynomial in the "top variable" lies at the core of much of the enumerative side of combinatorics, starting with (0.0.1)—the very first sample counting function in this book. Our proof also hints at the fact that the binomial coefficients $\binom{n}{m}$ form a basis for the space of polynomials. Much of what we will do in this chapter can be viewed as switching bases in one way or another. Here are the key players:

$$\begin{aligned}
\text{(M)-basis:} &\quad \{x^m \,:\, 0 \le m \le d\}\,,\\
(\gamma)\text{-basis:} &\quad \left\{x^m(1-x)^{d-m} \,:\, 0 \le m \le d\right\},\\
(\Delta)\text{-basis:} &\quad \left\{\tbinom{x}{m} \,:\, 0 \le m \le d\right\},\\
(h^*)\text{-basis:} &\quad \left\{\tbinom{x+m}{d} \,:\, 0 \le m \le d\right\}.
\end{aligned}$$

The (Δ)-basis is intimately related (via (4.1.3) below) to the Δ-operator which we will introduce momentarily. Our terminology for the (M)-basis simply stands for *monomials*; that of the (γ)- and (h^*)-bases are standard in the combinatorics literature. The latter will play a prominent role in Section 5.5.

Proposition 4.1.2. *The sets* (M), (γ), (Δ), *and* (h^*) *are bases for the vector space* $\mathbb{C}[x]_{\le d} := \{f \in \mathbb{C}[x] \,:\, \deg(f) \le d\}$.

Proof. The set (M) is the canonical basis of $\mathbb{C}[x]_{\leq d}$. An explicit change of basis from (Δ) to (M) is given in (1.3.1) and (4.1.2) below. For (γ), we observe that

$$1 \;=\; (x+(1-x))^d \;=\; \sum_{i=0}^{d} \binom{d}{i} x^i (1-x)^{d-i} \tag{4.1.1}$$

gives an explicit linear combination of x^0. You will pursue this idea further in Exercise 4.2. Finally, for (h^*), it is sufficient to show that its $d+1$ polynomials are linearly independent; see Exercise 4.2. □

Looking back once more to Chapter 1, a geometric passage from (h^*) to (Δ) is implicit in (1.4.4). This will be much clearer after Section 4.6.

Our proof of Proposition 4.1.1 is even closer to the principal structures of polynomials than one might think. To this end, we consider three linear operators on the vector space $\{(f(n))_{n\geq 0}\}$ of all complex-valued sequences:

$$\begin{aligned}
(If)(n) &:= f(n) && \textit{(identity operator)},\\
(\Delta f)(n) &:= f(n+1)-f(n) && \textit{(difference operator)},\\
(Sf)(n) &:= f(n+1) && \textit{(shift operator)}.
\end{aligned}$$

They are naturally related through $S = I + \Delta$. Aficionados of calculus will anticipate the following result; we would like to emphasize the analogy of its proof with that of Proposition 4.1.1.

Proposition 4.1.3. *A sequence $f(n)$ is given by a polynomial of degree $\leq d$ if and only if $(\Delta^m f)(0) = 0$ for all $m > d$.*

Proof. We start by noting that $f(n) = (S^n f)(0)$. If $(\Delta^m f)(0) = 0$ for all $m > d$, then

$$\begin{aligned}
f(n) &= \left((I+\Delta)^n f\right)(0)\\
&= \sum_{m=0}^{n} \binom{n}{m} (\Delta^m f)(0) \;=\; \sum_{m=0}^{d} \binom{n}{m} (\Delta^m f)(0)\,,
\end{aligned} \tag{4.1.2}$$

a polynomial of degree $\leq d$.

Conversely, if $f(n)$ is a polynomial of degree $\leq d$, then we can express it in terms of the (Δ)-basis, i.e., $f(n) = \sum_{m=0}^{d} \alpha_m \binom{n}{m}$ for some $\alpha_0, \alpha_1, \dots, \alpha_d \in \mathbb{C}$. Now, observe that

$$\Delta \binom{n}{m} \;=\; \binom{n}{m-1} \tag{4.1.3}$$

and hence $\Delta f = \sum_{m=1}^{d} \alpha_m \binom{n}{m-1}$ is a polynomial of degree $\leq d-1$. Induction shows that $\Delta^m f = 0$ for $m > d$. □

The numbers $(\Delta^m f)(0)$ play a central role in the above proof, namely, they are the coefficients of the polynomial $f(n)$ when expressed in terms of the (Δ)-basis. They will appear time and again in this section, and thus we define

$$f^{(m)} := (\Delta^m f)(0), \qquad \text{whence} \qquad f(n) = \sum_{m=0}^{d} f^{(m)} \binom{n}{m}. \tag{4.1.4}$$

We make one more observation, namely, that $f(n)$ has degree d if and only if $f^{(d)} \neq 0$.

We now introduce the main tool of this chapter, the **generating function** of the sequence $f(n)$: the formal power series

$$F(z) := \sum_{n \geq 0} f(n)\, z^n.$$

The set $\mathbb{C}[\![z]\!]$ of all such series is a vector space with the natural addition and scalar multiplication

$$\sum_{n\geq 0} f(n)\, z^n + c \sum_{n \geq 0} g(n)\, z^n := \sum_{n\geq 0} (f(n) + c\, g(n))\, z^n,$$

where $c \in \mathbb{C}$. Moreover, formal power series constitute a commutative ring with 1 under the multiplication

$$\left(\sum_{n\geq 0} f(n)\, z^n\right) \cdot \left(\sum_{n\geq 0} g(n)\, z^n\right) := \sum_{n\geq 0}\left(\sum_{k+l=n} f(k)\, g(l)\right) z^n,$$

and Exercise 4.3 says which formal power series have multiplicative inverses. Borrowing a leaf from calculus, we can also define the derivative

$$\frac{d}{dz} \sum_{n\geq 0} f(n)\, z^n := \sum_{n \geq 1} n\, f(n)\, z^{n-1}, \tag{4.1.5}$$

and Exercise 4.4 invites you to check that the usual differentiation rules hold in this setting.

Most of our use of generating functions is intuitively sensible but with the above definitions also on a solid algebraic foundation. For example, our favorite counting function comes with a generating function that is (or rather, should be) known to any calculus student: for $f(n) = \binom{n}{m}$, we compute

$$F(z) = \sum_{n\geq 0} \binom{n}{m} z^n = \frac{1}{m!}\, z^m \sum_{n \geq m} n(n-1)\cdots(n-m+1)\, z^{n-m}$$

$$= \frac{1}{m!}\, z^m \left(\frac{d}{dz}\right)^m \frac{1}{1-z} = \frac{z^m}{(1-z)^{m+1}}. \tag{4.1.6}$$

A generating function, such as the above example, that can be expressed as a proper rational function (i.e., the quotient $\frac{p(z)}{q(z)}$ of two polynomials where the degree of $p(z)$ is smaller than that of $q(z)$) is **rational**. Note that a proper rational function $\frac{p(z)}{q(z)}$ represents a power series if and only if $q(0) \neq 0$, by Exercise 4.3.

Thanks to (4.1.4), the sample generating function (4.1.6) gives us a way to compute the generating function for *every* polynomial $f(n)$: with the notation of (4.1.4), we obtain

$$F(z) = \sum_{n\geq 0} f(n)\, z^n = \sum_{m=0}^{d} f^{(m)} \sum_{n\geq 0} \binom{n}{m} z^n$$

$$(4.1.7) \qquad = \sum_{m=0}^{d} f^{(m)} \frac{z^m}{(1-z)^{m+1}}$$

$$(4.1.8) \qquad = \frac{\sum_{m=0}^{d} f^{(m)}\, z^m (1-z)^{d-m}}{(1-z)^{d+1}}.$$

The (equivalent) expressions (4.1.7) and (4.1.8) representing the rational generating function of the polynomial $f(n)$ allow us to develop several simple but powerful properties of polynomials and their generating functions.

Proposition 4.1.4. *A sequence $f(n)$ is given by a polynomial of degree $\leq d$ if and only if*

$$\sum_{n\geq 0} f(n)\, z^n = \frac{h(z)}{(1-z)^{d+1}}$$

for some polynomial $h(z)$ of degree $\leq d$. Furthermore, $f(n)$ has degree d if and only if $h(1) \neq 0$.

Proof. The first part of the proposition follows from (4.1.8) and by writing $f(n)$ in terms of the (Δ)-basis and $h(z)$ in terms of the (γ)-basis.

The second part follows from (4.1.4) which illustrates that $f(n)$ has degree d if and only if $f^{(d)} \neq 0$. This, in turn, holds (by (4.1.8)) if and only if $h(1) \neq 0$. □

Proposition 4.1.4 can be rephrased in terms of recursions for the sequence $(f(n))_{n\geq 0}$: given a polynomial $f(n)$, we can rewrite the rational generating function identity given in Proposition 4.1.4 as

$$h(z) = (1-z)^{d+1} \sum_{n\geq 0} f(n)\, z^n = \sum_{n\geq 0} f(n) \sum_{j=0}^{d+1} \binom{d+1}{j} (-1)^j\, z^{n+j}.$$

Since $h(z)$ has degree $\leq d$, the coefficients of z^n for $n > d$ on the right-hand side must be zero, that is,

$$\sum_{j=0}^{d+1} \binom{d+1}{j} (-1)^j f(n-j) \;=\; 0 \tag{4.1.9}$$

for all $n \geq d+1$.

Reflecting some more on (4.1.9) naturally leads one to see the full strength of rational generating functions: a sequence $(f(n))_{n\geq 0}$ satisfies a **linear recurrence** if for some $c_0, \ldots, c_d \in \mathbb{C}$ with $c_0, c_d \neq 0$

$$c_0\, f(n+d) + c_1\, f(n+d-1) + \cdots + c_d\, f(n) \;=\; 0 \tag{4.1.10}$$

for all $n \geq 0$. Thus, the sequence $(f(n))_{n\geq 0}$ is fully described by the coefficients of the linear recurrence $c_0, \ldots, c_d$ and the starting values $f(0), \ldots, f(d)$.

Proposition 4.1.5. *Let $(f(n))_{n\geq 0}$ be a sequence of numbers. Then $(f(n))_{n\geq 0}$ satisfies a linear recurrence of the form* (4.1.10) *(with $c_0, c_d \neq 0$) if and only if*

$$F(z) \;=\; \sum_{n\geq 0} f(n)\, z^n \;=\; \frac{p(z)}{c_d z^d + c_{d-1} z^{d-1} + \cdots + c_0}$$

for some polynomial $p(z)$ with $\deg(p) < d$.

The reasoning that led to (4.1.9) also yields Proposition 4.1.5; see Exercises 4.6 and 4.7.

Here is an example. The first recursion any student sees is likely that for the **Fibonacci numbers** defined through

$$\begin{aligned} &f(0) := 0\,, \\ &f(1) = f(2) := 1\,, \\ &f(n) := f(n-1) + f(n-2) \quad \text{for} \quad n \geq 3\,. \end{aligned}$$

Proposition 4.1.5 says that the generating function for the Fibonacci numbers is rational with denominator $1 - z - z^2$. The starting data give

$$\sum_{n\geq 0} f(n)\, z^n \;=\; \frac{z}{1 - z - z^2}\,.$$

Exercise 4.8 shows how we can concretely compute formulas from this rational generating function, but its use is not limited to that—in the next section we will illustrate how one can derive interesting identities from generating functions.

Coming back to our setup of $f(n)$ being a polynomial of degree d, we claim that (4.1.9) has to hold also for $n < 0$. Indeed, $f(-n)$ is, of course, a polynomial in n of degree d and so comes with a rational generating function. Towards computing it, we start by plugging negative arguments into (4.1.4):

$$f(-n) \;=\; \sum_{m=0}^{d} f^{(m)} \binom{-n}{m} \;=\; \sum_{m=0}^{d} f^{(m)} (-1)^m \binom{n+m-1}{m}, \tag{4.1.11}$$

by our very first combinatorial reciprocity instance (0.0.2). We define the accompanying generating function as

$$F^{\circ}(z) \;:=\; \sum_{n \ge 1} f(-n)\, z^n . \tag{4.1.12}$$

(There are several good reasons for starting this series only at $n = 1$, as we will see shortly.) Note that $F^{\circ}(z)$ is rational, by Proposition 4.1.4. Here is a blueprint for a formal reciprocity on the level of rational generating functions: by (4.1.11),

$$\begin{aligned}
F^{\circ}(z) \;&=\; \sum_{n \ge 1} \sum_{m=0}^{d} f^{(m)} (-1)^m \binom{n+m-1}{m} z^n \\
&=\; \sum_{m=0}^{d} f^{(m)} (-1)^m \sum_{n \ge 1} \binom{n+m-1}{m} z^n \\
&=\; \sum_{m=0}^{d} f^{(m)} (-1)^m \frac{z}{(1-z)^{m+1}} \;=\; -\sum_{m=0}^{d} f^{(m)} \frac{\left(\frac{1}{z}\right)^m}{\left(1-\frac{1}{z}\right)^{m+1}} \\
&=\; -F\!\left(\tfrac{1}{z}\right),
\end{aligned}$$

where the third equation follows from Exercise 4.5. This is a formal rather than a true combinatorial reciprocity, as the calculation does not tell us if we can think of $f(-n)$ as a genuine counting function.

If $(f(n))_{n \ge 0}$ is a sequence satisfying a linear recurrence of the form (4.1.10), then we can let the recurrence run *backwards*: for example, since we assume that $c_d \ne 0$, the value of $f(-1)$ is determined by $f(0), \dots, f(d-1)$ by way of (4.1.10). The sequence $f^{\circ}(n) := f(-n)$ satisfies the linear recurrence

$$c_d\, f^{\circ}(n+d) + c_{d-1}\, f^{\circ}(n+d-1) + \cdots + c_0\, f^{\circ}(n) \;=\; 0 . \tag{4.1.13}$$

For example, for the Fibonacci numbers, this gives

n	1	2	3	4	5	6	7	8	...
$f^{\circ}(n)$	1	-1	2	-3	5	-8	13	-21	...

The corresponding rational generating function is

$$F^{\circ}(z) \;=\; \frac{z}{1+z-z^2} .$$

In terms of general rational generating functions, this formal reciprocity reads as follows.

Theorem 4.1.6. *Let $F(z) = \sum_{n\geq 0} f(n)\, z^n$ be a rational generating function. Then $F^\circ(z)$, defined in (4.1.12), is also rational. The two rational functions are related as*

$$F^\circ(z) \;=\; -F(\tfrac{1}{z}).$$

Proof. Assume that

$$F(z) \;=\; \frac{p(z)}{c_d z^d + c_{d-1} z^{d-1} + \cdots + c_0},$$

where $c_0, c_d \neq 0$ and $p(z)$ is a polynomial of degree $< d$. It follows that

$$-F(\tfrac{1}{z}) \;=\; -\frac{z^d\, p(\frac{1}{z})}{c_0 z^d + c_1 z^{d-1} + \cdots + c_d},$$

also a rational generating function. The coefficients of $-F(\frac{1}{z})$ satisfy the linear recurrence (4.1.13). To prove Theorem 4.1.6, we thus have to only verify that the starting data of $f^\circ(n)$ is encoded in the numerator of $-F(\frac{1}{z})$, that is,

$$\left(c_0 z^d + c_1 z^{d-1} + \cdots + c_d\right) F^\circ(z) \;=\; -z^d\, p(\tfrac{1}{z}).$$

This is left to Exercise 4.9. □

The assumption that $F(z)$ is a *rational* generating function is essential—see Exercise 4.11. On the other hand, we can tweak things, up to a point, when $F(z)$ evaluates to an *improper* rational function, i.e., $F(z) = \frac{p(z)}{q(z)}$, where $\deg(p) \geq \deg(q)$. In this case, we can use long division to write $F(z)$ as a sum of a polynomial and a proper rational function. We give a sample result, paralleling Proposition 4.1.4, and invite you to check the details in Exercise 4.12.

Corollary 4.1.7. *A sequence $f(n)$ is eventually polynomial*[1] *of degree $\leq d$ if and only if*

$$\sum_{n\geq 0} f(n)\, z^n \;=\; g(z) + \frac{h(z)}{(1-z)^{d+1}}$$

for some polynomials $g(z)$ and $h(z)$ with $\deg(h) \leq d$. *Furthermore, $f(n)$ has degree d if and only if $h(1) \neq 0$.*

We finish this section with a reciprocity theorem, analogous to (and directly following from) Theorem 4.1.6, for improper rational generating functions. We will see it in action in Sections 4.9 and 5.6, where $g(z)$ will be a constant.

[1] A sequence $f(n)$ is **eventually polynomial** if there exist $k \in \mathbb{Z}_{\geq 0}$ and a polynomial $p(n)$ such that $f(n) = p(n)$ for $n > k$.

Corollary 4.1.8. *Let*

$$F(z) = \sum_{n \ge 0} f(n)\, z^n = g(z) + \frac{h(z)}{(1-z)^{d+1}}$$

be an improper rational generating function, for some polynomials $g(z) = g_k z^k + g_{k-1} z^{k-1} + \cdots + g_0$ *and* $h(z)$ *with* $g_k \neq 0$ *and* $\deg(h) \le d$. *Let*

$$\tilde{f}(n) \;:=\; \begin{cases} f(n) - g_n & \text{if } n \le k, \\ f(n) & \text{if } n > k, \end{cases}$$

so that $\tilde{F}(z) := \sum_{n \ge 0} \tilde{f}(n)\, z^n = \frac{h(z)}{(1-z)^{d+1}}$. *Then*

$$F\left(\tfrac{1}{z}\right) \;=\; g\left(\tfrac{1}{z}\right) - \tilde{F}^{\circ}(z)\,.$$

4.2. Compositions

A nifty interpretation of $\binom{n}{k}$ goes as follows: since

$$\binom{n}{k} \;=\; \left|\left\{(a_1, a_2, \dots, a_k) \in \mathbb{Z}^k \,:\, 0 < a_1 < a_2 < \cdots < a_k < n+1\right\}\right|,$$

we can set $a_0 := 0$, $a_{k+1} := n+1$ and define $b_j := a_j - a_{j-1}$ for $1 \le j \le k+1$, which gives

$$\binom{n}{k} \;=\; \left|\left\{(b_1, b_2, \dots, b_{k+1}) \in \mathbb{Z}_{>0}^{k+1} \,:\, b_1 + b_2 + \cdots + b_{k+1} = n+1\right\}\right|.$$

We call a vector $(b_1, b_2, \dots, b_{k+1}) \in \mathbb{Z}_{>0}^{k+1}$ such that $b_1 + b_2 + \cdots + b_{k+1} = n+1$ a **composition** of $n+1$; the $b_1, b_2, \dots, b_{k+1}$ are called the **parts** of the composition. So $\binom{n}{k}$ equals the number of compositions of $n+1$ with $k+1$ parts. This interpretation also gives an alternative reasoning for the generating function of $\binom{n}{k}$ given in (4.1.6). Namely,[2]

$$\begin{aligned}
\sum_{n \ge 0} \binom{n}{k} q^n \;&=\; \sum_{n \ge 0} \#(\text{compositions of } n+1 \text{ with } k+1 \text{ parts})\; q^n \\
&= \frac{1}{q} \sum_{n \ge 0} \#(\text{compositions of } n+1 \text{ with } k+1 \text{ parts})\; q^{n+1} \\
&= \frac{1}{q} \sum_{b_1, b_2, \dots, b_{k+1} \ge 1} q^{b_1 + b_2 + \cdots + b_{k+1}} \qquad (4.2.1) \\
&= \frac{1}{q} \left(\sum_{b_1 \ge 1} q^{b_1}\right)\left(\sum_{b_2 \ge 1} q^{b_2}\right) \cdots \left(\sum_{b_{k+1} \ge 1} q^{b_{k+1}}\right)
\end{aligned}$$

[2]You might wonder about our switching to the variable q starting with this generating function. The reason will be given in Section 4.7.

$$= \frac{1}{q}\left(\frac{q}{1-q}\right)^{k+1} = \frac{q^k}{(1-q)^{k+1}} .$$

This computation might be an overkill for the binomial coefficient, but it hints at the powerful machinery that generating functions provide. We give another sample of this machinery next.

Proposition 4.2.1. *Given a set $A \subseteq \mathbb{Z}_{>0}$, let $c_A(n)$ denote the number of compositions of n with parts in A. Then its generating function is*

$$C_A(q) := 1 + \sum_{n\ge 1} c_A(n)\, q^n = \frac{1}{1-\sum_{m\in A} q^m} .$$

Note that the last expression is *not* a rational generating function, unless A is finite. It is the multiplicative inverse of the formal power series $1 - \sum_{m\in A} q^m$, which exists by Exercise 4.3.

Proof. Essentially by repeating the argument in (4.2.1), the generating function for all compositions with exactly j parts (all in A) equals $\left(\sum_{m\in A} q^m\right)^j$. Thus

$$C_A(q) = 1 + \sum_{j\ge 1}\left(\sum_{m\in A} q^m\right)^j = \frac{1}{1-\sum_{m\in A} q^m} . \qquad \square$$

We show two pieces of art from the exhibition provided by Proposition 4.2.1. First, suppose A consists of all odd positive integers. Then

$$\frac{1}{1-\sum_{m\ge 1 \text{ odd}} q^m} = \frac{1}{1-\frac{q}{1-q^2}} = \frac{1-q^2}{1-q-q^2} = 1 + \frac{q}{1-q-q^2}$$

—smell familiar? Now let A consist of all integers ≥ 2. Then

$$\frac{1}{1-\sum_{m\ge 2} q^m} = \frac{1}{1-\frac{q^2}{1-q}} = \frac{1-q}{1-q-q^2} = 1 + \frac{q^2}{1-q-q^2} .$$

What these two generating functions prove, in two simple lines, is the following.

Theorem 4.2.2. *The number of compositions of n using only odd parts and the number of compositions of $n-1$ using only parts ≥ 2 are both given by the n-th Fibonacci number.*

4.3. Plane Partitions

Our next counting function concerns the simplest case of a **plane partition**; namely, we will count all ways of writing $n = a_1 + a_2 + a_3 + a_4$ such that the integers $a_1, a_2, a_3, a_4 \geq 0$ satisfy the inequalities[3]

$$\begin{array}{ccc} a_1 & \geq & a_2 \\ \text{IV} & & \text{IV} \\ a_3 & \geq & a_4 \,. \end{array} \tag{4.3.1}$$

(For a general plane partition, this array of inequalities can form a rectangle of any size.) Let $pl(n)$ denote the number of plane partitions of n of the form (4.3.1). We will compute its generating function

$$Pl(q) := \sum_{n \geq 0} pl(n)\, q^n = \sum q^{a_1+a_2+a_3+a_4},$$

where the last sum is over all integers $a_1, a_2, a_3, a_4 \geq 0$ satisfying (4.3.1), from first principles, using geometric series:

$$\begin{aligned} Pl(q) &= \sum_{a_4 \geq 0} q^{a_4} \sum_{a_3 \geq a_4} q^{a_3} \sum_{a_2 \geq a_4} q^{a_2} \sum_{a_1 \geq \max(a_2,a_3)} q^{a_1} \\ &= \sum_{a_4 \geq 0} q^{a_4} \sum_{a_3 \geq a_4} q^{a_3} \sum_{a_2 \geq a_4} q^{a_2} \frac{q^{\max(a_2,a_3)}}{1-q} \\ &= \frac{1}{1-q} \sum_{a_4 \geq 0} q^{a_4} \sum_{a_3 \geq a_4} q^{a_3} \left(q^{a_3} \sum_{a_2=a_4}^{a_3-1} q^{a_2} + \sum_{a_2 \geq a_3} q^{2a_2} \right) \\ &= \frac{1}{1-q} \sum_{a_4 \geq 0} q^{a_4} \sum_{a_3 \geq a_4} q^{a_3} \left(q^{a_3} \frac{q^{a_4} - q^{a_3}}{1-q} + \frac{q^{2a_3}}{1-q^2} \right) \\ &= \frac{1}{1-q} \sum_{a_4 \geq 0} \left(\frac{q^{4a_4}}{(1-q)(1-q^2)} - \frac{q^{4a_4}}{(1-q)(1-q^3)} + \frac{q^{4a_4}}{(1-q^2)(1-q^3)} \right) \\ &= \frac{1+q^2}{(1-q)(1-q^2)(1-q^3)} \sum_{a_4 \geq 0} q^{4a_4} \\ &= \frac{1+q^2}{(1-q)(1-q^2)(1-q^3)(1-q^4)} \\ &= \frac{1}{(1-q)(1-q^2)^2(1-q^3)} \,. \end{aligned} \tag{4.3.2}$$

(If this computation seems lengthy, or even messy, don't worry: we will shortly develop theory that will simplify computations tremendously.)

[3] If we relax our definition of *composition* just a tad by allowing parts that are 0, plane partitions of n are special cases of compositions of n.

Next we would like to compute an explicit formula for $pl(n)$ from this generating function. One way—namely, through a partial-fraction expansion of $Pl(q)$—is outlined in Exercise 4.19. A second approach is through expanding both numerator and denominator of $Pl(q)$ by the same factor to transform the denominator into a simple form:

(4.3.3)

$$Pl(q) = \frac{\left(\begin{array}{c} q^{16}+q^{15}+3\,q^{14}+4\,q^{13}+7\,q^{12}+9\,q^{11}+10\,q^{10}+13\,q^{9} \\ +12\,q^{8}+13\,q^{7}+10\,q^{6}+9\,q^{5}+7\,q^{4}+4\,q^{3}+3\,q^{2}+q+1 \end{array}\right)}{(1-q^6)^4} .$$

The reason for this transformation is that the new denominator yields an easy power series expansion, by Exercise 4.5:

$$\frac{1}{(1-q^6)^4} = \sum_{n\geq 0} \binom{n+3}{3} q^{6n}.$$

This suggests that the resulting plane-partition counting function $pl(n)$ has a certain periodic character, with period 6. Indeed, we can split up the rational generating function $Pl(q)$ into six parts as follows:

$$\begin{aligned} Pl(q) &= \frac{7\,q^{12}+10\,q^6+1}{(1-q^6)^4} + \frac{4\,q^{13}+13\,q^7+q}{(1-q^6)^4} + \frac{3\,q^{14}+12\,q^8+3\,q^2}{(1-q^6)^4} \\ &\quad + \frac{q^{15}+13\,q^9+4\,q^3}{(1-q^6)^4} + \frac{q^{16}+10\,q^{10}+7\,q^4}{(1-q^6)^4} + \frac{9\,q^{11}+9\,q^5}{(1-q^6)^4} \\ &= \sum_{n\geq 0} \left(7\tbinom{n+1}{3} + 10\tbinom{n+2}{3} + \tbinom{n+3}{3}\right) q^{6n} \\ &\quad + \sum_{n\geq 0} \left(4\tbinom{n+1}{3} + 13\tbinom{n+2}{3} + \tbinom{n+3}{3}\right) q^{6n+1} \\ &\quad + \sum_{n\geq 0} \left(3\tbinom{n+1}{3} + 12\tbinom{n+2}{3} + 3\tbinom{n+3}{3}\right) q^{6n+2} \\ &\quad + \sum_{n\geq 0} \left(\tbinom{n+1}{3} + 13\tbinom{n+2}{3} + 4\tbinom{n+3}{3}\right) q^{6n+3} \\ &\quad + \sum_{n\geq 0} \left(\tbinom{n+1}{3} + 10\tbinom{n+2}{3} + 7\tbinom{n+3}{3}\right) q^{6n+4} \\ &\quad + \sum_{n\geq 0} \left(9\tbinom{n+2}{3} + 9\tbinom{n+3}{3}\right) q^{6n+5}. \end{aligned}$$

From this data, a quick calculation yields

$$pl(n) = \begin{cases} \frac{1}{72}n^3 + \frac{1}{6}n^2 + \frac{2}{3}n + 1 & \text{if } n \equiv 0 \bmod 6, \\ \frac{1}{72}n^3 + \frac{1}{6}n^2 + \frac{13}{24}n + \frac{5}{18} & \text{if } n \equiv 1 \bmod 6, \\ \frac{1}{72}n^3 + \frac{1}{6}n^2 + \frac{2}{3}n + \frac{8}{9} & \text{if } n \equiv 2 \bmod 6, \\ \frac{1}{72}n^3 + \frac{1}{6}n^2 + \frac{13}{24}n + \frac{1}{2} & \text{if } n \equiv 3 \bmod 6, \\ \frac{1}{72}n^3 + \frac{1}{6}n^2 + \frac{2}{3}n + \frac{7}{9} & \text{if } n \equiv 4 \bmod 6, \\ \frac{1}{72}n^3 + \frac{1}{6}n^2 + \frac{13}{24}n + \frac{7}{18} & \text{if } n \equiv 5 \bmod 6. \end{cases} \tag{4.3.4}$$

The counting function $pl(n)$ is our first instance of a **quasipolynomial**, that is, a function $p : \mathbb{Z} \to \mathbb{C}$ of the form

$$p(n) = c_d(n)\, n^d + \cdots + c_1(n)\, n + c_0(n)\,,$$

where $c_0(n), c_1(n), \ldots, c_d(n)$ are periodic functions in n. In the case $p(n) = pl(n)$ of our plane-partition counting function, the coefficient $c_0(n)$ has period 6, the coefficient $c_1(n)$ has period 2, and $c_2(n)$ and $c_3(n)$ have period 1, i.e., they are constants. Of course, we can also think of the quasipolynomial $pl(n)$ as given by the list (4.3.4) of six polynomials, which we run through cyclically as n increases. We will say more about basic properties of quasipolynomials in Section 4.5.

We finish our study of plane partitions by observing a simple algebraic relation for $Pl(q)$, namely,

$$Pl\left(\frac{1}{q}\right) = q^8\, Pl(q)\,. \tag{4.3.5}$$

This suggests some relation to Theorem 4.1.6. Indeed, this theorem says

$$Pl\left(\frac{1}{q}\right) = -\sum_{n \ge 1} pl(-n)\, q^n. \tag{4.3.6}$$

But then equating the right-hand sides of (4.3.5) and (4.3.6) yields the following reciprocity relation for the plane-partition quasipolynomials:

$$pl(-n) = -pl(n-8)\,. \tag{4.3.7}$$

This is a combinatorial reciprocity that we are after: up to a shift by 8, the plane-partition counting function is **self-reciprocal**. (We have seen another example of self-reciprocal functions in Theorem 2.3.3.) We can think of this reciprocity also as follows: if we count the number of plane partitions of n for which all the inequalities in (4.3.1) are strict and all parts are strictly positive, then this is exactly the number of quadruples $(a_1, a_2, a_3, a_4) \in \mathbb{Z}^4_{\ge 0}$ such that

$$\begin{array}{ccc} a_1 + 3 & \ge & a_2 + 2 \\ \text{\rotatebox{90}{$\le$}} & & \text{\rotatebox{90}{$\le$}} \\ a_3 + 2 & \ge & a_4 + 1\,. \end{array} \tag{4.3.8}$$

But these are just plane partitions with $a_1 + \cdots + a_4 = n - 8$, which are counted on the right-hand side of (4.3.7).

There are many generalizations of $pl(n)$; one is given in Exercise 4.20.

4.4. Restricted Partitions

An **(integer) partition** of n is a sequence $(a_1 \geq a_2 \geq \cdots \geq a_k \geq 1)$ of non-increasing positive integers such that

$$n = a_1 + a_2 + \cdots + a_k . \tag{4.4.1}$$

The numbers $a_1, a_2, \ldots, a_k$ are the **parts** of this partition. For example, $(4,2,1)$ and $(3,2,1,1)$ are partitions of 7.

Our goal is to enumerate partitions with certain restrictions, which will allow us to prove a combinatorial reciprocity theorem not unlike that for the plane-partition counting function $pl(n)$. (The enumeration of unrestricted partitions is the subject of Exercise 4.27, but it will not yield a reciprocity theorem.) Namely, we restrict the parts of a partition to a finite set $A := \{a_1 > a_2 > \cdots > a_d\} \subset \mathbb{Z}_{>0}$, that is, we only allow partitions of the form

$$(a_1, \ldots, a_1, a_2, \ldots, a_2, \ldots, a_d, \ldots, a_d) .$$

(It is interesting—and related to topics to appear soon—to allow A to be a *multiset*; see Exercise 4.24.) The **restricted partition function** for A is

$$p_A(n) := \left| \left\{ (m_1, m_2, \ldots, m_d) \in \mathbb{Z}^d_{\geq 0} : m_1 a_1 + m_2 a_2 + \cdots + m_d a_d = n \right\} \right| .$$

By now it will not come as a surprise that we will approach the counting function $p_A(n)$ through its generating function

$$P_A(q) := \sum_{n \geq 0} p_A(n)\, q^n .$$

One advantage of this (and any other) generating function is that it allows us, in a sense, to manipulate the sequence $(p_A(n))_{n \geq 0}$ algebraically:

$$\begin{aligned} P_A(q) &= \sum_{m_1, m_2, \ldots, m_d \geq 0} q^{m_1 a_1 + m_2 a_2 + \cdots + m_d a_d} \\ &= \left(\sum_{m_1 \geq 0} q^{m_1 a_1} \right) \left(\sum_{m_2 \geq 0} q^{m_2 a_2} \right) \cdots \left(\sum_{m_d \geq 0} q^{m_d a_d} \right) \qquad (4.4.2) \\ &= \frac{1}{(1 - q^{a_1})(1 - q^{a_2}) \cdots (1 - q^{a_d})} , \end{aligned}$$

where the last identity comes from the geometric series. To see how the generating function of this counting function helps us understand the latter, we look at the simplest case when A contains only one positive integer a. In this case

$$P_{\{a\}}(q) = \frac{1}{1-q^a} = 1 + q^a + q^{2a} + \cdots,$$

the generating function for

$$p_{\{a\}}(n) = \begin{cases} 1 & \text{if } a|n, \\ 0 & \text{otherwise} \end{cases}$$

(as expected from the definition of $p_{\{a\}}(n)$). Note that the counting function $p_{\{a\}}(n)$ is a (simple) instance of a quasipolynomial, namely, $p_{\{a\}}(n) = c_0(n)$, where $c_0(n)$ is the periodic function (with period a) that returns 1 if n is a multiple of a and 0 otherwise.

Next we consider the case when A has two elements. The product structure of the accompanying generating function

$$P_{\{a,b\}}(q) = \frac{1}{(1-q^a)(1-q^b)}$$

means that we can compute

$$p_{\{a,b\}}(n) = \sum_{s=0}^{n} p_{\{a\}}(s)\, p_{\{b\}}(n-s).$$

Note that we are summing a quasipolynomial here, and so $p_{\{a,b\}}(n)$ is again a quasipolynomial by the next proposition, whose proof we leave as Exercise 4.21.

Proposition 4.4.1. *If $p(n)$ is a quasipolynomial, so is $r(n) := \sum_{s=0}^{n} p(s)$. More generally, if $f(n)$ and $g(n)$ are quasipolynomials, so is their* **convolution**

$$c(n) := \sum_{s=0}^{n} f(s)\, g(n-s).$$

We invite you to explicitly compute some examples of restricted partition functions, such as $p_{\{1,2\}}(n)$ (Exercise 4.23). Naturally, we can repeatedly apply Proposition 4.4.1 to deduce:

Corollary 4.4.2. *The restricted partition function $p_A(n)$ is a quasipolynomial in n.*

Since $p_A(n)$ is a quasipolynomial, we are free to evaluate it at negative integers. We define

$$p_A^\circ(n) := \left|\left\{(m_1, m_2, \ldots, m_d) \in \mathbb{Z}_{>0}^d : m_1 a_1 + m_2 a_2 + \cdots + m_d a_d = n\right\}\right|,$$

the number of restricted partitions of n such that every a_i is used at least once.

Theorem 4.4.3. *If $A = \{a_1, a_2, \ldots, a_d\} \subset \mathbb{Z}_{>0}$, then*

$$p_A(-n) = (-1)^{d-1} p_A(n - a_1 - a_2 - \cdots - a_d) = (-1)^{d-1} p_A^\circ(n) .$$

Proof. We first observe that the number of partitions of n in which a_i is used at least once equals $p_A(n - a_i)$. Thus, setting $a := a_1 + a_2 + \cdots + a_d$, we see that

$$p_A^\circ(n) = p_A(n-a) ,$$

and this gives the second equality. To prove the first, we use simple algebra on (4.4.2) to obtain

$$P_A\left(\frac{1}{q}\right) = \frac{q^{a_1+a_2+\cdots+a_d}}{(q^{a_1}-1)(q^{a_2}-1)\cdots(q^{a_d}-1)} = (-1)^d q^a P_A(q) . \tag{4.4.3}$$

Now we use Theorem 4.1.6, which gives

$$P_A\left(\frac{1}{q}\right) = -\sum_{n \ge 1} p_A(-n)\, q^n . \tag{4.4.4}$$

Equating the right-hand sides of (4.4.3) and (4.4.4) proves the claim. □

4.5. Quasipolynomials

We have defined a quasipolynomial $p(n)$ as a function $\mathbb{Z} \to \mathbb{C}$ of the form

$$p(n) = c_d(n)\, n^d + \cdots + c_1(n)\, n + c_0(n) , \tag{4.5.1}$$

where $c_0, c_1, \ldots, c_d$ are periodic functions in n. Assuming that c_d is not the zero function, the **degree** of $p(n)$ is d, and the least common period of $c_0(n), c_1(n), \ldots, c_d(n)$ is the **period** of $p(n)$. Alternatively, for a quasipolynomial $p(n)$, there exist a positive integer k and polynomials

$$p_0(n), p_1(n), \ldots, p_{k-1}(n)$$

such that

$$p(n) = \begin{cases} p_0(n) & \text{if } n \equiv 0 \bmod k, \\ p_1(n) & \text{if } n \equiv 1 \bmod k, \\ \vdots & \\ p_{k-1}(n) & \text{if } n \equiv k-1 \bmod k. \end{cases}$$

The minimal such k is the period of $p(n)$, and for this minimal k, the polynomials $p_0(n), p_1(n), \ldots, p_{k-1}(n)$ are the **constituents** of $p(n)$. Of course, when $k = 1$, we need only one constituent and the coefficient functions $c_0(n), c_1(n), \ldots, c_d(n)$ are constants, and so $p(n)$ is a *polynomial*. Yet another perspective on quasipolynomials is explored in Exercise 4.31.

As we have seen in (4.3.3) and (4.4.2), the quasipolynomials arising from plane partitions and restricted partitions can be encoded into rational generating functions with a particular denominator.

Proposition 4.5.1. *Let $p : \mathbb{Z} \to \mathbb{C}$ be a function with associated generating function*

$$P(z) := \sum_{n \geq 0} p(n)\, z^n .$$

Then $p(n)$ is a quasipolynomial of degree $\leq d$ and period dividing k if and only if

$$P(z) = \frac{h(z)}{(1 - z^k)^{d+1}} ,$$

where $h(z)$ is a polynomial of degree at most $k(d+1) - 1$.

This generalizes Proposition 4.1.4 from polynomials (i.e., the case $k = 1$) to quasipolynomials. Proposition 4.5.1 also explains, in some sense, why we rewrote the generating function $Pl(q)$ for plane partitions in (4.3.3). Apparently the generating function (4.4.2) for the restricted partition function is not of this form, but multiplying numerator and denominator by appropriate terms yields the denominator $(1 - z^k)^d$, where $k = \operatorname{lcm}(a_1, a_2, \ldots, a_d)$. For example,

$$P_{\{2,3\}}(z) = \frac{1}{(1 - z^2)(1 - z^3)} = \frac{1 + z^2 + z^3 + z^4 + z^5 + z^7}{(1 - z^6)^2} .$$

For a general recipe to convert (4.4.2) into a form fitting Proposition 4.5.1, we refer to Exercise 4.32. In particular, the presentation of $P(z)$ in Proposition 4.5.1 is typically not reduced, and by getting rid of common factors it can be seen that $\frac{h(z)}{g(z)}$ gives rise to a quasipolynomial if and only if each zero γ of $g(z)$ satisfies $\gamma^k = 1$ for some $k \in \mathbb{Z}_{>0}$. The benefit of Proposition 4.5.1 is that it gives a pretty effective way of showing that a function $q : \mathbb{Z} \to \mathbb{C}$ is a quasipolynomial.

Proof of Proposition 4.5.1. Suppose $p(n)$ is a quasipolynomial of degree $\leq d$ and period dividing k, so there are polynomials $p_0(n), p_1(n), \ldots, p_{k-1}(n)$ of degree $\leq d$ such that

$$p(n) = \begin{cases} p_0(n) & \text{if } n \equiv 0 \bmod k, \\ p_1(n) & \text{if } n \equiv 1 \bmod k, \\ \quad \vdots & \\ p_{k-1}(n) & \text{if } n \equiv k-1 \bmod k. \end{cases}$$

Thus

$$P(z) = \sum_{a \geq 0} \sum_{b=0}^{k-1} p(ak+b)\, z^{ak+b} = \sum_{b=0}^{k-1} z^b \sum_{a \geq 0} p_b(ak+b)\, z^{ak} ,$$

and since $p_b(ak+b)$ is a polynomial in a of degree $\le d$, we can use Proposition 4.1.4 to conclude that

$$P(z) = \sum_{b=0}^{k-1} z^b \frac{h_b(z^k)}{(1-z^k)^{d+1}}$$

for some polynomials $h_b(z)$ of degree $\le d$. Since $\sum_{b=0}^{k-1} z^b h_b(z^k)$ is a polynomial of degree $\le k(d+1)-1$, this proves the forward implication of Proposition 4.5.1.

For the converse implication, suppose $P(z) = \frac{h(z)}{(1-z^k)^{d+1}}$, where $h(z)$ is a polynomial of degree $\le k(d+1)-1$, say

$$h(z) = \sum_{m=0}^{k(d+1)-1} c_m \, z^m = \sum_{a=0}^{d} \sum_{b=0}^{k-1} c_{ak+b} \, z^{ak+b}.$$

Then

$$\begin{aligned} P(z) &= h(z) \sum_{j \ge 0} \binom{d+j}{d} z^{kj} = \sum_{j\ge 0} \sum_{b=0}^{k-1} \sum_{a=0}^{d} c_{ak+b} \binom{d+j}{d} z^{k(j+a)+b} \\ &= \sum_{j\ge 0} \sum_{b=0}^{k-1} \sum_{a=0}^{d} c_{ak+b} \binom{d+j-a}{d} z^{kj+b} = \sum_{j\ge 0} \sum_{b=0}^{k-1} p_b(kj+b) \, z^{kj+b}, \end{aligned}$$

where $p_b(kj+b) = \sum_{a=0}^{d} c_{ak+b} \binom{d+j-a}{d}$, a polynomial in j of degree $\le d$. In other words, $P(z)$ is the generating function of the quasipolynomial with constituents $p_0(n), p_1(n), \dots, p_{b-1}(n)$. □

4.6. Integer-point Transforms and Lattice Simplices

It is time to return to geometry. The theme of this chapter has been generating functions that we constructed from some (counting) function, with the hope of obtaining more information about this function. Now we will *start* with a generating function and try to interpret it geometrically as an object in its own right.

Our setting is that of Section 1.4 but in general dimensions: we consider a lattice polytope $\mathsf{P} \subset \mathbb{R}^d$ and its Ehrhart function

$$\text{(4.6.1)} \qquad \operatorname{ehr}_{\mathsf{P}}(n) := \left| n\,\mathsf{P} \cap \mathbb{Z}^d \right| = \left| \mathsf{P} \cap \tfrac{1}{n}\mathbb{Z}^d \right| ;$$

that is, $\operatorname{ehr}_{\mathsf{P}}(n)$ counts the lattice points as we dilate P or, equivalently, shrink the integer lattice $\mathbb{Z}^d$. We call the accompanying generating function

$$\text{(4.6.2)} \qquad \operatorname{Ehr}_{\mathsf{P}}(z) := 1 + \sum_{n \ge 1} \operatorname{ehr}_{\mathsf{P}}(n) \, z^n$$

the **Ehrhart series** of P. Our promised geometric interpretation of this generating function uses a technique from Chapter 3; namely, we consider the homogenization of P, defined in (3.1.9):

$$\begin{aligned}\operatorname{hom}(\mathsf{P}) &= \left\{(\mathbf{p},t) \in \mathbb{R}^{d+1} \,:\, t \geq 0,\ \mathbf{p} \in t\,\mathsf{P}\right\} \\ &= \operatorname{cone}\left\{(\mathbf{v},1) \,:\, \mathbf{v} \in \operatorname{vert}(\mathsf{P})\right\}.\end{aligned}$$

One advantage of hom(P) is that we can see a copy of the dilate $n\mathsf{P}$ as the intersection of hom(P) with the hyperplane $t = x_{d+1} = n$, as illustrated in Figure 4.1; we will say that points on this hyperplane are **at height** n and that the integer points in hom(P) are **graded** by the last coordinate. In

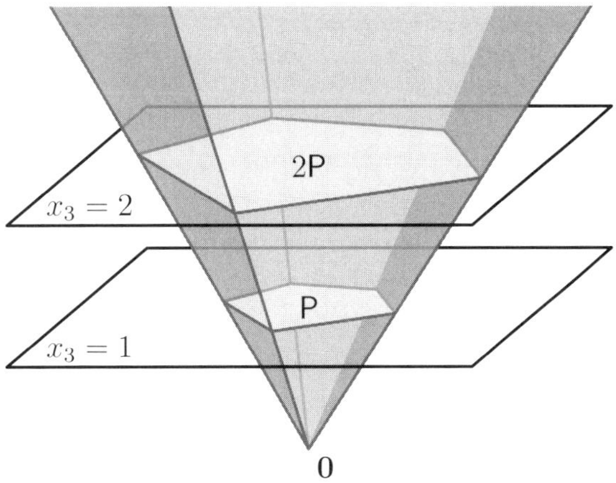

Figure 4.1. Recovering dilates of P in hom(P).

other words, the Ehrhart series of P can be computed through

$$\operatorname{Ehr}_{\mathsf{P}}(z) = \sum_{n \geq 0} \#\,(\text{lattice points in } \operatorname{hom}(\mathsf{P}) \text{ at height } n)\, z^n \tag{4.6.3}$$

(and this also explains our convention of starting $\operatorname{Ehr}_{\mathsf{P}}(z)$ with constant term 1). This motivates the study of the arithmetic of a set $S \subseteq \mathbb{R}^{d+1}$ by computing a *multivariate* generating function, namely, its **integer-point transform**

$$\sigma_S(z_1, \dots, z_{d+1}) := \sum_{\mathbf{m} \in S \cap \mathbb{Z}^{d+1}} \mathbf{z}^{\mathbf{m}},$$

where we abbreviated $\mathbf{z}^{\mathbf{m}} := z_1^{m_1} z_2^{m_2} \cdots z_{d+1}^{m_{d+1}}$; we will often write $\sigma_S(\mathbf{z})$ to shorten the notation involving the variables. In the language of integer-point transforms, (4.6.3) can be rewritten as

$$\operatorname{Ehr}_{\mathsf{P}}(z) = \sigma_{\operatorname{hom}(\mathsf{P})}(1, \dots, 1, z). \tag{4.6.4}$$

An integer-point transform is typically not a formal power series but a formal *Laurent series* $\sum_{\mathbf{m}\in\mathbb{Z}^{d+1}} c_{\mathbf{m}}\, \mathbf{z}^{\mathbf{m}} \in \mathbb{C}[\![z_1^{\pm 1}, \ldots, z_{d+1}^{\pm 1}]\!]$. If we want to multiply formal Laurent series (which we do), we have to be a little careful for which sets S we want to consider integer-point transforms; see Exercise 4.34. In particular, we are safe if we work with convex, line free sets such as $\hom(\mathsf{P})$.

We illustrate the above train of thoughts with the down-to-earth example from Section 1.4, the lattice triangle

$$\triangle := \left\{ (x_1, x_2) \in \mathbb{R}^2 : 0 \le x_1 \le x_2 \le 1 \right\}.$$

Its vertices are $(0,0)$, $(0,1)$, and $(1,1)$, and so

$$\hom(\triangle) = \mathbb{R}_{\ge 0} \begin{pmatrix} 0 \\ 0 \\ 1 \end{pmatrix} + \mathbb{R}_{\ge 0} \begin{pmatrix} 0 \\ 1 \\ 1 \end{pmatrix} + \mathbb{R}_{\ge 0} \begin{pmatrix} 1 \\ 1 \\ 1 \end{pmatrix}. \tag{4.6.5}$$

The three generators of $\hom(\triangle)$—call them $\mathbf{v}_1, \mathbf{v}_2, \mathbf{v}_3$—form a **lattice basis** of $\mathbb{Z}^3$, i.e., every point in $\mathbb{Z}^3$ can be uniquely expressed as an integral linear combination of $\mathbf{v}_1, \mathbf{v}_2, \mathbf{v}_3$ (see Exercise 4.33). In particular, every lattice point in $\hom(\triangle)$ can be uniquely written as a nonnegative integral linear combination of $\mathbf{v}_1, \mathbf{v}_2, \mathbf{v}_3$. Hence

$$\begin{aligned} \sigma_{\hom(\triangle)}(\mathbf{z}) &= \sum_{\mathbf{m}\in\hom(\triangle)\cap\mathbb{Z}^3} \mathbf{z}^{\mathbf{m}} = \sum_{k_1,k_2,k_3\ge 0} \mathbf{z}^{k_1\mathbf{v}_1 + k_2\mathbf{v}_2 + k_3\mathbf{v}_3} \\ &= \frac{1}{(1-\mathbf{z}^{\mathbf{v}_1})(1-\mathbf{z}^{\mathbf{v}_2})(1-\mathbf{z}^{\mathbf{v}_3})} \\ &= \frac{1}{(1-z_3)(1-z_2z_3)(1-z_1z_2z_3)}. \end{aligned} \tag{4.6.6}$$

With (4.6.4),

$$\mathrm{Ehr}_{\triangle}(z) = \sigma_{\hom(\triangle)}(1,1,z) = \frac{1}{(1-z)^3} = \sum_{n\ge 0} \binom{n+2}{2} z^n \tag{4.6.7}$$

and thus we recover the Ehrhart polynomial of $\triangle$ which we computed in Section 1.4.

Our computation of $\sigma_{\hom(\triangle)}(\mathbf{z})$ in (4.6.6) is somewhat misleading when thinking about the integer-point transform of the homogenization of a general lattice simplex, because the cone $\hom(\triangle)$ is **unimodular**: it is generated by a lattice basis of $\mathbb{Z}^3$.

We now change our example slightly and consider the lattice triangle $\mathsf{P} := \operatorname{conv}\{(0,0),(0,1),(2,1)\}$; note that

$$\hom(\mathsf{P}) = \mathbb{R}_{\ge 0} \begin{pmatrix} 0 \\ 0 \\ 1 \end{pmatrix} + \mathbb{R}_{\ge 0} \begin{pmatrix} 0 \\ 1 \\ 1 \end{pmatrix} + \mathbb{R}_{\ge 0} \begin{pmatrix} 2 \\ 1 \\ 1 \end{pmatrix} \tag{4.6.8}$$

is *not* unimodular. We can still implement the basic idea behind our computation of $\sigma_{\hom(\triangle)}(\mathbf{z})$, but we have to take into account some extra data.

We write $\mathbf{w}_1, \mathbf{w}_2, \mathbf{w}_3$ for the generators of $\hom(\mathsf{P})$ given in (4.6.8). We observe that the lattice point $\mathbf{q} = (1,1,1)$ is contained in $\hom(\mathsf{P})$ but is not an *integer* linear combination of the generators; see Figure 4.2 (mind you, the generators *are* linearly independent). Hence, we will miss $\mathbf{z}^{\mathbf{q}}$ if we exercise (4.6.6) with P instead of $\triangle$. Stronger yet, the linear independence of $\mathbf{w}_1, \mathbf{w}_2, \mathbf{w}_3$ yields that none of the points

$$\mathbf{q} + k_1\mathbf{w}_1 + k_2\mathbf{w}_2 + k_3\mathbf{w}_3 \qquad \text{for } k_1, k_2, k_3 \in \mathbb{Z}_{\geq 0}$$

can be expressed as a nonnegative integer linear combination of $\mathbf{w}_1, \mathbf{w}_2, \mathbf{w}_3$. A little bit of thought (see Exercise 4.38 for more) brings the following insight: we define the half-open parallelepiped

$$\Box := \left\{ \lambda_1 \begin{pmatrix} 0 \\ 0 \\ 1 \end{pmatrix} + \lambda_2 \begin{pmatrix} 0 \\ 1 \\ 1 \end{pmatrix} + \lambda_3 \begin{pmatrix} 2 \\ 1 \\ 1 \end{pmatrix} : 0 \leq \lambda_1, \lambda_2, \lambda_3 < 1 \right\};$$

then for each $\mathbf{p} \in \hom(\mathsf{P}) \cap \mathbb{Z}^3$ there are unique $k_1, k_2, k_3 \in \mathbb{Z}_{\geq 0}$ and $\mathbf{r} \in \Box \cap \mathbb{Z}^3$ such that

$$\mathbf{p} = k_1\mathbf{w}_1 + k_2\mathbf{w}_2 + k_3\mathbf{w}_3 + \mathbf{r}.$$

Figure 4.2 illustrates a geometric reason behind our claim, namely, we can

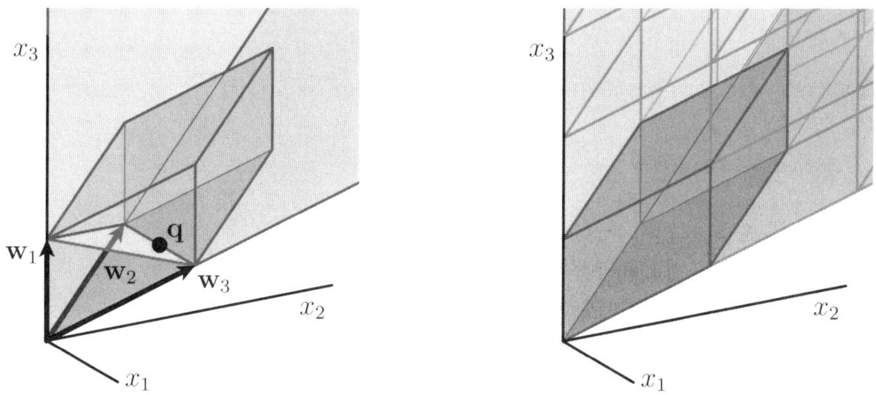

Figure 4.2. Tiling $\hom(\mathsf{P})$ with translates of its fundamental parallelepiped.

tile the cone $\hom(\mathsf{P})$ with translates of $\Box$. In our specific case you can check (Exercise 4.37) that

$$\Box \cap \mathbb{Z}^3 = \{\mathbf{0}, \mathbf{q}\}.$$

Thus, we compute

$$\begin{aligned} \sigma_{\hom(\mathsf{P})}(\mathbf{z}) &= \sum_{k_1,k_2,k_3 \geq 0} \mathbf{z}^{k_1\mathbf{w}_1 + k_2\mathbf{w}_2 + k_3\mathbf{w}_3} + \sum_{k_1,k_2,k_3 \geq 0} \mathbf{z}^{\mathbf{q} + k_1\mathbf{w}_1 + k_2\mathbf{w}_2 + k_3\mathbf{w}_3} \\ &= \frac{1}{(1 - \mathbf{z}^{\mathbf{w}_1})(1 - \mathbf{z}^{\mathbf{w}_2})(1 - \mathbf{z}^{\mathbf{w}_3})} + \frac{\mathbf{z}^{\mathbf{q}}}{(1 - \mathbf{z}^{\mathbf{w}_1})(1 - \mathbf{z}^{\mathbf{w}_2})(1 - \mathbf{z}^{\mathbf{w}_3})} \end{aligned}$$

$$= \frac{1 + z_1 z_2 z_3}{(1 - z_3)\,(1 - z_2 z_3)\,\left(1 - z_1^2 z_2 z_3\right)}\,.$$

This implies, again with (4.6.4),

$$\mathrm{Ehr}_{\mathsf{P}}(z) \;=\; \sigma_{\mathrm{hom}(\mathsf{P})}(1,1,z) \;=\; \frac{1+z}{(1-z)^3}\,.$$

Note that the form of this rational generating function, as that of (4.6.7), ensures that $\mathrm{ehr}_{\mathsf{P}}(n)$ is a polynomial, by Proposition 4.1.4. Of course, this merely confirms Theorem 1.4.1.

We can interpret the coefficients of the numerator polynomial of $\mathrm{Ehr}_{\mathsf{P}}(z)$: the constant is 1 (stemming from the origin of $\mathrm{hom}(\mathsf{P})$), and the remaining term in the example above comes from the fact that the fundamental parallelepiped of $\mathrm{hom}(\mathsf{P})$ contains precisely one integer point at height 1 and none of larger height. Exercise 4.42 gives a few general interpretations of the Ehrhart series coefficients.

The general case of the integer-point transform of a simplicial cone is not much more complicated than our computations in this section; we will treat it in Theorem 4.8.1 below. One of its consequences will be the following result, which gives a simplicial version of Theorem 1.4.1 in all dimensions. We recall that $\triangle^\circ$ denotes the relative interior of $\triangle$.

Theorem 4.6.1. *Suppose $\triangle \subset \mathbb{R}^d$ is a lattice simplex.*

(a) *For positive integers n, the counting function $\mathrm{ehr}_\triangle(n)$ agrees with a polynomial in n of degree $\dim(\triangle)$ whose constant term equals 1.*

(b) *When this polynomial is evaluated at negative integers, we obtain*

$$(-1)^{\dim(\triangle)}\,\mathrm{ehr}_\triangle(-n) \;=\; \left|n\,\triangle^\circ \cap \mathbb{Z}^d\right|.$$

In other words, the Ehrhart polynomials of $\triangle$ and $\triangle^\circ$ are related as

$$(-1)^{\dim(\triangle)}\,\mathrm{ehr}_\triangle(-n) \;=\; \mathrm{ehr}_{\triangle^\circ}(n)\,. \tag{4.6.9}$$

This theorem holds for every lattice polytope (not just simplices), as we will show in Chapter 5. The general version of Theorem 4.6.1(a) is known as *Ehrhart's theorem* and that of Theorem 4.6.1(b) as *Ehrhart–Macdonald reciprocity*.

As a first example, we discuss the Ehrhart polynomial of a simplex

$$\triangle \;=\; \mathrm{conv}\{\mathbf{v}_0, \mathbf{v}_1, \dots, \mathbf{v}_d\} \;\subset\; \mathbb{R}^d$$

that is **unimodular**, i.e., the vectors $\mathbf{v}_1 - \mathbf{v}_0, \mathbf{v}_2 - \mathbf{v}_0, \dots, \mathbf{v}_d - \mathbf{v}_0$ form a $\mathbb{Z}$-basis for $\mathbb{Z}^d$ (which, by Exercise 4.33, is equivalent to the condition $\det(\mathbf{v}_1 - \mathbf{v}_0, \mathbf{v}_2 - \mathbf{v}_0, \dots, \mathbf{v}_\mathrm{d} - \mathbf{v}_0) = \pm 1$). The following result, which can be proved *without* assuming Theorem 4.6.1, is subject to Exercise 4.35.

Proposition 4.6.2. *Let $\triangle$ be the convex hull of the origin and the d unit vectors in $\mathbb{R}^d$. Then* $\mathrm{ehr}_\triangle(n) = \binom{n+d}{d}$, *and this polynomial satisfies the conclusions of Theorem* 4.6.1. *More generally,* $\mathrm{ehr}_\triangle(n) = \binom{n+d}{d}$ *for every unimodular simplex* $\triangle$.

Before proving (a generalization of) Theorem 4.6.1, we will revisit the partition and composition counting functions that appeared earlier in this chapter. Again we can use a cone setting to understand these functions geometrically, but instead of grading integer points by their last coordinate, i.e., along $(0, 0, 1)$, we will introduce a grading along a different direction.

4.7. Gradings of Cones and Rational Polytopes

In the last section, when considering the cone $\mathsf{C} = \hom(\mathsf{P})$, we said that C is *graded by height*; this suggests that there are other ways to grade a cone.

Let $\mathsf{C} \subset \mathbb{R}^d$ be a pointed, rational d-dimensional cone. A **grading** of C is a vector $\mathbf{a} \in \mathbb{Z}^d$ such that $\langle \mathbf{a}, \mathbf{p} \rangle > 0$ for all $\mathbf{p} \in \mathsf{C} \setminus \{\mathbf{0}\}$. (This is where our demanding C to be pointed is essential.) For a grading $\mathbf{a} \in \mathbb{Z}^d$, we define the **Hilbert function** of C as

$$h_{\mathsf{C}}^{\mathbf{a}}(n) \ := \ \left|\left\{\mathbf{m} \in \mathsf{C} \cap \mathbb{Z}^d \ : \ \langle \mathbf{a}, \mathbf{m} \rangle = n\right\}\right| .$$

Since C is pointed, it follows that $h_{\mathsf{C}}^{\mathbf{a}}(n) < \infty$ for all $n \in \mathbb{Z}_{\geq 1}$ and thus we can define the generating function

$$H_{\mathsf{C}}^{\mathbf{a}}(z) \ := \ 1 + \sum_{n \geq 1} h_{\mathsf{C}}^{\mathbf{a}}(n)\, z^n ,$$

the **Hilbert series** of C with respect to $\mathbf{a}$. One benefit of integer-point transforms is the following.

Proposition 4.7.1. *Let $\mathsf{C} \subset \mathbb{R}^d$ be a pointed, rational cone and $\mathbf{a} \in \mathbb{Z}^d$ a grading of* C. *Then*

$$H_{\mathsf{C}}^{\mathbf{a}}(z) \ = \ \sigma_{\mathsf{C}}\left(z^{a_1}, z^{a_2}, \dots, z^{a_d}\right).$$

Proof. We compute

$$\sigma_{\mathsf{C}}\left(z^{a_1}, \dots, z^{a_d}\right) \ = \sum_{\mathbf{m} \in \mathsf{C} \cap \mathbb{Z}^d} z^{a_1 m_1 + \cdots + a_d m_d} \ = \ 1 + \sum_{\mathbf{m} \in \mathsf{C} \cap \mathbb{Z}^d \setminus \{\mathbf{0}\}} z^{\langle \mathbf{a}, \mathbf{m} \rangle}$$
$$= \ 1 + \sum_{n \geq 1} h_{\mathsf{C}}^{\mathbf{a}}(n)\, z^n . \qquad \square$$

Consider again the cone defined by (4.6.5),

$$\mathsf{C} \ := \ \mathbb{R}_{\geq 0} \begin{pmatrix} 0 \\ 0 \\ 1 \end{pmatrix} + \mathbb{R}_{\geq 0} \begin{pmatrix} 0 \\ 1 \\ 1 \end{pmatrix} + \mathbb{R}_{\geq 0} \begin{pmatrix} 1 \\ 1 \\ 1 \end{pmatrix} .$$

In the last section we computed its integer-point transform

$$\sigma_{\mathsf{C}}(\mathbf{z}) = \frac{1}{(1-z_3)(1-z_2z_3)(1-z_1z_2z_3)},$$

and from the grading $(0,0,1)$ we obtained the Ehrhart series $\frac{1}{(1-z)^3}$ of the triangle $\triangle := \{(x_1,x_2) \in \mathbb{R}^2 : 0 \le x_1 \le x_2 \le 1\}$. If we instead grade the points in C along $(1,1,1)$, we meet another friendly face:[4]

$$H_{\mathsf{C}}^{(1,1,1)}(q) = \sigma_{\mathsf{C}}(q,q,q) = \frac{1}{(1-q)(1-q^2)(1-q^3)},$$

the generating function for the restricted partition function with parts 1, 2, and 3, as well as (by Exercise 4.29) for the number of partitions into at most three parts. We remark the obvious, namely, that this generating function is quite different from the Ehrhart series (4.6.7) of $\triangle$; yet they both come from the same cone and the same integer-point transform, specialized along two different gradings.

Next we address some of the other composition/partition counting functions that appeared earlier in this chapter. The compositions into $k+1$ parts, which—up to a factor of $\frac{1}{q}$—we enumerated in (4.2.1), live in the open (unimodular) cone $\mathbb{R}_{>0}^{k+1}$ which comes with the integer-point transform

$$\sigma_{\mathbb{R}_{>0}^{k+1}}(\mathbf{z}) = \frac{z_1}{1-z_1}\frac{z_2}{1-z_2}\cdots\frac{z_{k+1}}{1-z_{k+1}},$$

and indeed, $\sigma_{\mathbb{R}_{>0}^{k+1}}(q,q,\dots,q) = \frac{q^{k+1}}{(1-q)^{k+1}}$ confirms (4.2.1).

The plane partitions of Section 4.3 are the integer points in the cone

$$\mathsf{C} := \left\{ \mathbf{x} \in \mathbb{R}_{\ge 0}^4 : \begin{array}{ccc} x_1 & \ge & x_2 \\ \text{\rotatebox{90}{$\le$}} & & \text{\rotatebox{90}{$\le$}} \\ x_3 & \ge & x_4 \end{array} \right\}.$$

The computation of its integer-point transform is less trivial, as C is not simplicial. However, thinking about possible orderings of x_2 and x_3 in the definition of C gives the inclusion–exclusion formula

$$\sigma_{\mathsf{C}}(\mathbf{z}) = \sigma_{\mathsf{C}_{2\ge3}}(\mathbf{z}) + \sigma_{\mathsf{C}_{3\ge2}}(\mathbf{z}) - \sigma_{\mathsf{C}_{2=3}}(\mathbf{z}), \tag{4.7.1}$$

where

$$\begin{aligned} \mathsf{C}_{2\ge3} &:= \left\{\mathbf{x} \in \mathbb{R}_{\ge0}^4 : x_1 \ge x_2 \ge x_3 \ge x_4\right\}, \\ \mathsf{C}_{3\ge2} &:= \left\{\mathbf{x} \in \mathbb{R}_{\ge0}^4 : x_1 \ge x_3 \ge x_2 \ge x_4\right\}, \quad \text{and} \\ \mathsf{C}_{2=3} &:= \left\{\mathbf{x} \in \mathbb{R}_{\ge0}^4 : x_1 \ge x_2 = x_3 \ge x_4\right\}. \end{aligned}$$

[4]We follow the convention of using the variable q when grading points along $(1,1,\dots,1)$. This also explains our mysterious shift from z to q in Section 4.2.

Each of these cones is unimodular (and, in fact, can be interpreted as containing partitions featured in Exercise 4.29), and so

$$\begin{aligned} \sigma_{\mathsf{C}_{2\geq 3}}(\mathbf{z}) &= \frac{1}{(1-z_1z_2z_3z_4)(1-z_1z_2z_3)(1-z_1z_2)(1-z_1)}\,, \\ \sigma_{\mathsf{C}_{3\geq 2}}(\mathbf{z}) &= \frac{1}{(1-z_1z_2z_3z_4)(1-z_1z_2z_3)(1-z_1z_3)(1-z_1)}\,, \\ \sigma_{\mathsf{C}_{2=3}}(\mathbf{z}) &= \frac{1}{(1-z_1z_2z_3z_4)(1-z_1z_2z_3)(1-z_1)}\,. \end{aligned} \tag{4.7.2}$$

Combining these integer-point transforms according to (4.7.1) gives

$$\sigma_{\mathsf{C}}(\mathbf{z}) = \frac{1-z_1^2z_2z_3}{(1-z_1z_2z_3z_4)(1-z_1z_2z_3)(1-z_1z_2)(1-z_1z_3)(1-z_1)} \tag{4.7.3}$$

and $\sigma_{\mathsf{C}}(q,q,q,q) = \frac{1}{(1-q)(1-q^2)^2(1-q^3)}$ confirms the generating function (4.3.2) for the plane partitions.

Finally, we interpret the restricted partition function from Section 4.4 geometrically. There are two ways to go about that: first, we can interpret

$$p_A(n) = \left|\left\{(m_1,m_2,\dots,m_d)\in\mathbb{Z}_{\geq 0}^d \,:\, m_1a_1+m_2a_2+\cdots+m_da_d=n\right\}\right|$$

as counting lattice points in the cone $\mathbb{R}_{\geq 0}^d$ along the grading $(a_1,a_2,\dots,a_d)$; this comes with the integer-point transform

$$\sigma_{\mathbb{R}_{\geq 0}^d}(\mathbf{z}) = \frac{1}{(1-z_1)(1-z_2)\cdots(1-z_d)}$$

which specializes to

$$\sigma_{\mathbb{R}_{\geq 0}^d}\left(q^{a_1},q^{a_2},\dots,q^{a_d}\right) = \frac{1}{(1-q^{a_1})(1-q^{a_2})\cdots(1-q^{a_d})}\,,$$

confirming (4.4.2). Possibly the more insightful geometric interpretation of $p_A(n)$ goes back to the viewpoint through Ehrhart series. Namely, restricted partitions are integer points in the cone

$$\mathsf{C} := \left\{(x_1,x_2,\dots,x_d,x_{d+1})\in\mathbb{R}_{\geq 0}^{d+1} \,:\, x_1a_1+x_2a_2+\cdots+x_da_d=x_{d+1}\right\}$$

and now we are interested in the "Ehrhart grading" along $(0,0,\dots,0,1)$. In fact, C is the homogenization of a polytope, but this polytope is not a lattice polytope anymore: the generators of C can be chosen as

$$\begin{pmatrix} \frac{1}{a_1} \\ 0 \\ 0 \\ \vdots \\ 0 \\ 1 \end{pmatrix}, \begin{pmatrix} 0 \\ \frac{1}{a_2} \\ 0 \\ \vdots \\ 0 \\ 1 \end{pmatrix}, \dots, \begin{pmatrix} 0 \\ 0 \\ \vdots \\ 0 \\ \frac{1}{a_d} \\ 1 \end{pmatrix}.$$

Indeed, from the look of the generating function (4.4.2) of $p_A(n)$ we should not expect the Ehrhart series of a lattice polytope, as the denominator of $P_A(q)$ is not simply a power of $(1-q)$.

These considerations suggest a variant of Theorem 4.6.1 for **rational polytopes**—those with vertices in $\mathbb{Q}^d$. We leave its proof as Exercise 4.43 (with which you should wait until understanding our proof of Theorem 4.8.1 below).

Theorem 4.7.2. *If $\triangle$ is a rational simplex, then for positive integers n, the counting function $\operatorname{ehr}_\triangle(n)$ is a quasipolynomial in n whose period divides the least common multiple of the denominators of the vertex coordinates of $\triangle$. When this quasipolynomial is evaluated at negative integers, we obtain*

$$(-1)^{\dim(\triangle)} \operatorname{ehr}_\triangle(-n) = \left| n\,\triangle^\circ \cap \mathbb{Z}^d \right| .$$

In other words, the Ehrhart quasipolynomials of $\triangle$ and $\triangle^\circ$ are related as

$$(-1)^{\dim(\triangle)} \operatorname{ehr}_\triangle(-n) = \operatorname{ehr}_{\triangle^\circ}(n) .$$

The different types of grading motivate the study of integer-point transforms as multivariate generating functions. Our next goal is to prove structural results for them, in the case that the underlying polyhedron is a simplicial rational cone.

4.8. Stanley Reciprocity for Simplicial Cones

We return once more to the plane partitions of Section 4.3, and again we think of them as integer points in the 4-dimensional cone

$$\mathsf{C} := \left\{ \mathbf{x} \in \mathbb{R}^4_{\geq 0} : \begin{array}{ccc} x_1 & \geq & x_2 \\ \text{\rotatebox{90}{$\leq$}} & & \text{\rotatebox{90}{$\leq$}} \\ x_3 & \geq & x_4 \end{array} \right\} .$$

In (4.7.1) we computed the integer-point transform of C via inclusion–exclusion, through the integer-point transforms of three unimodular cones (two 4-dimensional and one 3-dimensional cone). Now we will play a variation of the same theme; however, we will use only full-dimensional cones. Namely, if we let

$$\mathsf{C}_{2\geq 3} := \left\{ \mathbf{x} \in \mathbb{R}^4_{\geq 0} : x_1 \geq x_2 \geq x_3 \geq x_4 \right\} \quad \text{and}$$
$$\mathsf{C}_{3>2} := \left\{ \mathbf{x} \in \mathbb{R}^4_{\geq 0} : x_1 \geq x_3 > x_2 \geq x_4 \right\} ,$$

then $\mathsf{C}_{2\geq 3} \uplus \mathsf{C}_{3>2} = \mathsf{C}$ (recall that this symbol denotes a *disjoint* union), and thus

$$\sigma_{\mathsf{C}}(\mathbf{z}) = \sigma_{\mathsf{C}_{2\geq 3}}(\mathbf{z}) + \sigma_{\mathsf{C}_{3>2}}(\mathbf{z}) , \tag{4.8.1}$$

from which we can compute $Pl(q)$ once more by specializing $\mathbf{z} = (q, q, q, q)$. Thus, compared to (4.7.1) we have fewer terms to compute, but we are paying a price by having to deal with *half-open* cones. There is a general philosophy surfacing here, which we will see in action time and again: it is quite useful to decompose a polyhedron in different ways (in this case, giving rise to (4.7.1) and (4.8.1)), each having its own advantages.

We have computed the integer-point transform of $\mathsf{C}_{2\ge3}$ already in (4.7.2):

$$\sigma_{\mathsf{C}_{2\ge3}}(\mathbf{z}) = \frac{1}{(1 - z_1z_2z_3z_4)(1 - z_1z_2z_3)(1 - z_1z_2)(1 - z_1)}.$$

The analogous computation for $\mathsf{C}_{3>2}$ has a little twist stemming from the fact that $\mathsf{C}_{3>2}$ is half open: in terms of generators,

$$\mathsf{C}_{3>2} = \mathbb{R}_{\ge0}\begin{pmatrix}1\\1\\1\\1\end{pmatrix} + \mathbb{R}_{\ge0}\begin{pmatrix}1\\1\\1\\0\end{pmatrix} + \mathbb{R}_{>0}\begin{pmatrix}1\\0\\1\\0\end{pmatrix} + \mathbb{R}_{\ge0}\begin{pmatrix}1\\0\\0\\0\end{pmatrix}.$$

The integer-point transform of its closure $\mathsf{C}_{3\ge2}$ we also computed in (4.7.2), from which we deduce

$$\sigma_{\mathsf{C}_{3>2}}(\mathbf{z}) = z_1z_3\,\sigma_{\mathsf{C}_{3\ge2}}(\mathbf{z}) = \frac{z_1z_3}{(1 - z_1z_2z_3z_4)(1 - z_1z_2z_3)(1 - z_1z_3)(1 - z_1)}.$$

Putting it all together, we recover with (4.8.1) the integer-point transform for C which we computed in (4.7.3):

$$\sigma_{\mathsf{C}}(\mathbf{z}) = \frac{1 - z_1^2z_2z_3}{(1 - z_1z_2z_3z_4)(1 - z_1z_2z_3)(1 - z_1z_2)(1 - z_1z_3)(1 - z_1)}.$$

There are two features that made the "geometric computations" of $\sigma_{\mathsf{C}}(\mathbf{z})$ through (4.7.1) and (4.8.1) possible: the fact that we could decompose C into simplicial cones (in the first instance $\mathsf{C}_{2\ge3}$, $\mathsf{C}_{3\ge2}$, and $\mathsf{C}_{2=3}$, in the second $\mathsf{C}_{2\ge3}$ and $\mathsf{C}_{3>2}$), and the fact that these cones were unimodular. We will see in Chapter 5 that *every* pointed cone can be similarly decomposed into simplicial cones (this is called a *triangulation* of the cone), and the following important theorem shows that the integer-point transform of every half-open simplicial cone (not just a unimodular one) is nice. Even better, it gives our first reciprocity theorem for a multivariate generating function.

Theorem 4.8.1. *Fix linearly independent vectors* $\mathbf{v}_1, \mathbf{v}_2, \dots, \mathbf{v}_k \in \mathbb{Z}^d$ *and* $1 \le m \le k$ *and define the two half-open cones*[5]

$$\widehat{\mathsf{C}} := \mathbb{R}_{\ge0}\mathbf{v}_1 + \cdots + \mathbb{R}_{\ge0}\mathbf{v}_{m-1} + \mathbb{R}_{>0}\mathbf{v}_m + \cdots + \mathbb{R}_{>0}\mathbf{v}_k$$

and

$$\check{\mathsf{C}} := \mathbb{R}_{>0}\mathbf{v}_1 + \cdots + \mathbb{R}_{>0}\mathbf{v}_{m-1} + \mathbb{R}_{\ge0}\mathbf{v}_m + \cdots + \mathbb{R}_{\ge0}\mathbf{v}_k\,.$$

[5]Technically, the cones are only half open for $m > 1$.

Then $\sigma_{\widehat{C}}(\mathbf{z})$ *and* $\sigma_{\breve{C}}(\mathbf{z})$ *are rational generating functions in* $z_1, z_2, \ldots, z_d$ *which are related via*

$$\sigma_{\breve{C}}\left(\tfrac{1}{\mathbf{z}}\right) = (-1)^k \, \sigma_{\widehat{C}}(\mathbf{z}) \,, \tag{4.8.2}$$

where $\frac{1}{\mathbf{z}} := \left(\frac{1}{z_1}, \frac{1}{z_2}, \ldots, \frac{1}{z_d}\right)$.

This is the simplicial version of *Stanley reciprocity*, the general case of which we will prove in Theorem 5.4.2. We remark that (4.8.2) is an identity of rational functions (not formal Laurent series), just as, e.g., (4.6.9) is an identity of polynomials (not counting functions).

Proof. We start with $\widehat{C}$ and use a tiling argument, which we hinted at already in Figure 4.2, to compute its generating function. Namely, let

$$\widehat{\Pi} := [0,1)\,\mathbf{v}_1 + \cdots + [0,1)\,\mathbf{v}_{m-1} + (0,1]\,\mathbf{v}_m + \cdots + (0,1]\,\mathbf{v}_k \,, \tag{4.8.3}$$

the **fundamental parallelepiped** of $\widehat{C}$; see Figure 4.3 for an example. Then

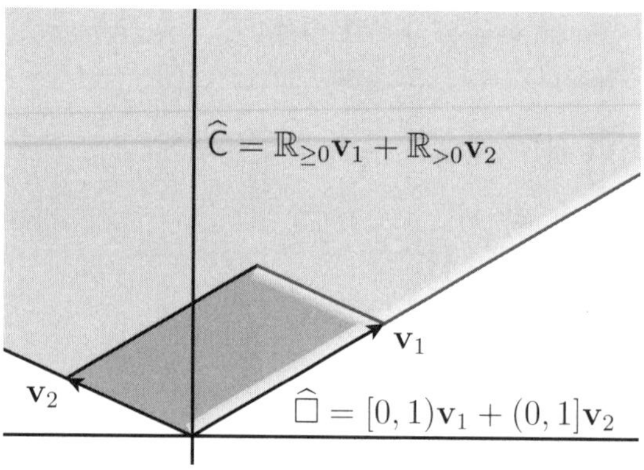

Figure 4.3. The fundamental parallelogram of a 2-cone.

we can tile $\widehat{C}$ by translates of $\widehat{\Pi}$, as we invite you to prove in Exercise 4.38:

$$\widehat{C} = \biguplus_{j_1,\ldots,j_k \geq 0} \left(j_1\mathbf{v}_1 + \cdots + j_k\mathbf{v}_k + \widehat{\Pi}\right) \tag{4.8.4}$$

(see Figure 4.4). This can be translated into the language of generating functions as

$$\begin{aligned} \sigma_{\widehat{C}}(\mathbf{z}) &= \left(\sum_{j_1 \geq 0} \mathbf{z}^{j_1\mathbf{v}_1}\right) \cdots \left(\sum_{j_k \geq 0} \mathbf{z}^{j_k\mathbf{v}_k}\right) \sigma_{\widehat{\Pi}}(\mathbf{z}) \\ &= \frac{\sigma_{\widehat{\Pi}}(\mathbf{z})}{(1-\mathbf{z}^{\mathbf{v}_1})\cdots(1-\mathbf{z}^{\mathbf{v}_k})} \,. \end{aligned} \tag{4.8.5}$$

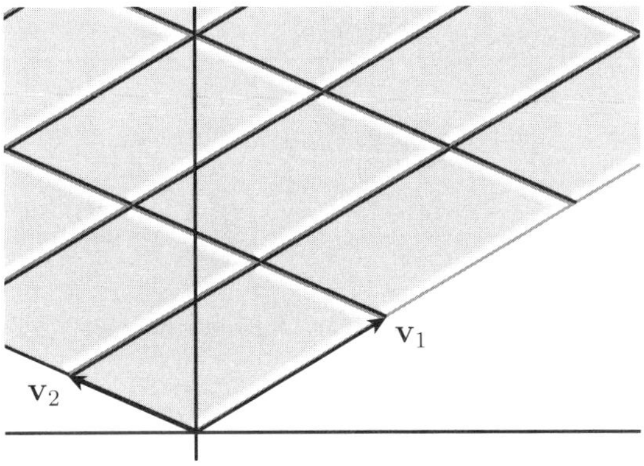

Figure 4.4. Translates of the fundamental parallelogram tile the cone.

Completely analogously, we compute

$$\sigma_{\check{\mathsf{C}}}(\mathbf{z}) = \frac{\sigma_{\check{\square}}(\mathbf{z})}{(1-\mathbf{z}^{\mathbf{v}_1})\cdots(1-\mathbf{z}^{\mathbf{v}_k})},$$

where

$$\check{\square} := (0,1]\,\mathbf{v}_1 + \cdots + (0,1]\,\mathbf{v}_{m-1} + [0,1)\,\mathbf{v}_m + \cdots + [0,1)\,\mathbf{v}_k\,.$$

These two fundamental parallelepipeds are intimately related: if $\mathbf{x} \in \check{\square}$, then we can write $\mathbf{x} = \lambda_1\mathbf{v}_1 + \cdots + \lambda_k\mathbf{v}_k$ for some $0 < \lambda_1, \ldots, \lambda_{m-1} \leq 1$ and $0 \leq \lambda_m, \ldots, \lambda_k < 1$. But then

$$\mathbf{v}_1 + \mathbf{v}_2 + \cdots + \mathbf{v}_k - \mathbf{x} = (1-\lambda_1)\mathbf{v}_1 + \cdots + (1-\lambda_k)\mathbf{v}_k$$

is a point in $\hat{\square}$. Since $\mathbf{v}_1, \mathbf{v}_2, \ldots, \mathbf{v}_k$ are linearly independent, this proves

$$\hat{\square} = \mathbf{v}_1 + \mathbf{v}_2 + \cdots + \mathbf{v}_k - \check{\square} \tag{4.8.6}$$

(illustrated in Figure 4.5) or, in terms of generating functions,

$$\sigma_{\hat{\square}}(\mathbf{z}) = \mathbf{z}^{\mathbf{v}_1+\mathbf{v}_2+\cdots+\mathbf{v}_k}\,\sigma_{\check{\square}}\left(\frac{1}{\mathbf{z}}\right).$$

This yields

$$\sigma_{\check{\mathsf{C}}}\left(\frac{1}{\mathbf{z}}\right) = \frac{\sigma_{\check{\square}}(\frac{1}{\mathbf{z}})}{(1-\mathbf{z}^{-\mathbf{v}_1})\cdots(1-\mathbf{z}^{-\mathbf{v}_k})} = \frac{\mathbf{z}^{-\mathbf{v}_1-\mathbf{v}_2-\cdots-\mathbf{v}_k}\,\sigma_{\hat{\square}}(\mathbf{z})}{(1-\mathbf{z}^{-\mathbf{v}_1})\cdots(1-\mathbf{z}^{-\mathbf{v}_k})}$$
$$= (-1)^k\frac{\sigma_{\hat{\square}}(\mathbf{z})}{(1-\mathbf{z}^{\mathbf{v}_1})\cdots(1-\mathbf{z}^{\mathbf{v}_k})} = (-1)^k\,\sigma_{\hat{\mathsf{C}}}(\mathbf{z})\,. \qquad \square$$

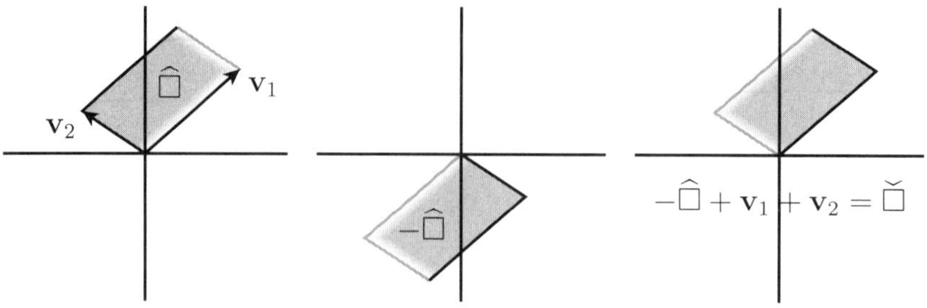

Figure 4.5. The geometry of (4.8.6) in dimension 2.

Theorem 4.8.1 is at the heart of this book. Once we have developed the machinery of decomposing polyhedra in Chapter 5, we will be able to extend it to every pointed rational cone (Theorem 5.4.2). For now, we limit ourselves to consequences for simplicial cones. For $m = 1$, we observe that in Theorem 4.8.1

$$\widehat{\mathsf{C}} = \mathsf{C}^\circ \quad \text{and} \quad \check{\mathsf{C}} = \mathsf{C} .$$

This extreme case yields a reciprocity for simplicial cones equipped with an arbitrary grading.

Corollary 4.8.2. *Let $\mathsf{C} \subset \mathbb{R}^{d+1}$ be a rational simplicial cone with generators $\mathbf{v}_1, \dots, \mathbf{v}_k$ and fundamental parallelepiped $\square$. For a grading $\mathbf{a} \in \mathbb{Z}^{d+1}$,*

$$H_{\mathsf{C}}^{\mathbf{a}}(z) \;=\; \sum_{n \geq 0} h_{\mathsf{C}}^{\mathbf{a}}(n)\, z^n \;=\; \frac{H_{\square}^{\mathbf{a}}(z)}{\left(1 - z^{\langle \mathbf{a}, \mathbf{v}_1 \rangle}\right) \cdots \left(1 - z^{\langle \mathbf{a}, \mathbf{v}_k \rangle}\right)} \tag{4.8.7}$$

and

$$H_{\mathsf{C}}^{\mathbf{a}}\left(\frac{1}{z}\right) \;=\; (-1)^k\, H_{\mathsf{C}^\circ}^{\mathbf{a}}(z) .$$

One interesting fact to note is that the fundamental parallelepiped depends on a choice of generators and hence is unique only up to scaling of the generators. Technically, this means that the right-hand side of (4.8.7) depends on a choice of generators, whereas the left-hand side does not; see Exercise 4.39.

Corollary 4.8.2 also furnishes a proof of Theorem 4.6.1.

Proof of Theorem 4.6.1. Recall our setup for this result, stated at the beginning of Section 4.6: to a given r-dimensional lattice simplex $\triangle \subset \mathbb{R}^d$ with vertices $\mathbf{v}_1, \dots, \mathbf{v}_{r+1}$, we associate its homogenization $\mathsf{C} = \operatorname{hom}(\triangle)$. This is a simplicial cone with generators $\mathbf{w}_i = (\mathbf{v}_i, 1)$ for $i = 1, \dots, r+1$ and associated parallelepiped

$$\square \;:=\; [0,1)\mathbf{w}_1 + [0,1)\mathbf{w}_2 + \cdots + [0,1)\mathbf{w}_{r+1} .$$

For the grading $\mathbf{a} = (0, \dots, 0, 1)$, we can write the Ehrhart series of $\triangle$ as

$$\operatorname{Ehr}_{\triangle}(z) = 1 + \sum_{n \ge 1} \operatorname{ehr}_{\triangle}(n)\, z^n = H^{\mathbf{a}}_{\mathsf{C}}(z) = \frac{H^{\mathbf{a}}_{\square}(z)}{(1-z)^{r+1}},$$

where the last equality stems from Corollary 4.8.2. The numerator polynomial

$$h^*(z) := H^{\mathbf{a}}_{\square}(z) = h_0^* + h_1^* z + \cdots + h_r^* z^r$$

enumerates the number of lattice points in $\square$ according to their height. Note that the h_j^*s are nonnegative integers. We also observe that there are no lattice points of height $> r$ and that the origin $\mathbf{0}$ is the unique lattice point in $\square$ of height zero; hence $h_i^* \ge 0$ for $i = 0, \dots, r$ and $h_0^* = 1$. Using Proposition 4.1.4, this shows that $\operatorname{ehr}_{\triangle}(n)$ is a polynomial of degree r, as $h^*(1) > 0$. By the same token, we observe that $\operatorname{ehr}_{\triangle}(0) = H^{\mathbf{a}}_{\mathsf{C}}(0) = h^*(0) = 1$ which proves part (a).

For part (b) we combine Theorem 4.1.6 with Corollary 4.8.2 to infer

$$\begin{aligned}(-1)^r \sum_{n \ge 1} \operatorname{ehr}_{\triangle}(-n)\, z^n &= (-1)^{r+1} \sum_{n \ge 0} \operatorname{ehr}_{\triangle}(n) \left(\frac{1}{z}\right)^n \\ &= (-1)^{r+1} \operatorname{Ehr}_{\triangle}\left(\frac{1}{z}\right) = \operatorname{Ehr}_{\triangle^\circ}(z) = \sum_{n \ge 1} \operatorname{ehr}_{\triangle^\circ}(n)\, z^n,\end{aligned}$$

which gives the combinatorial reciprocity that we were after. $\square$

We remark that the penultimate line in the above proof essentially forced us to define the Ehrhart series of an *open* polytope to start at $n = 1$:

$$\operatorname{Ehr}_{\mathsf{P}^\circ}(z) := \sum_{n \ge 1} \operatorname{ehr}_{\mathsf{P}^\circ}(n)\, z^n.$$

One reason to do so is to make reciprocity work (specifically, Theorem 4.1.6), but our definition also gives the correct open analogue to (4.6.4), as

$$\operatorname{Ehr}_{\mathsf{P}^\circ}(z) = \sigma_{\operatorname{hom}(\mathsf{P})^\circ}(1, \dots, 1, z).$$

4.9. Chain Partitions and the Dehn–Sommerville Relations

We finish this chapter with the study of a class of partition functions defined via posets, which are hybrids of sorts between the restricted partition functions from Section 4.4 and the zeta polynomials from Chapter 2. Let Π be a finite poset with $\hat{0}$ and $\hat{1}$ and let $\phi : \Pi \setminus \{\hat{0}, \hat{1}\} \to \mathbb{Z}_{>0}$ be an order-preserving map. A (Π, ϕ)**-chain partition** of $n \in \mathbb{Z}_{>0}$ is a partition of the form

$$n = \phi(c_m) + \phi(c_{m-1}) + \cdots + \phi(c_1)$$

for some multichain

$$\hat{1} \succ c_m \succeq c_{m-1} \succeq \cdots \succeq c_1 \succ \hat{0}$$

in Π. We define $cp_{\Pi,\phi}(n)$ to be the number of chain partitions of n and we set $cp_{\Pi,\phi}(0) := 1$. The connection to restricted partition functions is immediate: for $A = \{b_d > \cdots > b_1 > 0\} \subset \mathbb{Z}_{\geq 0}$, we set $\phi(i) := b_i$ for $i \in [d]$. Then

$$cp_{\Pi,\phi}(n) \;=\; p_A(n)\,,$$

where $\Pi = [d] \cup \{\hat{0}, \hat{1}\}$ is the $(d+2)$-chain. On the other extreme, if Π is an antichain together with $\hat{0}$ and $\hat{1}$, then (for the same ϕ) we are counting the compositions from Section 4.2:

$$cp_{\Pi,\phi}(n) \;=\; c_A(n)\,.$$

Our counting function $cp_{\Pi,\phi}(n)$ for a general pair (Π, ϕ) comes, naturally, with a generating function; we will first construct a multivariate generating function—analogous to an integer-point transform—, from which the univariate version follows by specialization. Let $\mathbf{z} := (z_a : a \in \Pi)$ be a set of variables indexed by the elements of Π, and set

$$\Upsilon_\Pi(\mathbf{z}) \;:=\; \sum_{\hat{0} \prec c_1 \preceq c_2 \preceq \cdots \preceq c_m \prec \hat{1}} z_{c_1} z_{c_2} \cdots z_{c_m}\,.$$

In words, the monomials in $\Upsilon_\Pi(\mathbf{z})$ encode the multichains in $\Pi \setminus \{\hat{0}, \hat{1}\}$. In particular,

$$CP_{\Pi,\phi}(q) \;:=\; \sum_{n \geq 0} cp_{\Pi,\phi}(n)\, q^n \;=\; \Upsilon\big(z_a = q^{\phi(a)} \,:\, a \in \Pi\big)\,, \tag{4.9.1}$$

the univariate generating function corresponding to our chain-partition counting function $cp_{\Pi,\phi}(n)$. Not unlike the ideas used in Section 2.2, the generating function $\Upsilon_\Pi(\mathbf{z})$ can be developed by accounting for the possible repetitions in a multichain stemming from a chain:

$$\begin{aligned} \Upsilon_\Pi(\mathbf{z}) \;&=\; \sum_{\hat{0} \prec c_1 \prec c_2 \prec \cdots \prec c_m \prec \hat{1}} \; \sum_{j_1, j_2, \ldots, j_m \geq 1} z_{c_1}^{j_1} z_{c_2}^{j_2} \cdots z_{c_m}^{j_m} \\ &=\; \sum_{\hat{0} \prec c_1 \prec c_2 \prec \cdots \prec c_m \prec \hat{1}} \frac{z_{c_1}}{1 - z_{c_1}} \frac{z_{c_2}}{1 - z_{c_2}} \cdots \frac{z_{c_m}}{1 - z_{c_m}}\,. \end{aligned} \tag{4.9.2}$$

As a first milestone, we compute $cp_{\Pi,\phi}$ for a particular class of pairs (Π, ϕ). We recall that a poset Π is graded if every maximal chain in Π has the same length $d+1$. In this case, we define the **rank** of $a \in \Pi$, denoted by $\operatorname{rk}_\Pi(a)$, as the length of any maximal chain ending in a. (This definition made a brief appearance in Exercise 2.15.) Thus the rank of Π is the rank of any maximal element in Π, for which we write $\operatorname{rk}(\Pi) = d + 1$.

For $S = \{0 \leq s_1 < s_2 < \cdots < s_m \leq d+1\}$, let

$$\alpha_\Pi(S) \;:=\; \left|\{c_1 \prec c_2 \prec \cdots \prec c_m \,:\, \operatorname{rk}(c_j) = s_j \text{ for } j = 1, \ldots, m\}\right|;$$

in words, $\alpha_\Pi(S)$ counts all chains with ranks in S. The collection

$$(\alpha_\Pi(S) \,:\, S \subseteq \{0, 1, \ldots, d+1\})$$

is the **flag f-vector** of Π. If Π has a minimum $\hat{0}$ and maximum $\hat{1}$, then $\alpha_\Pi(S) = \alpha_\Pi(S \cup \{0\}) = \alpha_\Pi(S \cup \{d+1\})$ for any S. Hence we may restrict the flag f-vector to $(\alpha_\Pi(S) : S \subseteq [d])$.

The order-preserving map ϕ is **ranked** if $\phi(a) = \phi(b)$ whenever $\mathrm{rk}_\Pi(a) = \mathrm{rk}_\Pi(b)$. For example, $\phi(a) := \mathrm{rk}_\Pi(a)$ is a ranked order-preserving map. If ϕ is ranked, we set $\phi_i := \phi(a)$, where $a \in \Pi$ is any element of rank $1 \le i \le d$.

Theorem 4.9.1. *Suppose Π is a graded poset of rank $d+1$ with $\hat{0}$ and $\hat{1}$, and $\phi : \Pi \setminus \{\hat{0}, \hat{1}\} \to \mathbb{Z}_{\ge 0}$ is a ranked order-preserving map. Then*

$$CP_{\Pi,\phi}(q) \;=\; \frac{\sum_{S \subseteq [d]} \alpha(S) \prod_{s \in S} q^{\phi_s} \prod_{s \notin S} (1 - q^{\phi_s})}{(1-q^{\phi_1})(1-q^{\phi_2}) \cdots (1-q^{\phi_d})} \,.$$

In particular, $cp_{\Pi,\phi}(n)$ is a quasipolynomial in n.

Proof. Consider a summand in the right-hand side of (4.9.2). Under the specialization $z_a \mapsto q^{\phi(a)}$, it takes on the form

$$\frac{q^{\phi(c_1)}}{1-q^{\phi(c_1)}} \cdots \frac{q^{\phi(c_m)}}{1-q^{\phi(c_m)}} \,,$$

which only depends on $S = \{0 < \mathrm{rk}_\Pi(c_1) < \cdots < \mathrm{rk}_\Pi(c_m) < d+1\}$ and which thus occurs exactly $\alpha_\Pi(S)$ times in (4.9.1). We can move the summands to the common denominator $(1-q^{\phi_1}) \cdots (1-q^{\phi_d})$ by realizing that in each chain there are no two elements with the same rank. $\square$

The form of (4.9.2) suggests that there is a reciprocity theorem for $\Upsilon_\Pi(\mathbf{z})$ waiting in the wings. Indeed,

$$\text{(4.9.3)} \qquad \Upsilon_\Pi\left(\tfrac{1}{\mathbf{z}}\right) \;=\; \sum_{\hat{0} \prec c_1 \prec \cdots \prec c_m \prec \hat{1}} (-1)^m \frac{1}{1-z_{c_1}} \frac{1}{1-z_{c_2}} \cdots \frac{1}{1-z_{c_m}}$$

is a weighted sum of monomials $z_{b_1}^{k_1} z_{b_2}^{k_2} \cdots z_{b_j}^{k_j}$ corresponding to the multichain

$$\hat{0} \prec \underbrace{b_1 \preceq \cdots \preceq b_1}_{k_1} \prec \underbrace{b_2 \preceq \cdots \preceq b_2}_{k_2} \prec \cdots \prec \underbrace{b_j \preceq \cdots \preceq b_j}_{k_j} \prec \hat{1} \,.$$

We observe that the coefficient with which this multichain occurs depends only on the underlying chain $b_1 \prec b_2 \prec \cdots \prec b_j$. As in Section 3.5, we will abbreviate this chain by $\mathbf{b}$ and recall that $l(\mathbf{b}) = j - 1$ is the length of $\mathbf{b}$. Any summand in the right-hand side of (4.9.3) can be expanded as a sum over multichains supported on *subchains* of $c_1 \prec c_2 \prec \cdots \prec c_m$. To be more precise, the coefficient of $z_{b_1} z_{b_2} \cdots z_{b_j}$ in (4.9.3) is

$$\sum_{\mathbf{c} \supseteq \mathbf{b}} (-1)^{l(\mathbf{c})-1}, \tag{4.9.4}$$

where this sum is over all chains $\mathbf{c}$ containing $\mathbf{b}$ as a subchain.

There is a natural construction that will help us with the bookkeeping. Let Π be a general poset. For $a, b \in \Pi$, we define the **order complex** $\Delta_\Pi(a,b)$ as the collection of sets $\sigma = \{c_1, c_2, \ldots, c_m\}$ such that $a \prec c_1 \prec c_2 \prec \cdots \prec c_m \prec b$. This is a partially ordered set under inclusion with minimum $\varnothing$, the empty chain. The order complex of Π itself is $\Delta(\Pi)$, the collection of *all* chains in Π.

Coming back to the coefficient of $z^{b_1} \cdots z^{b_j}$, setting $b_0 := \hat{0}$ and $b_{j+1} :- \hat{1}$ allows us to rewrite

$$\sum_{\mathbf{c} \supseteq \mathbf{b}} (-1)^{l(\mathbf{c})} = \prod_{h=0}^{j} \sum_{\sigma \in \Delta_\Pi(b_h, b_{h+1})} (-1)^{|\sigma|-1} = \prod_{h=0}^{j} \sum_{k \geq 1} (-1)^k c_k(b_h, b_{h+1}),$$

where $c_k(b_h, b_{h+1})$ was introduced in Section 2.4 as the number of chains of length k in the open interval $(b_h, b_{h+1}) \subseteq \Pi$. Bringing Theorem 2.4.6 into the mix of (4.9.3) and (4.9.4) thus yields the following.

Lemma 4.9.2. *Let Π be a finite poset with $\hat{0}$ and $\hat{1}$ and Möbius function μ. Then*

$$-\Upsilon_\Pi\left(\frac{1}{\mathbf{z}}\right) = \sum_{\hat{0} \prec b_1 \prec \cdots \prec b_j \prec \hat{1}} \mu(\hat{0}, b_1)\mu(b_1, b_2) \cdots \mu(b_j, \hat{1}) \frac{z_{b_1}}{1 - z_{b_1}} \cdots \frac{z_{b_j}}{1 - z_{b_j}}.$$

This is surprisingly close to (4.9.2). We can get even closer if Π is Eulerian.

Theorem 4.9.3. *Suppose Π is an Eulerian poset with $\operatorname{rk}(\Pi) = d+1$. Then*

$$\Upsilon_\Pi\left(\tfrac{1}{\mathbf{z}}\right) = (-1)^d \, \Upsilon_\Pi(\mathbf{z}).$$

Proof. We have done most of the leg work by deriving Lemma 4.9.2. Since Π is Eulerian, we have $\mu(a,b) = (-1)^{l(a,b)} = (-1)^{\operatorname{rk}(b)-\operatorname{rk}(a)}$ for any $a \preceq b$ and hence

$$\begin{aligned} \mu(\hat{0}, b_1)\, \mu(b_1, b_2) \cdots \mu(b_j, \hat{1}) &= (-1)^{l(\hat{0}, b_1)} (-1)^{l(b_1, b_2)} \cdots (-1)^{l(b_j, \hat{1})} \\ &= (-1)^{d+1}, \end{aligned}$$

and so Lemma 4.9.2 yields

$$-\Upsilon_\Pi\left(\tfrac{1}{\mathbf{z}}\right) = \sum_{\hat{0} \prec b_1 \prec \cdots \prec b_j \prec \hat{1}} (-1)^{d+1} \frac{z_{b_1}}{1 - z_{b_1}} \cdots \frac{z_{b_j}}{1 - z_{b_j}} = (-1)^{d+1} \, \Upsilon_\Pi(\mathbf{z}). \qquad \square$$

Specializing at $\mathbf{z} = (q^{\phi(a)} : \hat{0} \prec a \prec \hat{1})$ gives the following generalization of Theorem 4.4.3.

Corollary 4.9.4. *Let* Π *be an Eulerian poset of rank* $d+1$. *If* $\phi : \Pi \setminus \{\hat{0}, \hat{1}\} \to \mathbb{Z}_{\geq 0}$ *is a ranked order-preserving map with values* $\phi_1 \leq \cdots \leq \phi_d$ *then*

$$(-1)^d cp_{\Pi,\phi}(-n) \;=\; cp_{\Pi,\phi}(n - \phi_1 - \phi_2 - \cdots - \phi_d)\,.$$

The counting function $cp_{\Pi,\phi}(n)$ is particularly nice for the following class of (particularly nice) posets. An **(abstract) simplicial complex** is a nonempty collection Γ of subsets of some finite set V that is closed under taking subsets, that is, if $\sigma \in \Gamma$ and $\sigma' \subseteq \sigma$, then $\sigma' \in \Gamma$. For example, if Π is any poset, then its order complex $\Delta(\Pi)$ is a simplicial complex. A simplicial complex is **pure** if all inclusion-maximal sets $\sigma \in \Gamma$ have the same cardinality. In this case, Γ is a graded poset with rank function $\mathrm{rk}_\Gamma(\sigma) = |\sigma|$. In particular, Γ has the unique minimal element $\varnothing$.

Another class of examples for simplicial complexes can be derived from simplicial polytopes: if P is a simplicial polytope with vertex set V, then every proper face $\mathsf{F} \prec \mathsf{P}$ is a simplex, and so any subset of $\mathrm{vert}(\mathsf{F})$ is the set of vertices of a face of P. This means that

$$\Gamma_{\mathsf{P}} \;:=\; \{\mathrm{vert}(\mathsf{F}) \,:\, \mathsf{F} \prec \mathsf{P} \text{ proper face}\}$$

is a simplicial complex.[6] The elements $\sigma \in \Gamma$ are called **faces** of Γ and, for consistency with simplicial polytopes, we set $\dim \sigma := |\sigma| - 1$ and $\dim \Gamma := \max\{\dim \sigma : \sigma \in \Gamma\}$.

For a simplicial complex, there is a convenient way to encode multichains. Each $\mathbf{a} \in \mathbb{Z}^V_{\geq 0}$ represents a multisubset of V by interpreting a_v as the multiplicity of the element v. We define the **support** of $\mathbf{a} \in \mathbb{Z}^V_{\geq 0}$ as

$$\mathrm{supp}(\mathbf{a}) \;:=\; \{v \in V \,:\, a_v > 0\}\,,$$

the subset underlying the multisubset $\mathbf{a}$, and we write $|\mathbf{a}| := \sum_v a_v$ for the cardinality of $\mathbf{a}$.

Lemma 4.9.5. *Let* Γ *be a simplicial complex on* V. *Then there is a one-to-one correspondence between multisubsets* $\mathbf{a} \in \mathbb{Z}^V_{\geq 0} \setminus \{\mathbf{0}\}$ *with* $\mathrm{supp}(\mathbf{a}) \in \Gamma$ *and* $|\mathbf{a}| = n$ *and multichains*

$$\varnothing \subset \sigma_1 \subseteq \sigma_2 \subseteq \cdots \subseteq \sigma_m \;=\; \mathrm{supp}(\mathbf{a})$$

with $n = |\sigma_1| + \cdots + |\sigma_m|$.

[6] It is a short step to identify each vertex set in Γ with the accompanying face of P, and the resulting simplices form a simplicial complex for which we drop the attribute *abstract*; we will have much more to say about this in the next chapter.

Proof. We will actually prove a refined statement: there is a one-to-one correspondence between multisubsets $\mathbf{a} \in \mathbb{Z}_{\geq 0}^V \setminus \{\mathbf{0}\}$ with $\mathrm{supp}(\mathbf{a}) \in \Gamma$ and $|\mathbf{a}| = n$ and multichains

$$\varnothing \subset \sigma_1 \subseteq \sigma_2 \subseteq \cdots \subseteq \sigma_m = \mathrm{supp}(\mathbf{a}),$$

where $m = \max\{a_v : v \in V\}$ and $n = |\sigma_1| + \cdots + |\sigma_m|$. If $m = 1$, then $\mathbf{a} \in \{0,1\}^V$ and setting $\sigma_1 := \mathrm{supp}(\mathbf{a})$ does the trick. Otherwise ($m > 1$), consider $\mathbf{a}' \in \mathbb{Z}^V$ with

$$a'_v := \begin{cases} a_v - 1 & \text{if } v \in \mathrm{supp}(\mathbf{a}), \\ 0 & \text{otherwise.} \end{cases}$$

By construction, $\mathbf{a}' \geq 0$, $\mathrm{supp}(\mathbf{a}') \subseteq \mathrm{supp}(\mathbf{a})$, and $\max\{a'_v : v \in V\} = m-1$. By induction on m, there is a chain $\varnothing \subset \sigma_1 \subseteq \cdots \subseteq \sigma_{m-1} = \mathrm{supp}(\mathbf{a}')$ and we simply set $\sigma_m := \mathrm{supp}(\mathbf{a})$.

Our construction also suggests what the inverse map should be. For a subset $A \subseteq V$, let $\mathbf{e}_A \in \{0,1\}^V$ with

$$(\mathbf{e}_A)_v = \begin{cases} 1 & \text{if } v \in A, \\ 0 & \text{if } v \notin A. \end{cases}$$

Then, given a multichain $\varnothing \subset \sigma_1 \subseteq \cdots \subseteq \sigma_m \in \Gamma$ with $|\sigma_1| + \cdots + |\sigma_m| = n$, we define $\mathbf{a} = \mathbf{e}_{\sigma_1} + \cdots + \mathbf{e}_{\sigma_m} \in \mathbb{Z}_{\geq 0}^V$. □

The above way to encode multichains is the key to the following result. To every simplicial complex Γ, we can associate an ***f*-vector** $f(\Gamma) = (f_{-1}, f_0, \ldots, f_{d-1})$, where $f_i = f_i(\Gamma)$ is the number of faces $\sigma \in \Gamma$ of dimension i or, equivalently, of rank $i+1$.

Theorem 4.9.6. *Let* Γ *be a simplicial complex of dimension* $d-1$ *with rank function* $\mathrm{rk}(\sigma) = |\sigma|$. *Then*

$$cp_{\Gamma \cup \{\hat{1}\}, \mathrm{rk}}(n) = \sum_{k=0}^{d} f_{k-1}(\Gamma) \binom{n}{k}.$$

Proof. Unravelling the definitions, $cp_{\Gamma \cup \{\hat{1}\}, \mathrm{rk}}(n)$ counts the number of multichains $\varnothing \subset \sigma_1 \subseteq \sigma_2 \subseteq \cdots \subseteq \sigma_m$ in Γ with $n = |\sigma_1| + \cdots + |\sigma_m|$. Using Lemma 4.9.5, we compute for $n > 0$

$$\begin{aligned} cp_{\Gamma \cup \{\hat{1}\}, \mathrm{rk}}(n) &= \left|\left\{\mathbf{a} \in \mathbb{Z}_{\geq 0}^V : \mathrm{supp}(\mathbf{a}) \in \Gamma,\ |\mathbf{a}| = n\right\}\right| \\ &= \sum_{\sigma \in \Gamma} \left|\left\{\mathbf{a} \in \mathbb{Z}_{\geq 0}^V : \mathrm{supp}(\mathbf{a}) = \sigma,\ |\mathbf{a}| = n\right\}\right| \\ &= \sum_{\sigma \in \Gamma} \left|\left\{\mathbf{a} \in \mathbb{Z}_{\geq 0}^V : \mathrm{supp}(\mathbf{a}) \subseteq \sigma,\ |\mathbf{a}| = n - |\sigma|\right\}\right| \end{aligned}$$

$$= \sum_{\sigma \in \Gamma} \binom{n}{|\sigma|} = \sum_{k=0}^{d} f_{k-1} \binom{n}{k} . \qquad \square$$

If you get the feeling that this formula looks familiar, you might want to revisit (3.5.6) in Section 3.5.

For any two $F \subseteq G \in \Gamma$, the interval $[F, G]$ is isomorphic to the Boolean lattice consisting of all subsets of $G \setminus F$ (partially ordered by $\subseteq$), and hence $\mu_\Gamma(F, G) = (-1)^{l(F,G)}$. In particular, if $\Gamma = \Gamma_{\mathsf{P}}$ for some d-dimensional simplicial polytope P, then $\Gamma_{\mathsf{P}} \cup \{\hat{1}\}$ is isomorphic to the face lattice $\Phi(\mathsf{P})$, which is an Eulerian poset by Theorem 3.5.1. We can then deduce from Theorem 4.9.6 that

$$CP_{\Phi(\mathsf{P}),\mathrm{rk}}(q) = \frac{h_0(\mathsf{P}) + h_1(\mathsf{P})\, q + \cdots + h_{d-1}(\mathsf{P})\, q^{d-1}}{(1-q)^d}$$

for some integers $h_0(\mathsf{P}), h_1(\mathsf{P}), \ldots, h_{d-1}(\mathsf{P})$ depending on P. Corollary 4.9.4 now implies that

$$h_i(\mathsf{P}) = h_{d-1-i}(\mathsf{P}) \tag{4.9.5}$$

for $i = 0, \ldots, d-1$. At this stage, this is where our story takes a break, as we do not have an interpretation for the numbers $h_i(\mathsf{P})$. In fact, we do not even know whether $h_i(\mathsf{P}) \geq 0$ (in which case there is a chance that they count something). We will develop the necessary machinery in the next chapter and shed more light on the $h_i(\mathsf{P})$ in Section 5.6.

Notes

We have barely started to touch on the useful and wonderful world of generating functions. We heartily recommend [**114**] and [**186**] if you'd like to explore more.

Partitions and compositions have been around since at least Leonard Euler's time. They provide a fertile ground for famous theorems (see, e.g., the work of Hardy, Ramanujan, and Rademacher) and open problems (e.g., nobody understands exactly how the *parity* of the number of partitions of n behaves), and they provide a just-as-fertile ground for connections to other areas in mathematics and physics (e.g., Young tableaux, which open a window to representation theory). Again we barely scratched the surface in this chapter; see, e.g., [**4, 5, 82**] for further study. The earliest reference for Proposition 4.2.1 we are aware of is William Feller's book [**63**, p. 311] on probability theory, which was first published in 1950. The earliest combinatorics paper that includes Proposition 4.2.1 seems to be [**125**]. A two-variable generalization of Proposition 4.2.1, which appeared in [**87**], is described in Exercise 4.18. It is not clear who first proved Theorem 4.2.2. The earliest reference we are aware of is [**87**] but we suspect that the theorem has been known earlier. Arthur Cayley's collected works [**45**, p. 16] contain the result that $c_{\{j\in\mathbb{Z}\,:\,j\geq 2\}}(n)$ equals the n-th Fibonacci number, but establishing a bijective proof of the fact that $c_{\{j\in\mathbb{Z}\,:\,j\geq 2\}}(n) = c_{\{2j+1\,:\,j\geq 0\}}(n)$ (which follows from Theorem 4.2.2) is nontrivial [**157**]. Other applications of Proposition 4.2.1 include recent theorems, e.g., [**23, 144, 145, 189**]; see Exercises 4.16 and 4.17.

Plane partitions were introduced by Percy MacMahon about a century ago, who proved a famous generating-function formula for the general case of an $m \times n$ plane partition [**117**]. There are various generalizations of plane partitions, for example, the *plane partition diamonds* given in Exercise 4.20, due to George Andrews, Peter Paule, and Axel Riese [**6**].

Restricted partition functions are closely related to a famous problem in combinatorial number theory: namely, what is the largest integer root of $p_A(n)$ (the *Frobenius number* associated with the set A)?[7] This problem, first raised by Georg Frobenius in the 19th century, is often called the *coin-exchange problem*—it can be phrased in layman terms as looking for the largest amount of money that we cannot change given coin denominations in the set A. Exercise 4.25 (which gives a formula for the restricted partition function in the case that A contains two elements; it goes back to an 1811 book on elementary number theory by Peter Barlow [**13**, pp. 323–325]) suggests that the Frobenius problem is easy for $|A| = 2$ (and you may use Exercise 4.25 to find a formula for the Frobenius number in this case), but this is deceiving: the Frobenius problem is much harder for $|A| = 3$ (though there exist formulas of sorts [**54**]) and certainly wide open for $|A| \geq 4$. The Frobenius problem is also interesting from a computational perspective: while the Frobenius number is known to be polynomial-time computable for fixed $|A|$ [**98**], implementable algorithms are harder to come by (see, e.g., [**27**]). For much more on the Frobenius problem, we refer to [**139**].

Theorem 4.4.3 is due to Eugène Ehrhart [**58**], who also proved Theorem 4.7.2, which gives a vast generalization of the reciprocity featured in Theorem 4.4.3, from one to an arbitrary (finite) number of linear constraints. We will have more to say about Ehrhart's work in the next chapter. Our development of Ehrhart polynomials follows Eugène Ehrhart's original *ansatz* [**57**]. We'll have (much) more to say about this in the next chapter. The same goes for the multivariate analogues of Ehrhart series in Section 4.8, which were initiated by Richard Stanley.

For a rational polyhedral cone $\mathsf{C} \subset \mathbb{R}^{d+1}$, the set $\mathsf{C} \cap \mathbb{Z}^{d+1}$ is a finitely generated semigroup, called an *affine semigroup*. The corresponding semigroup algebra $\mathbb{C}[\mathsf{C} \cap \mathbb{Z}^{d+1}]$ is the subalgebra of $S := \mathbb{C}[x_1^{\pm 1}, x_2^{\pm 1}, \dots, x_{d+1}^{\pm 1}]$

[7] For this question to make sense, we need to assume that the elements of A are relatively prime.

generated by the monomials $\mathbf{x}^{\mathbf{u}}$ for $\mathbf{u} \in \mathsf{C} \cap \mathbb{Z}^{d+1}$. Any $\mathbf{a} \in \mathbb{Z}^{d+1}$ such that $\langle \mathbf{a}, \mathbf{u} \rangle > 0$ for $\mathbf{u} \in \mathsf{C} \cap \mathbb{Z}^{d+1}$ and $\mathbf{u} \neq 0$ defines a grading of S and the Hilbert functions of Section 4.7 are in fact the Hilbert functions of S in the given grading. Affine semigroup algebras for cones of the form $\mathsf{C} = \hom(\mathsf{P})$, where P is a lattice polytope, are of particular interest in connection with toric algebraic geometry and combinatorial commutative algebra. See, e.g., [**122**] for more on this.

Flag f-vectors of posets, particularly, of Eulerian posets, have been studied extensively; see, for example, [**19, 29**]. In particular, flag f-vectors of Eulerian posets satisfy the *generalized* Dehn–Sommerville relations; see Exercise 4.45. If Π is an Eulerian poset of rank $d+1$, then the flag f-vector has 2^d entries but the complete flag f-vector can be recovered from the knowledge of $f(n)$ many entries, where $f(n)$ is the n-th Fibonacci number. Note that not any $f(n)$ entries will do the job. Which subsets work, however, is still open; see [**46**] for the easier case of f-, respectively, h-vectors of Eulerian posets. General chain partitions are, to the best of our knowledge, new. The chain-generating function (4.9.2) and its relative (4.9.3) are very natural and have appeared in different guises. If Γ is a simplicial complex on $[n]$, then its *Stanley–Reisner ring* is $\mathbb{C}[\Gamma] := \mathbb{C}[x_1, \dots, x_n]/I_\Gamma$, where I_Γ is the ideal generated by all $\mathbf{x}^{\mathbf{a}}$ for which $\operatorname{supp}(\mathbf{a}) \notin \Gamma$. The ideal I_Γ is homogeneous and $\mathbb{C}[\Gamma]$ inherits the natural grading of polynomials. The corresponding Hilbert function is exactly $cp_{\Gamma \cup \{\hat{1}\}, \mathrm{rk}}(n)$. For the algebraic perspective on our derivation of the Dehn–Sommerville relations, see [**122**].

Exercises

4.1 Prove the following extension of Proposition 4.1.1: Let $\mathbf{A} \in \mathbb{C}^{d \times d}$, fix indices $i, j \in [d]$, and consider the sequence $a(n) := (A^n)_{ij}$ formed by the (i,j)-entries of the n-th powers of $\mathbf{A}$. Then $a(n)$ agrees with a polynomial in n if and only if $\mathbf{A}$ is unipotent. (*Hint:* Consider the Jordan normal form of $\mathbf{A}$.)

4.2 ○ Complete the proof of Proposition 4.1.2 that (M), (γ), (Δ), and (h^*) are bases for the vector space $\mathbb{C}[z]_{\le d} = \{f \in \mathbb{C}[z] : \deg(f) \le d\}$:

(a) Give an explicit change of bases from (γ) to (M) in the spirit of (4.1.1). For example, consider $z^j \frac{\mathrm{d}}{\mathrm{d}z}(z+w)^d$ and set $w = 1 - z$.

(b) Assume that there are numbers $\alpha_0, \dots, \alpha_d$ such that

$$\alpha_0 \binom{z}{d} + \alpha_0 \binom{z+1}{d} + \cdots + \alpha_d \binom{z+d}{d} = 0 .$$

Argue, by specializing z, that $\alpha_j = 0$ for all j.

(c) Can you find explicit changes of bases for the sets (M), (γ), (Δ), and (h^*) and give them combinatorial meaning?

4.3 ○ Show that for $F(z) = \sum_{n\geq 0} f(n)z^n$ there exists a power series $G(z)$ such that $F(z)\,G(z) = 1$ if and only if $f(0) \neq 0$. (This explains why we do not allow A to contain the number 0 in Proposition 4.2.1.)

4.4 ○ Check that our definition (4.1.5) for the derivative of a formal power series satisfies the following properties: given $F(z) := \sum_{n\geq 0} f(n)\,z^n$ and $G(z) := \sum_{n\geq 0} g(n)\,z^n$, define $F'(z)$ and $G'(z)$ via (4.1.5). Then:
(a) If $\lambda \in \mathbb{C}$, then $(F(z) + \lambda\, G(z))' = F'(z) + \lambda\, G'(z)$.
(b) $(F(z)\,G(z))' = F'(z)\,G(z) + F(z)\,G'(z)$.
(c) If $G(z)$ has a multiplicative inverse, then

$$\left(\frac{F(z)}{G(z)}\right)' = \frac{F'(z)\,G(z) - F(z)\,G'(z)}{G(z)^2}\,.$$

4.5 Show that for integers $m \geq k \geq 0$,

$$\sum_{n\geq k} \binom{n+m-k}{m} z^n = \frac{z^k}{(1-z)^{m+1}}\,.$$

4.6 ○ Prove Proposition 4.1.5: Let $(f(n))_{\geq 0}$ be a sequence of numbers. Then $(f(n))_{n\geq 0}$ satisfies a linear recurrence of the form (4.1.10) (with $c_0, c_d \neq 0$) if and only if

$$F(z) = \sum_{n\geq 0} f(n) z^n = \frac{p(z)}{c_d z^d + c_{d-1} z^{d-1} + \cdots + c_0}$$

for some polynomial $p(z)$ of degree $< d$.

4.7 ○ Let $(f(n))_{n\geq 0}$ be a sequence of numbers. Show that $f(n)$ satisfies a linear recursion (with nonzero constant term) for sufficiently large $n > 0$ if and only if $\sum_{n\geq 0} f(n) z^n = \frac{p(z)}{q(z)}$ for some polynomials $p(z)$ and $q(z)$, with no restriction on the degree of $p(z)$.

4.8 Show that, if $f(n)$ is the sequence of Fibonacci numbers, then

$$\sum_{n\geq 1} f(n)\, z^n = \frac{z}{1-z-z^2}\,.$$

Expand this rational function into partial fractions to give a closed formula for $f(n)$.

4.9 ○ Let $f^\circ(n) := f(-n)$ be the sequence satisfying the recurrence (4.1.13) with starting values $f(0), f(1), \ldots, f(d-1)$. Compute the numerator for $F^\circ(z) = \sum_{n\geq 1} f^\circ(n) z^n$ and thereby complete the proof of Theorem 4.1.6.

4.10 Let $\mathbf{A} \in \mathbb{C}^{d\times d}$, and for $i, j \in [d]$ define $f(n) := (\mathbf{A}^n)_{ij}$.
(a) Show that $f(n)$ satisfies a linear recurrence.

(b) If $\mathbf{A}$ is invertible, show that $f^\circ(n) = (\mathbf{A}^{-n})_{ij}$ (in the language of Exercise 4.9). What is $f^\circ(n)$ if $\mathbf{A}$ is not invertible?

(c) If $\mathbf{A}$ is diagonalizable with eigenvalues $\lambda_1, \dots, \lambda_d$, show that there are $a_1, \dots, a_d \in \mathbb{C}$ such that

$$f(n) \;=\; a_1 \lambda_1^n + \cdots + a_d \lambda_d^n \,.$$

4.11 ⬠ Consider the formal power series $F(z) = \sum_{n \ge 0} \frac{z^n}{n!}$.

(a) Prove that $F(z)$ is not rational.

(b) Show that $F(\frac{1}{z})$ is not a generating function. (*Hint:* This time you might want to think about $F(\frac{1}{z})$ as a function.)

4.12 ⬠ Prove Corollary 4.1.7: A sequence $f(n)$ is eventually polynomial of degree $\le d$ if and only if

$$\sum_{n \ge 0} f(n)\, z^n \;=\; g(z) + \frac{h(z)}{(1-z)^{d+1}}$$

for some polynomials $g(z)$ and $h(z)$ with $\deg(h) \le d$. Furthermore, $f(n)$ has degree d if and only if $h(1) \ne 0$.

4.13 The **Bernoulli polynomials** $B_k(n)$ are defined through the generating function

$$\frac{z\, e^{nz}}{e^z - 1} \;=\; \sum_{k \ge 0} \frac{B_k(n)}{k!}\, z^k. \tag{4.9.6}$$

The **Bernoulli numbers** are $B_k := B_k(0)$. Prove the following properties of Bernoulli polynomials and numbers.

(a) $\displaystyle\sum_{j=0}^{n-1} j^{k-1} = \tfrac{1}{k}\left(B_k(n) - B_k\right).$

(b) $\displaystyle B_k(n) = \sum_{m=0}^{k} \binom{k}{m} B_{k-m}\, n^m.$

(c) $B_k(1-n) = (-1)^k B_k(n)$.

(d) $B_k = 0$ for all odd $k \ge 3$.

4.14 Prove that, if P is a simplicial d-polytope, then

$$Z_{\Phi(\mathsf{P})}(n) = \sum_{m=0}^{d} \frac{n^{m+1}}{m+1} \sum_{k=m}^{d} \binom{k}{m} B_{k-m}\, f_{k-1}$$

and conclude with Theorem 2.3.3 and Corollary 3.5.4 the following alternative version of the Dehn–Sommerville relations (Theorem 3.5.5):

$$\sum_{k=m}^{d} \binom{k}{m} B_{k-m}\, f_{k-1} \;=\; 0$$

for $m = d-1, d-3, \dots$.

4.15 Give an alternative proof of Proposition 4.2.1 by utilizing the fact that $c_A(n) = \sum_{m \in A} c_A(n-m)$.

4.16 Let $A := \{n \in \mathbb{Z}_{>0} : 3 \nmid n\}$. Compute the generating function for $c_A(n)$ and derive from it both a recursion and closed form for $c_A(n)$. Generalize.

4.17 Fix positive integers a and b. Prove that the number of compositions of n with parts a and b equals the number of compositions of $n+a$ with parts in $\{a + bj : j \geq 0\}$ (and thus, by symmetry, also the number of compositions of $n+b$ with parts in $\{aj + b : j \geq 0\}$).

4.18 Towards a two-variable generalization of Proposition 4.2.1, let $c_A(n,m)$ denote the number of compositions of n with precisely m parts in the set A, and let

$$C_A(x,y) \ := \ \sum_{n,m \geq 0} c_A(n,m)\, x^n y^m,$$

where we set $c_A(0,0) := 1$. Prove that

$$C_A(x,y) \ = \ \frac{1}{1 - y \sum_{m \in A} x^m} \, .$$

Compute a formula for $c_A(n,m)$ when A is the set of all positive odd integers.

4.19 Compute the quasipolynomial $pl(n)$ through a partial-fraction expansion of $Pl(q)$ of the form $\frac{a(q)}{(1-q)^4} + \frac{b(q)}{(1-q^2)^2} + \frac{c(q)}{1-q^3}$. Compare your formula for $pl(n)$ with (4.3.4).

4.20 Show that the generating function for **plane partition diamonds**

$$\begin{array}{ccccccc}
a_1 & \geq & a_2 & & & & \\
\rotatebox{-90}{$\geq$} & & \rotatebox{-90}{$\geq$} & & & & \\
a_3 & \geq & a_4 & \geq & a_5 & & \\
 & & \rotatebox{-90}{$\geq$} & & \rotatebox{-90}{$\geq$} & & \\
 & & a_6 & \geq & a_7 & & \\
 & & & & & \ddots & \\
 & & & & & & a_{3n-2} \ \geq \ a_{3n-1} \\
 & & & & & & \rotatebox{-90}{$\geq$} \qquad\quad \rotatebox{-90}{$\geq$} \\
 & & & & & & a_{3n} \ \ \geq \ \ a_{3n+1}
\end{array}$$

is

$$\frac{(1+q^2)(1+q^5)(1+q^8)\cdots(1+q^{3n-1})}{(1-q)(1-q^2)\cdots(1-q^{3n+1})} \, .$$

Derive a reciprocity theorem for the associated plane-partition-diamond counting function.

4.21 ○ Prove Proposition 4.4.1: If $p(n)$ is a quasipolynomial, so is $r(n) := \sum_{s=0}^{n} p(s)$. More generally, if $f(n)$ and $g(n)$ are quasipolynomials, then so is their convolution

$$c(n) = \sum_{s=0}^{n} f(s)\, g(n-s)\,.$$

4.22 Continuing Exercise 4.21, let $c(n)$ be the convolution of the quasipolynomials $f(n)$ and $g(n)$. What can you say about the degree and the period of $c(n)$, given the degrees and periods of $f(n)$ and $g(n)$?

4.23 Compute the quasipolynomial $p_A(n)$ for the case $A = \{1, 2\}$.

4.24 How does your computation of both the generating function and the quasipolynomial $p_A(n)$ change when we switch from Exercise 4.23 to the case of the *multiset* $A = \{1, 2, 2\}$?

4.25 Suppose a and b are relatively prime positive integers. Define the integers α and β through

$$b\beta \equiv 1 \bmod a \qquad \text{and} \qquad a\alpha \equiv 1 \bmod b\,,$$

and denote by $\{x\}$ the **fractional part** of x, defined through

$$x = \lfloor x \rfloor + \{x\}\,,$$

where $\lfloor x \rfloor$ is the largest integer $\leq x$. Prove that

$$p_{\{a,b\}}(n) = \frac{n}{ab} - \left\{ \frac{\beta n}{a} \right\} - \left\{ \frac{\alpha n}{b} \right\} + 1\,.$$

4.26 The **(unrestricted) partition function** $p(n)$ counts all partitions of n. Show that its generating function is

$$1 + \sum_{n \geq 1} p(n)\, q^n = \prod_{k \geq 1} \frac{1}{1 - q^k}\,.$$

4.27 Let $d(n)$ denote the number of partitions of n into distinct parts (i.e., no part is used more than once), and let $o(n)$ denote the number of partitions of n into odd parts (i.e., each part is an odd integer). Compute the generating functions of $d(n)$ and $o(n)$, and prove that they are equal (and thus $d(n) = o(n)$ for all positive integers n).

4.28 In this exercise we consider the problem of counting partitions of n with an arbitrary but finite number of parts, restricting the maximal size of each part. That is, let

$$p_{\leq m}(n) := \left| \left\{ \begin{array}{l} (m \geq a_1 \geq a_2 \geq \cdots \geq a_k \geq 1) : \\ k \in \mathbb{Z}_{>0} \text{ and } a_1 + a_2 + \cdots + a_k = n \end{array} \right\} \right|.$$

Prove that

$$p_{\leq m}(n) = p_{\{1,2,\ldots,m\}}(n)\,.$$

4.29 Let $p_k(n)$ denote the number of partitions of n into at most k parts.

(a) Show that

$$1 + \sum_{n \ge 1} p_k(n)\, q^n \;=\; \frac{1}{(1-q)(1-q^2)\cdots(1-q^k)}$$

and conclude that $p_k(n)$ is a quasipolynomial in n.

(b) Prove that $(-1)^{k-1} p_k(-n)$ equals the number of partitions of n into exactly k distinct parts.

4.30 Compute the constituents and the rational generating function of the quasipolynomial $p(n) = n + (-1)^n$.

4.31 Recall that $\zeta \in \mathbb{C}$ is a root of unity if $\zeta^m = 1$ for some $m \in \mathbb{Z}_{>0}$.

(a) Prove that if $c : \mathbb{Z} \to \mathbb{C}$ is a periodic function with period k, then there are roots of unity $\zeta_0, \zeta_1, \dots, \zeta_{k-1} \in \mathbb{C}$ and coefficients $c_0, c_1, \dots, c_{k-1} \in \mathbb{C}$ such that

$$c(n) \;=\; c_0\, \zeta_0^n + c_1\, \zeta_1^n + \cdots + c_{k-1}\, \zeta_{k-1}^n .$$

(b) Show that $p : \mathbb{Z}_{\ge 0} \to \mathbb{C}$ is a quasipolynomial if only if

$$p(n) \;=\; \sum_{i=1}^{m} c_i\, \zeta_i^n\, n^{k_i},$$

where $c_i \in \mathbb{C}$, $k_i \in \mathbb{Z}_{\ge 0}$, and ζ_i are roots of unity.

4.32 ○ For $A = \{a_1, a_2, \dots, a_d\} \subset \mathbb{Z}_{>0}$ let $k = \operatorname{lcm}(a_1, a_2, \dots, a_d)$ be the least common multiple of the elements of A. Provide an explicit polynomial $h_A(q)$ such that the generating function $P_A(q)$ for the restricted partitions with respect to A is

$$P_A(q) \;=\; \sum_{n \ge 0} p_A(n)\, q^n \;=\; \frac{h_A(q)}{(1-q^k)^d} .$$

4.33 ○ Recall that $\mathbf{v}_1, \mathbf{v}_2, \dots, \mathbf{v}_d \in \mathbb{Z}^d$ form a lattice basis of $\mathbb{Z}^d$ if every point in $\mathbb{Z}^d$ can be uniquely expressed as an integral linear combination of $\mathbf{v}_1, \mathbf{v}_2, \dots, \mathbf{v}_d$. Let $\mathbf{A}$ be the matrix with columns $\mathbf{v}_1, \mathbf{v}_2, \dots, \mathbf{v}_d$. Show that $\mathbf{v}_1, \mathbf{v}_2, \dots, \mathbf{v}_d$ is a lattice basis if and only if $\det(\mathbf{A}) = \pm 1$. (*Hint:* Use Cramer's rule and the fact that each unit vector can be written as an integral linear combination of the lattice basis.)

4.34 ○ For $\mathbf{w} \in \mathbb{Z}^{d+1} \setminus \{0\}$, let $T_{\mathbf{w}} : \mathbb{C}[\![z_1^{\pm 1}, \dots, z_{d+1}^{\pm 1}]\!] \to \mathbb{C}[\![z_1^{\pm 1}, \dots, z_{d+1}^{\pm 1}]\!]$ be given by

$$T_{\mathbf{w}}(f) \;:=\; (1 - \mathbf{z}^{\mathbf{w}}) f .$$

(a) Show that $T_{\mathbf{w}}$ is an invertible linear transformation.

(b) For $\mathbf{w} \in \mathbb{Z}^{d+1} \setminus \{0\}$ show that $f := \sum_{t \in \mathbb{Z}} \mathbf{z}^{t\mathbf{w}}$ equals zero.

(c) More generally, prove that if Q is a rational polyhedron that contains a line, then $\sigma_{\mathsf{Q}}(\mathbf{z}) = 0$.

(d) Let $S \subset \mathbb{R}^{d+1}$ and $\mathbf{a} \in \mathbb{Z}^{d+1}$ such that for each $\delta \in \mathbb{Z}$ the sets

$$S_\delta := \left\{ \mathbf{m} \in S \cap \mathbb{Z}^{d+1} : \langle \mathbf{a}, \mathbf{m} \rangle = \delta \right\}$$

are all finite. Show that the specialization $\sigma_S(t\,\mathbf{a})$ is a well-defined element of $\mathbb{C}[\![t^{\pm 1}]\!]$.

(e) Let $S \subset \mathbb{R}^{d+1}$ be a convex, line-free set. Prove that $\sigma_S(\mathbf{z})$ is nonzero. Moreover, if $S_1, S_2 \subset \mathbb{R}^{d+1}$ are line-free convex sets, show that $\sigma_{S_1}(\mathbf{z})\,\sigma_{S_2}(\mathbf{z})$ is well defined.

4.35 ⬠ Prove Proposition 4.6.2 (without assuming Theorem 4.6.1): Let $\triangle$ be the convex hull of the origin and the d unit vectors in $\mathbb{R}^d$. Then $\mathrm{ehr}_\triangle(n) = \binom{n+d}{d}$, and this polynomial satisfies Theorem 4.6.1. More generally, $\mathrm{ehr}_\triangle(n) = \binom{n+d}{d}$ for every unimodular simplex $\triangle$.

4.36 Show that the cone $\mathsf{C}_{2\geq 3} = \left\{ \mathbf{x} \in \mathbb{R}^4_{\geq 0} : x_1 \geq x_2 \geq x_3 \geq x_4 \right\}$ has generators

$$\begin{pmatrix} 1 \\ 1 \\ 1 \\ 1 \end{pmatrix}, \begin{pmatrix} 1 \\ 1 \\ 1 \\ 0 \end{pmatrix}, \begin{pmatrix} 1 \\ 1 \\ 0 \\ 0 \end{pmatrix}, \begin{pmatrix} 1 \\ 0 \\ 0 \\ 0 \end{pmatrix}.$$

4.37 ⬠ Let $\mathsf{P} := \mathrm{conv}\{(0,0),(0,1),(2,1)\}$. Show that every lattice point in $\mathrm{hom}(\mathsf{P})$ can be uniquely written as either

$$k_1 \begin{pmatrix} 0 \\ 0 \\ 1 \end{pmatrix} + k_2 \begin{pmatrix} 0 \\ 1 \\ 1 \end{pmatrix} + k_3 \begin{pmatrix} 2 \\ 1 \\ 1 \end{pmatrix}$$

or

$$\begin{pmatrix} 1 \\ 0 \\ 0 \end{pmatrix} + k_1 \begin{pmatrix} 0 \\ 0 \\ 1 \end{pmatrix} + k_2 \begin{pmatrix} 0 \\ 1 \\ 1 \end{pmatrix} + k_3 \begin{pmatrix} 2 \\ 1 \\ 1 \end{pmatrix}$$

for some nonnegative integers k_1, k_2, k_3.

4.38 ⬠ Prove (4.8.4), i.e., fix linearly independent vectors $\mathbf{v}_1, \mathbf{v}_2, \dots, \mathbf{v}_k \in \mathbb{Z}^d$ and define

$$\widehat{\mathsf{C}} := \mathbb{R}_{\geq 0}\mathbf{v}_1 + \cdots + \mathbb{R}_{\geq 0}\mathbf{v}_{m-1} + \mathbb{R}_{>0}\mathbf{v}_m + \cdots + \mathbb{R}_{>0}\mathbf{v}_k$$

and

$$\widehat{\square} := [0,1)\,\mathbf{v}_1 + \cdots + [0,1)\,\mathbf{v}_{m-1} + (0,1]\,\mathbf{v}_m + \cdots + (0,1]\,\mathbf{v}_k \,.$$

Then

$$\widehat{\mathsf{C}} = \biguplus_{j_1,\dots,j_k \geq 0} \left(j_1\mathbf{v}_1 + \cdots + j_k\mathbf{v}_k + \widehat{\square} \right).$$

4.39 Let C be a rational, simplicial cone with generators $\mathbf{v}_1, \dots, \mathbf{v}_k \in \mathbb{Z}^d$ and corresponding parallelepiped $\square$. Let $\square'$ be the parallelepiped for $\mathbf{v}'_i = \lambda_i \mathbf{v}_i$, $i \in [k]$ for some $\lambda_i \in \mathbb{Z}_{>0}$. Write $\sigma_{\square'}(\mathbf{z})$ in terms of $\sigma_\square(\mathbf{z})$ and verify that the right-hand side of (4.8.7) is independent of a particular choice of generators.

4.40 Let $\widehat{\mathsf{C}}$ and $\widehat{\square}$ be as in (the proof of) Theorem 4.8.1. Prove that if $\widehat{\mathsf{C}}$ is unimodular, then $\widehat{\square}$ contains precisely one lattice point (namely, the origin).

4.41 Pick four concrete points in $\mathbb{Z}^3$ and compute the Ehrhart polynomial of their convex hull.

4.42 Let $\triangle$ be a lattice d-simplex and write

$$\operatorname{Ehr}_\triangle(z) = \frac{h_d^* z^d + h_{d-1}^* z^{d-1} + \cdots + h_0^*}{(1-z)^{d+1}} .$$

Prove that:
(a) $h_d^* = \left|\triangle^\circ \cap \mathbb{Z}^d\right|$.
(b) $h_1^* = \left|\triangle \cap \mathbb{Z}^d\right| - d - 1$.
(c) $h_0^* + h_1^* + \cdots + h_d^* = d! \operatorname{vol}(\triangle)$.

4.43 ⬠ Prove Theorem 4.7.2: If $\triangle$ is a rational simplex, then for positive integers n, the counting function $\operatorname{ehr}_\triangle(n)$ is a quasipolynomial in n whose period divides the least common multiple of the denominators of the vertex coordinates of $\triangle$. When this quasipolynomial is evaluated at negative integers, we obtain

$$\operatorname{ehr}_\triangle(-n) = (-1)^{\dim(\triangle)} \operatorname{ehr}_{\triangle^\circ}(n) .$$

4.44 Let S be an m-dimensional subset of $\mathbb{R}^d$ (i.e., the affine span of S has dimension m). Then we define the **relative volume** of S to be

$$\operatorname{vol} S := \lim_{n \to \infty} \frac{1}{n^m} \left| n\, S \cap \mathbb{Z}^d \right| .$$

(a) Convince yourself that $\operatorname{vol} S$ is the usual volume if $m = d$.
(b) Show that, if $\triangle \subset \mathbb{R}^d$ is an m-dimensional rational simplex, then the leading coefficient of $\operatorname{ehr}_\triangle(n)$ (i.e., the coefficient of n^m) equals $\operatorname{vol} \triangle$.

4.45 Let Π be an Eulerian poset of rank $d+1$.
(a) For $0 \le i \le d+1$, let α_i be the number of elements of rank i. Show that

$$\alpha_0 - \alpha_1 + \cdots + (-1)^{d+1} \alpha_{d+1} = 0 .$$

(b) For $1 \le r < s \le d$, let $x, y \in \Pi$ be of ranks r and s, respectively. Let $\alpha_i(x,y)$ be the number of elements of rank i in $[x,y]$. This is equal to 0 if $x \not\preceq y$. Use the previous part to show that

$$\sum_{i=r}^{s} (-1)^i \alpha_\Pi(\{r,i,s\}) = 0 .$$

(c) Let $S \subseteq [d]$ and $0 \leq r < s-1 \leq d+1$ such that $S \cap \{r, r+1, \dots, s\} = \varnothing$. Show that

$$\sum_{i=r+1}^{s-1} (-1)^{i-r-1} \alpha(S \cup \{j\}) \;=\; \alpha(S)(1 + (-1)^{s-r}) \,.$$

Chapter 5

Subdivisions

A heavy warning used to be given that pictures are not rigorous; this has never had its bluff called and has permanently frightened its victims into playing for safety.
John Edensor Littlewood (1885–1977)

An idea that prevails in many parts of mathematics is to decompose a complex object into simpler ones, do computations on the simple pieces, and then put together the local information to get the global picture. This was our strategy, e.g., in Section 1.4 where we proved Ehrhart's theorem and Ehrhart–Macdonald reciprocity for lattice polygons (Theorem 1.4.1). In the plane we could appeal to your intuition that every lattice polygon can be triangulated into lattice triangles. This technique of hand waving fails in dimensions ≥ 3—it is no longer obvious that every (lattice) polytope can be decomposed into (lattice) simplices. Our goal for this chapter is to show it can be done, elegantly, and in more than one way.

5.1. Decomposing a Polyhedron

The first challenge in asserting that every polytope can be decomposed into simpler polytopes is to make precise what we actually mean by that. It is clear that counting lattice points through $E(S) := |S \cap \mathbb{Z}^d|$ is a *valuation* in the sense of (1.4.2) and (3.4.1): for bounded sets $S, T \subset \mathbb{R}^d$,

$$E(S \cup T) \;=\; E(S) + E(T) - E(S \cap T)\,.$$

We recall from Sections 1.4 and 4.6 that for a lattice polytope $\mathsf{P} \subset \mathbb{R}^d$, the Ehrhart function of P is given by

$$\operatorname{ehr}_{\mathsf{P}}(n) \;:=\; E(n\mathsf{P})$$

for integers $n \geq 1$. For a lattice simplex $\triangle$, Theorem 4.6.1 asserts that $\mathrm{ehr}_{\triangle}(n)$ agrees with a polynomial of degree $\dim \triangle$. Thus, for general lattice polytopes we could wish for the following: if for a lattice polytope $\mathsf{P} \subset \mathbb{R}^d$, we can find lattice simplices $\triangle_1, \dots, \triangle_m$ such that $\mathsf{P} = \triangle_1 \cup \cdots \cup \triangle_m$, then by the inclusion–exclusion principle from Section 2.4,

$$(5.1.1) \qquad \mathrm{ehr}_{\mathsf{P}}(n) \;=\; \sum_{\varnothing \neq I \subseteq [m]} (-1)^{|I|-1} \, \mathrm{ehr}_{\triangle_{\cap I}}(n) \,,$$

where $\triangle_{\cap I} := \bigcap_{i \in I} \triangle_i$. This attempt is bound to fail as the polytopes $\triangle_{\cap I}$ are in general not simplices and, even worse, not lattice polytopes. We need more structure.

A **dissection** of a polyhedron $\mathsf{Q} \subset \mathbb{R}^d$ is a collection of polyhedra $\mathsf{Q}_1, \dots, \mathsf{Q}_m$ of the same dimension such that

$$(5.1.2) \qquad \mathsf{Q} = \mathsf{Q}_1 \cup \cdots \cup \mathsf{Q}_m \quad \text{and} \quad \mathsf{Q}_i^\circ \cap \mathsf{Q}_j^\circ = \varnothing \text{ whenever } i \neq j \,.$$

This definition, at least, opens the door to induction on the dimension, although our problem with $\triangle_{\cap I}$ not being a lattice simplex prevails. As we will see in Section 5.3, a dissection into lattice simplices is indeed sufficient to extend Theorem 4.6.1 to arbitrary lattice polytopes, but there are alternatives. We will pursue a more combinatorial (and conservative) approach in this section.

A **polyhedral complex** is a nonempty finite collection $\mathcal{S}$ of polyhedra in $\mathbb{R}^d$ (which we call **cells** of $\mathcal{S}$) such that $\mathcal{S}$ satisfies the

containment property: if F is a face of $\mathsf{G} \in \mathcal{S}$ then $\mathsf{F} \in \mathcal{S}$, and the
intersection property: if $\mathsf{F}, \mathsf{G} \in \mathcal{S}$ then $\mathsf{F} \cap \mathsf{G}$ is a face of both F and G.

We call $\mathcal{S}$ a **polytopal complex** if all of its cells are polytopes, and $\mathcal{S}$ is a **fan** if all of its cells are polyhedral cones. A **(geometric) simplicial complex**[1] is a polytopal complex all of whose cells are simplices. The **support** of a polyhedral complex $\mathcal{S}$ is

$$|\mathcal{S}| \;:=\; \bigcup_{\mathsf{F} \in \mathcal{S}} \mathsf{F} \,,$$

the point set underlying $\mathcal{S}$. The **vertices** of $\mathcal{S}$ are the zero-dimensional polytopes contained in $\mathcal{S}$. We already know two seemingly trivial instances of polyhedral complexes, illustrated in Figure 5.1.

[1] This notion of *simplicial complex* is, naturally, not disjoint from that of an abstract simplicial complex defined in Section 4.9. In Section 5.6 we will witness the two notions simultaneously in action.

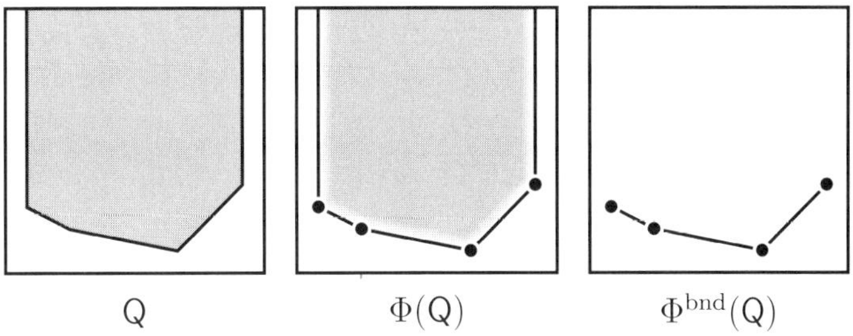

Figure 5.1. The polyhedral complexes given by the faces/bounded faces of a polyhedron.

Proposition 5.1.1. *If* Q *is a polyhedron, then the collection of faces* $\Phi(\mathsf{Q})$ *is a polyhedral complex. Moreover, the collection of bounded faces*

$$\Phi^{\mathrm{bnd}}(\mathsf{Q}) \ := \ \{\mathsf{F} \in \Phi(\mathsf{Q}) \ : \ \mathsf{F} \ \textit{bounded}\}$$

is a polyhedral complex.

Proof. By Proposition 3.3.1, a face of a face of Q is a face of Q, and the intersection of faces of Q is again a face. These are exactly the containment and intersection properties, respectively. As every face of a bounded face is necessarily bounded, the second claim follows. □

A **subdivision** of a polyhedron $\mathsf{P} \subset \mathbb{R}^d$ is a polyhedral complex $\mathcal{S}$ such that $\mathsf{P} = |\mathcal{S}|$. Proposition 5.1.1 yields that $\Phi(\mathsf{P})$ is trivially a subdivision of P. We call a subdivision $\mathcal{S}$ **proper** if $\mathcal{S} \neq \Phi(\mathsf{P})$. A subdivision $\mathcal{S}$ of a polytope P is a **triangulation** if all cells in $\mathcal{S}$ are simplices—in other words, $\mathcal{S}$ is a simplicial complex. All but the top right dissections in Figure 5.2 are subdivisions, the two middle ones are even triangulations.

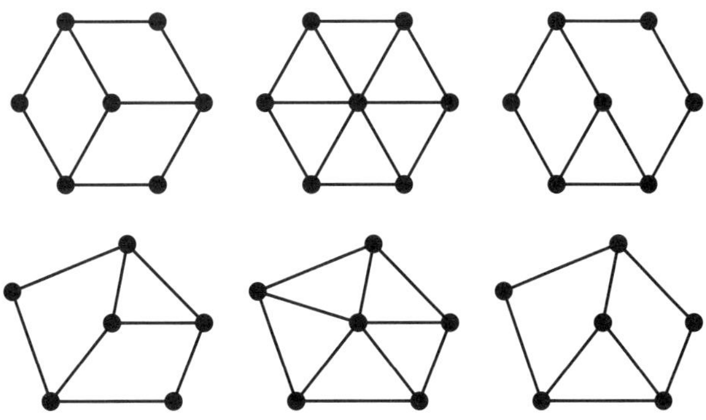

Figure 5.2. Various dissections.

The benefit of the intersection property of a subdivision is evident: if $\mathsf{P}_1, \ldots, \mathsf{P}_m$ are the inclusion-maximal cells in a subdivision $\mathcal{S}$, then for every

$I \subseteq [m]$, the polytope $\bigcap_{i \in I} \mathsf{P}_i$ is a lattice polytope whenever $\mathsf{P}_1, \dots, \mathsf{P}_m$ are.[2] For **lattice triangulations**, that is, triangulations into lattice simplices, Theorem 4.6.1 gives the following.

Corollary 5.1.2. *Let* $\mathsf{P} \subset \mathbb{R}^d$ *be a lattice polytope that admits a lattice triangulation. Then* $\operatorname{ehr}_\mathsf{P}(n)$ *agrees with a polynomial of degree* $\dim \mathsf{P}$.

Proof. Let $\triangle_1, \dots, \triangle_m$ be the inclusion-maximal cells of a lattice triangulation $\mathcal{T}$ of P. By the intersection property, each cell in $\mathcal{T}$ is a lattice simplex and Theorem 4.6.1 guarantees that (5.1.1) is an alternating sum of polynomials. Exercise 5.1 states that $\dim \triangle_i = \dim \mathsf{P}$ for $1 \le i \le m$, and since $\triangle_i \cap \triangle_j$ is a proper face of $\triangle_i$ whenever $i \neq j$, we conclude from Theorem 4.6.1 that $\deg(\operatorname{ehr}_\triangle(n)) < \dim \mathsf{P}$ for every cell $\triangle \in \mathcal{T}$ that is not one of $\triangle_1, \dots, \triangle_m$. Together with the fact that all $\operatorname{ehr}_{\triangle_i}(n)$ are polynomials of degree $\dim \mathsf{P}$ with positive leading coefficient, this proves our claim about the degree. □

In Corollary 5.1.6 below, we will show that every lattice polytope admits a lattice triangulation, from which we can (finally!) deduce that the Ehrhart counting function of every lattice polytope is a polynomial.

The motivation for requiring the containment property when defining polyhedral complexes is that we can appeal to Chapter 2 to express an Ehrhart function. Given a proper subdivision $\mathcal{S}$ of a lattice polytope P, we form the poset $\widehat{\mathcal{S}} := \mathcal{S} \cup \{\mathsf{P}\}$ with respect to inclusion; it has maximal element P.

Corollary 5.1.3. *Let* $\mathcal{S}$ *be a proper subdivision of a lattice polytope* P. *Then*

$$\operatorname{ehr}_\mathsf{P}(n) \;=\; \sum_{\mathsf{F} \in \mathcal{S}} -\mu_{\widehat{\mathcal{S}}}(\mathsf{F}, \mathsf{P}) \operatorname{ehr}_\mathsf{F}(n)\,. \tag{5.1.3}$$

Proof. We define a function $f_=$ on $\widehat{\mathcal{S}}$ by

$$f_=(\mathsf{F}) \;:=\; \begin{cases} \operatorname{ehr}_{\mathsf{F}^\circ}(n) & \text{if } \mathsf{F} \in \mathcal{S}\,, \\ 0 & \text{if } \mathsf{F} = \mathsf{P}\,. \end{cases}$$

Thus

$$f_\le(\mathsf{F}) \;=\; \sum_{\mathsf{G} \preceq \mathsf{F}} f_=(\mathsf{G}) \;=\; \operatorname{ehr}_\mathsf{F}(n)\,, \tag{5.1.4}$$

using Lemma 3.3.8 when $\mathsf{F} \in \mathcal{S}$, and the definition of a subdivision when $\mathsf{F} = \mathsf{P}$. By Möbius inversion (Theorem 2.4.2),

$$0 \;=\; f_=(\mathsf{P}) \;=\; f_\le(\mathsf{P}) + \sum_{\mathsf{F} \in \mathcal{S}} \mu_{\widehat{\mathcal{S}}}(\mathsf{F}, \mathsf{P}) \operatorname{ehr}_\mathsf{F}(n)\,.$$ □

[2] The empty set is, by definition, a lattice polytope.

In Theorem 5.2.1 we will determine the Möbius function of a subdivision concretely.

We still need to show that every (lattice) polytope has a (lattice) triangulation. In some sense, we actually already know how to do this—here is the setup: we fix a finite set $V \subset \mathbb{R}^d$ such that $\mathsf{P} := \operatorname{conv}(V)$ is full dimensional. (It would suffice to assume that V is the vertex set of P but as we will see, it will pay off to have some flexibility). For a function $\omega : V \to \mathbb{R}$, we denote its graph by

$$V^\omega := \left\{ (\mathbf{v}, \omega(\mathbf{v})) \in \mathbb{R}^{d+1} : \mathbf{v} \in V \right\}.$$

Let

$$\uparrow_{\mathbb{R}} := \left\{ (\mathbf{0}, t) \in \mathbb{R}^{d+1} : t \geq 0 \right\}$$

and denote by $\pi : \mathbb{R}^{d+1} \to \mathbb{R}^d$ the coordinate projection

$$\pi(x_1, \dots, x_d, x_{d+1}) := (x_1, \dots, x_d),$$

so that $\pi(\uparrow_{\mathbb{R}}) = \{\mathbf{0}\}$. Finally, we define

$$\mathsf{E}^\omega(V) := \operatorname{conv}(V^\omega) + \uparrow_{\mathbb{R}},$$

which we call the **convex epigraph** of ω, for reasons that will become clear in a moment. Figure 5.3 illustrates our construction. By the Minkowski–Weyl Theorem 3.2.5, $\mathsf{E}^\omega(V)$ is a genuine polyhedron in $\mathbb{R}^{d+1}$.

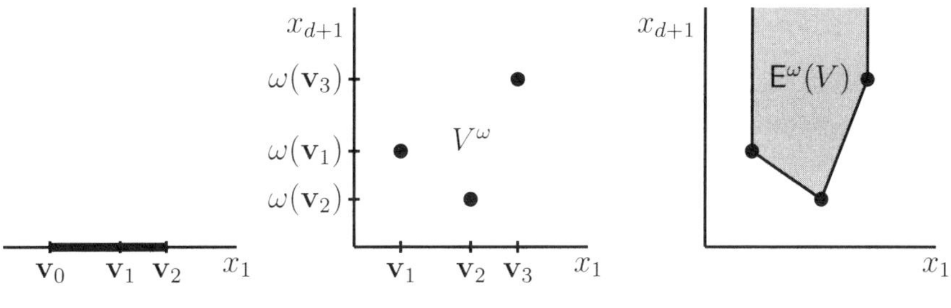

Figure 5.3. Constructing the convex epigraph of a polytope.

Proposition 5.1.4. *Fix a finite set $V \subset \mathbb{R}^d$ and a hyperplane*

$$\mathsf{H} = \left\{ (\mathbf{x}, x_{d+1}) \in \mathbb{R}^{d+1} : \langle \mathbf{a}, \mathbf{x} \rangle + a_{d+1} x_{d+1} = b \right\}.$$

If $\mathsf{E}^\omega(V) \subseteq \mathsf{H}^{\leq}$, then $a_{d+1} \leq 0$. If H is a supporting hyperplane, then the face $\mathsf{E}^\omega(V) \cap \mathsf{H}$ is bounded if and only if $a_{d+1} < 0$.

Proof. For every $\mathbf{p} \in \mathsf{E}^\omega(V)$, we have $\mathbf{p} + \uparrow_{\mathbb{R}} \subseteq \mathsf{E}^\omega(V)$. Hence, if $\mathsf{E}^\omega(V) \subseteq \mathsf{H}^{\leq}$, then also $\mathbf{p} + \uparrow_{\mathbb{R}} \subseteq \mathsf{H}^{\leq}$ which implies $a_{d+1} \leq 0$.

The nonempty face $\mathsf{F} = \mathsf{E}^\omega(V) \cap \mathsf{H}$ is unbounded if and only if $\mathsf{F} + \uparrow_{\mathbb{R}} \subseteq \mathsf{F}$. This happens if and only if for every point $(\mathbf{q}, q_{d+1}) \in \mathsf{H}$, we have $(\mathbf{q}, q_{d+1} + t) \in \mathsf{H}$ for all $t \geq 0$, a condition that is satisfied if and only if $a_{d+1} = 0$. □

By construction, $\pi(\mathsf{E}^\omega(V)) = \mathsf{P}$. The crucial insight now is that P is already the image under π of the collection of *bounded* faces of $\mathsf{E}^\omega(V)$. The statement in the following theorem is even better.

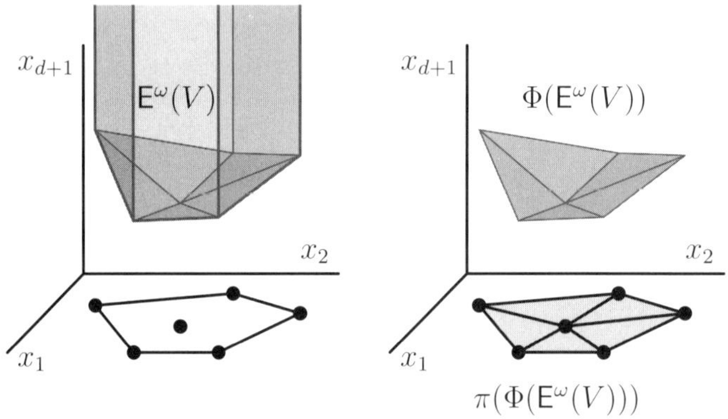

Figure 5.4. Two examples of a subdivision construction from the convex epigraph of a polytope.

Theorem 5.1.5. *Let* $\mathsf{P} = \operatorname{conv}(V)$ *and* $\omega : V \to \mathbb{R}$. *Then*

$$\mathcal{S}^\omega(V) \ := \ \left\{\pi(\mathsf{F}) \, : \, \mathsf{F} \in \Phi^{\text{bnd}}(\mathsf{E}^\omega(V))\right\}$$

is a subdivision of P *with vertices in* V. *Moreover,* $\mathcal{S}^\omega(V) \cong \Phi^{\text{bnd}}(\mathsf{E}^\omega(V))$ *as posets.*

An illustration of this theorem is given in Figure 5.4.

Proof. For a point $\mathbf{p} \in \mathsf{P}$, we denote by $\overline{\omega}(\mathbf{p})$ the smallest real number h such that $\hat{\mathbf{p}} := (\mathbf{p}, h) \in \mathsf{E}^\omega(V)$. Then $\hat{\mathbf{p}}$ is contained in the boundary of $\mathsf{E}^\omega(V)$ (see Figure 5.3), and so there is a unique face F of $\mathsf{E}^\omega(V)$ with $\hat{\mathbf{p}} \in \mathsf{F}^\circ$. We claim that F is bounded. Indeed, if F is unbounded, then $\mathsf{F} + \uparrow_{\mathbb{R}} \subseteq \mathsf{F}$ and so $(\mathbf{p}, \overline{\omega}(\mathbf{p}) + \varepsilon) \in \mathsf{F}^\circ$ and, consequently, $(\mathbf{p}, \overline{\omega}(\mathbf{p}) - \varepsilon) \in \mathsf{F}^\circ$ for some $\varepsilon > 0$. But this contradicts the minimality of $\overline{\omega}(\mathbf{p})$.

We further claim that π restricted to $|\Phi^{\text{bnd}}(\mathsf{E}^\omega(V))|$, the union of the bounded faces of $\mathsf{E}^\omega(V)$, is injective. Since for each $\mathbf{p} \in \mathsf{P}$ there is exactly one h such that $(\mathbf{p}, h)$ is contained in a bounded face of $\mathsf{E}^\omega(V)$, it suffices to show that π is injective when restricted to the relative interior of a bounded face F of $\mathsf{E}^\omega(V)$. This will also show that $\mathcal{S}^\omega(V)$ is isomorphic to $\Phi^{\text{bnd}}(\mathsf{E}^\omega(V))$.

Again by Proposition 5.1.4, we know that $\mathsf{F} = \mathsf{E}^\omega(V) \cap \mathsf{H}$ for some supporting hyperplane

$$\mathsf{H} = \left\{(\mathbf{x}, x_{d+1}) \in \mathbb{R}^{d+1} : \langle \mathbf{a}, \mathbf{x}\rangle - x_{d+1} = b\right\}.$$

Now we observe that the map $s : \mathbb{R}^d \to \mathsf{H}$ given by $s(\mathbf{x}) = (\mathbf{x}, \langle \mathbf{a}, \mathbf{x}\rangle - b)$ is a linear inverse to $\pi|_{\mathsf{H}} : \mathsf{H} \to \mathbb{R}^d$. Thus, $\pi|_{\mathsf{H}}$ is an isomorphism and, in particular, injective on F°. The geometric idea is captured in Figure 5.5. This shows that $\mathcal{S}^\omega(V)$ is a polyhedral complex with $|\mathcal{S}^\omega(V)| = \mathsf{P}$, and with vertices in V. □

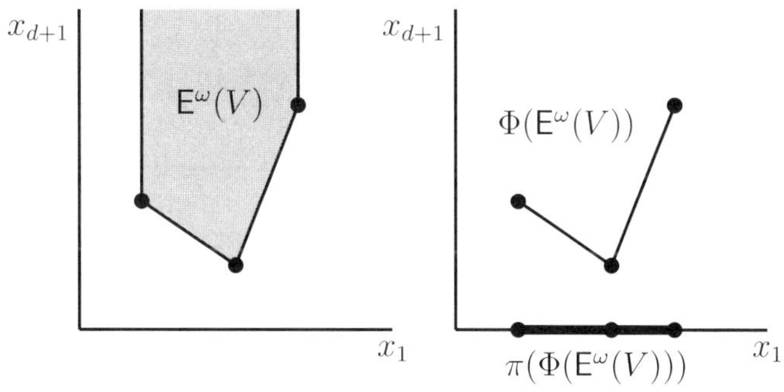

Figure 5.5. The complex of bounded faces and the induced subdivision.

We call a subdivision $\mathcal{S}$ of a polytope P **regular** (or **coherent**) if there is a finite subset $V \subset \mathsf{P}$ and a function $\omega : V \to \mathbb{R}$ such that $\mathcal{S} = \mathcal{S}^\omega(V)$. In Exercise 5.2 you will check that not all polytopal subdivisions are regular. Nevertheless, this construction proves the main result of this section.

Corollary 5.1.6. *Every (lattice) polytope has a (lattice) triangulation.*

Proof. Let $V \subset \mathbb{R}^d$ be a finite set containing the vertices of P. For each $\omega : V \to \mathbb{R}$, Theorem 5.1.5 yields a subdivision $\mathcal{S}^\omega(V)$ whose cells are polytopes with vertices in V. To make such a subdivision into a triangulation, we appeal to Exercise 5.4 for a suitable choice of ω. □

As we announced already, Corollaries 5.1.2 and 5.1.6 imply the following general variant of Theorem 4.6.1(a).

Theorem 5.1.7. *Suppose* P *is a lattice polytope. For positive integers* n*, the counting function* $\mathrm{ehr}_{\mathsf{P}}(n)$ *agrees with a polynomial in* n *of degree* $\dim \mathsf{P}$.

This is *Ehrhart's theorem* and we call the counting function $\mathrm{ehr}_{\mathsf{P}}(n)$ the *Ehrhart polynomial* of the lattice polytope P. Theorem 4.6.1 also said that the constant term of the Ehrhart polynomial of a lattice simplex is 1. Our next result extends this.

Theorem 5.1.8. *Let* P *be a lattice polytope and* $\mathrm{ehr}_{\mathsf{P}}(n)$ *its Ehrhart polynomial. Then* $\mathrm{ehr}_{\mathsf{P}}(0) = \chi(\mathsf{P}) = 1$.

Proof. Fix a lattice triangulation $\mathcal{S}$ of P. By (5.1.4),

$$\mathrm{ehr}_{\mathsf{P}}(n) = \sum_{\triangle \in \mathcal{S}} \mathrm{ehr}_{\triangle^\circ}(n) . \tag{5.1.5}$$

For every lattice simplex $\triangle$, the constant term of $\mathrm{ehr}_{\triangle^\circ}(n)$ is $(-1)^{\dim \triangle}$ (by Theorem 4.6.1), and so setting $n = 0$ in (5.1.5) computes the Euler characteristic of P. □

Theorem 5.1.5 yields an elegant way to obtain subdivisions. What the technique hides is that it is in general difficult to determine the actual subdivision $\mathcal{S}^\omega(V)$ from V and ω. The following example illustrates the power of Theorem 5.1.5 and, at the same time, gives an impression of potential difficulties.

Let $\mathsf{P} = [0,1]^d$, the d-dimensional cube with vertex set $V = \{0,1\}^d$. For $\mathbf{v} \in \{0,1\}^d$, let $n(\mathbf{v}) := \sum_i v_i$, the number of nonzero entries, and define $\omega : V \to \mathbb{R}$ by

$$\omega(\mathbf{v}) := n(\mathbf{v})\,(d - n(\mathbf{v})) . \tag{5.1.6}$$

This gives a subdivision $\mathcal{S}^\omega(V)$ of P into lattice polytopes but what is the subdivision exactly? To answer this, we have to determine which subsets of V correspond to the sets of vertices of bounded faces of $\mathsf{E}^\omega(V)$. In general, this is tantamount to computing an inequality description of a polyhedron that is given as a polytope plus a cone.

In our example, we can approach this as follows: given a point $\mathbf{p} \in [0,1]^d$, we seek to find $\overline{\omega}(\mathbf{p})$, that is, the smallest h such that $(\mathbf{p}, h) \in \mathsf{E}^\omega(V)$. In particular, $(\mathbf{p}, \overline{\omega}(\mathbf{p}))$ is in the boundary of $\mathsf{E}^\omega(V)$ and hence in the relative interior of a unique bounded face. Let $\mathbf{p}$ be a **generic point** relative to $[0,1]^d$ by which we mean for now that $p_i \neq p_j$ for all $i \neq j$. Thus there is a unique permutation σ of $[d]$ such that

$$0 \le p_{\sigma(1)} < p_{\sigma(2)} < \cdots < p_{\sigma(d)} \le 1 .$$

For this permutation σ, we set

$$\mathbf{u}_i^\sigma := \mathbf{e}_{\sigma(i+1)} + \mathbf{e}_{\sigma(i+2)} + \cdots + \mathbf{e}_{\sigma(d)} \in \{0,1\}^d \tag{5.1.7}$$

for $0 \le i \le d$. That is, $(\mathbf{u}_i^\sigma)_j = 1$ if and only if $\sigma^{-1}(j) > i$. In particular $\mathbf{u}_d^\sigma = \mathbf{0}$ and $\mathbf{u}_0^\sigma = (1,1,\dots,1)$. By writing

$$\mathbf{p} = p_{\sigma(1)}\mathbf{u}_0^\sigma + (p_{\sigma(2)} - p_{\sigma(1)})\mathbf{u}_1^\sigma + \cdots + (p_{\sigma(d)} - p_{\sigma(d-1)})\mathbf{u}_{d-1}^\sigma + (1 - p_{\sigma(d)})\mathbf{u}_d^\sigma ,$$

we see that $\mathbf{p}$ is a point in

$$\triangle_\sigma := \mathrm{conv}(\mathbf{u}_0^\sigma, \dots, \mathbf{u}_d^\sigma) ,$$

a d-dimensional simplex spanned by a subset of the vertices of $[0,1]^d$. (We have seen a precursor of this construction in (3.2.4).) Conversely, $\triangle_\sigma$ is exactly the set of points $\mathbf{p} \in \mathbb{R}^d$ that satisfy

$$0 \le p_{\sigma(1)} \le p_{\sigma(2)} \le \cdots \le p_{\sigma(d)} \le 1. \tag{5.1.8}$$

We claim that $\triangle_\sigma$ and its siblings yield our subdivision. As in Exercise 2.18, we denote by $\mathfrak{S}_d$ the set of all permutations of $[d]$.

Proposition 5.1.9. *Let $\mathcal{S}$ consist of $\triangle_\sigma$, for all $\sigma \in \mathfrak{S}_d$, and their faces. Then $\mathcal{S}$ is the regular triangulation of $[0,1]^d$ corresponding to (5.1.6). Moreover, $\mathcal{S}$ is a unimodular triangulation of $[0,1]^d$.*

Proof. Our claim is that $\triangle_\sigma$ is the projection of a bounded face of $\mathsf{E}^\omega(V)$. To show this, we construct a (linear) hyperplane $\mathsf{H} \subset \mathbb{R}^{d+1}$ such that $V^\omega \subseteq \mathsf{H}^{\le}$ and

$$V^\omega \cap \mathsf{H} = \{(\mathbf{u}_i^\sigma, \omega(\mathbf{u}_i^\sigma)) : 0 \le i \le d\}.$$

We claim that the linear function $\ell^\sigma : \mathbb{R}^d \to \mathbb{R}$ defined by

$$\ell^\sigma(\mathbf{x}) := \sum_{1 \le i < j \le d} \left(x_{\sigma(j)} - x_{\sigma(i)}\right)$$

satisfies

$$\ell^\sigma(\mathbf{v}) \le \omega(\mathbf{v})$$

for all $\mathbf{v} \in \{0,1\}^d$ with equality if and only if $\mathbf{v} = \mathbf{u}_k^\sigma$ for some $0 \le k \le d$. Indeed, for $i < j$ and $\mathbf{v} \in \{0,1\}^d$,

$$v_{\sigma(j)} - v_{\sigma(i)} \le 1$$

with equality if and only if $v_{\sigma(i)} = 0$ and $v_{\sigma(j)} = 1$. Hence $\ell^\sigma(\mathbf{v})$ is at most the number of pairs $(s < t)$ with $v_{\sigma(s)} = 0$ and $v_{\sigma(t)} = 1$, that is, at most

$$n(\mathbf{v})(d - n(\mathbf{v})) = \omega(\mathbf{v}).$$

Equality holds when σ sorts the entries of $\mathbf{v}$. Hence

$$\mathsf{H}^\sigma := \left\{(\mathbf{x}, t) \in \mathbb{R}^{d+1} : \ell^\sigma(\mathbf{x}) - t = 0\right\}$$

is supporting for $\mathsf{E}^\omega(V)$ and $\pi(\mathsf{E}^\omega(V) \cap \mathsf{H}^\sigma) = \triangle_\sigma$.

It remains to see that every maximal cell of $\mathcal{S}$ is of the form $\triangle_\sigma$ for some permutation $\sigma \in \mathfrak{S}_d$. However, *most* points in $[0,1]^d$ are generic and so the simplices $\triangle_\sigma$ cover $[0,1]^d$. That the simplices $\triangle_\sigma$ are unimodular is subject to Exercise 5.6. □

Next, we will live up to our promise and explain why we call $\mathsf{E}^\omega(V)$ the convex epigraph of $\mathsf{P} = \operatorname{conv}(V)$. In the proof of Theorem 5.1.5, we defined for each $\mathbf{p} \in \mathsf{P}$ the number $\overline{\omega}(\mathbf{p})$ as the smallest h such that $(\mathbf{p}, h) \in \mathsf{E}^\omega(V)$. This gives rise to a continuous function $\overline{\omega} : \mathsf{P} \to \mathbb{R}$.

Proposition 5.1.10. *Let* $\mathsf{P} = \operatorname{conv}(V)$ *be a polytope. For each* $\omega : V \to \mathbb{R}$, *the function* $\overline{\omega} : \mathsf{P} \to \mathbb{R}$ *is convex, and* $\overline{\omega}(\mathbf{v}) \le \omega(\mathbf{v})$ *for all* $\mathbf{v} \in V$.

Proof. Let $\mathbf{p}, \mathbf{q} \in \mathsf{P}$ and $0 \le \lambda \le 1$. Since $\mathsf{E}^\omega(V) \subset \mathbb{R}^{d+1}$ is a convex polyhedron, we know that

$$(1-\lambda)(\mathbf{p}, \overline{\omega}(\mathbf{p})) + \lambda(\mathbf{q}, \overline{\omega}(\mathbf{q})) \ \in \ \mathsf{E}^\omega(V)$$

and hence

$$(1-\lambda)\,\overline{\omega}(\mathbf{p}) + \lambda\,\overline{\omega}(\mathbf{q}) \ \ge \ \overline{\omega}((1-\lambda)\mathbf{p} + \lambda\,\mathbf{q})\,.$$

(See Figure 5.6 for an illustration.) Moreover, by definition $(\mathbf{v}, \omega(\mathbf{v})) \in \mathsf{E}^\omega(V)$ and therefore $\overline{\omega}(\mathbf{v}) \le \omega(\mathbf{v})$ for every $\mathbf{v} \in V$. □

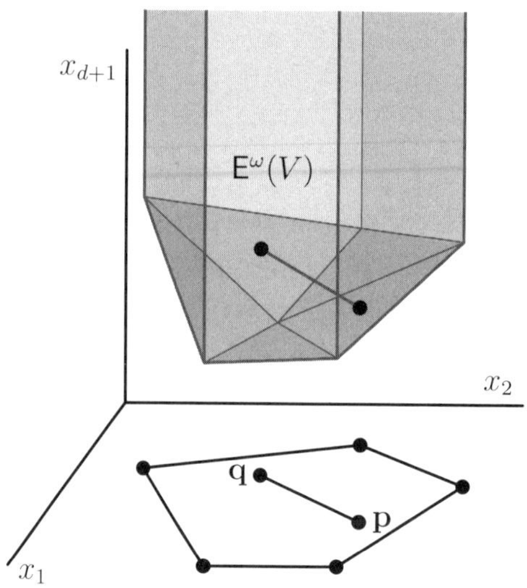

Figure 5.6. Illustrating the proof of Proposition 5.1.10.

In other words, $\mathsf{E}^\omega(V)$ is the epigraph $\{(\mathbf{p}, t) \,:\, \mathbf{p} \in \mathsf{P},\ t \ge \overline{\omega}(\mathbf{p})\}$ of the function $\overline{\omega}$. We note that there are many convex functions $f : \mathsf{P} \to \mathbb{R}$ such that $f(\mathbf{v}) \le \omega(\mathbf{v})$. However, $\overline{\omega}$ is special—it can be shown that $\overline{\omega}$ is the unique convex function with $f(v) \le \omega(v)$ for all $v \in V$ that minimizes the volume of the convex body

$$\{(\mathbf{p}, t) \,:\, \mathbf{p} \in \mathsf{P},\ f(\mathbf{p}) \le t \le M\}\,,$$

where $M := \max\{\omega(\mathbf{v}) \,:\, \mathbf{v} \in V\}$; see the Notes at the end of this chapter.

5.2. Möbius Functions of Subdivisions

Let $\mathsf{P} \subset \mathbb{R}^d$ be a full-dimensional lattice polytope and $\mathcal{T}$ a lattice triangulation of P (whose existence is vouched for by Corollary 5.1.6). Our next goal is a more refined statement about the representation (5.1.3) for $\mathrm{ehr}_{\mathsf{P}}(n)$ via the Möbius function $\mu_{\widehat{\mathcal{T}}}$ of $\widehat{\mathcal{T}} := \mathcal{T} \cup \{\mathsf{P}\}$.

If $\mathcal{T} = \mathcal{S}^\omega(\mathsf{P})$ is a regular triangulation, then $\mathcal{T}$ is isomorphic to the subposet $\Phi^{\mathrm{bnd}}(\mathsf{E}^\omega(V))$ of the bounded faces of $\mathsf{E}^\omega(V)$, and so our knowledge about Möbius functions of polyhedra (Theorem 3.5.1) gives almost everything; see Exercise 5.7. However, as we know from Exercise 5.2, not all subdivisions are regular. At any rate, it turns out that the methods developed in Section 3.5 help us to determine $\mu_{\widehat{\mathcal{T}}}$.

Theorem 5.2.1. *Let $\mathcal{S}$ be a proper subdivision of the polytope* P *and set $\widehat{\mathcal{S}} := \mathcal{S} \cup \{\mathsf{P}\}$. Then for $\mathsf{F}, \mathsf{G} \in \widehat{\mathcal{S}}$ with $\mathsf{F} \subseteq \mathsf{G}$,*

$$\mu_{\widehat{\mathcal{S}}}(\mathsf{F},\mathsf{G}) = \begin{cases} (-1)^{\dim \mathsf{G} - \dim \mathsf{F}} & \text{if } \mathsf{G} \neq \mathsf{P}, \\ (-1)^{\dim \mathsf{P} - \dim \mathsf{F} + 1} & \text{if } \mathsf{G} = \mathsf{P} \text{ and } \mathsf{F} \not\subseteq \partial \mathsf{P}, \\ 0 & \text{if } \mathsf{G} = \mathsf{P} \text{ and } \mathsf{F} \subseteq \partial \mathsf{P}. \end{cases}$$

If $\mathsf{G} \neq \mathsf{P}$, then the interval $[\mathsf{F},\mathsf{G}]$ in $\widehat{\mathcal{S}}$ is contained in the face lattice $\Phi(\mathsf{G})$ and hence the first case is subsumed by Theorem 3.5.1.

For $\mathsf{G} = \mathsf{P}$, we need to ponder the structure of the collection of cells $\mathsf{G}' \in \mathcal{S}$ that contain F. We recall from Section 3.5 that for a polyhedron Q and a point $\mathbf{q} \in \mathsf{Q}$, the tangent cone of Q at $\mathbf{q}$ is

$$\mathrm{T}_{\mathbf{q}}(\mathsf{Q}) = \{\mathbf{q} + \mathbf{u} : \mathbf{q} + \varepsilon\mathbf{u} \in \mathsf{Q} \text{ for all } \varepsilon > 0 \text{ sufficiently small}\}.$$

We saw in Proposition 3.5.2 that $\mathrm{T}_{\mathbf{q}}(\mathsf{Q})$ is the translate of a polyhedral cone which depends only on the (unique) face $\mathsf{F} \preceq \mathsf{Q}$ that contains $\mathbf{q}$ in its relative interior. In Chapter 3, tangent cones helped us to understand the intervals $[\mathsf{F},\mathsf{G}]$ in $\Phi(\mathsf{Q})$. The following is a strengthening of Lemma 3.5.3 from polyhedra to polyhedral subdivisions, illustrated in Figures 5.7 and 5.8.

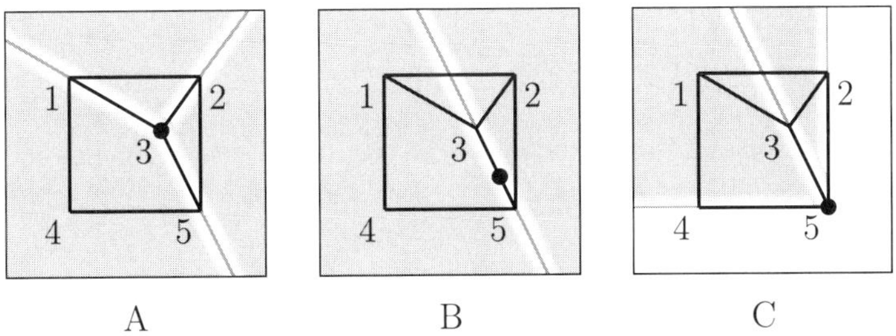

Figure 5.7. Various complexes of tangent cones.

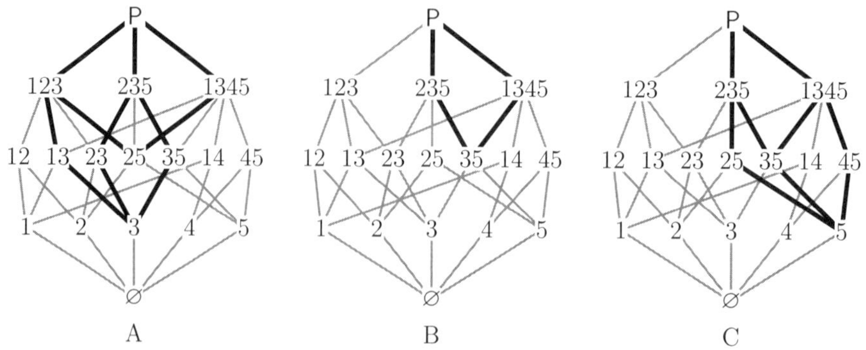

Figure 5.8. Intervals in $\widehat{\mathcal{S}}$ corresponding to the tangent cones in Figure 5.7.

Lemma 5.2.2. *Let $\mathcal{S}$ be a proper subdivision of a polytope P. For $\mathbf{q} \in \mathsf{P}$,*

$$\mathrm{T}_{\mathbf{q}}(\mathcal{S}) \ := \ \{\mathrm{T}_{\mathbf{q}}(\mathsf{G}) \,:\, \mathsf{G} \in \mathcal{S}\}$$

is a polyhedral complex with support $|\mathrm{T}_{\mathbf{q}}(\mathcal{S})| = \mathrm{T}_{\mathbf{q}}(\mathsf{P})$. Moreover, if $\mathsf{F} \in \mathcal{S}$ is the unique face with $\mathbf{q}$ in its relative interior, then the interval $[\mathsf{F}, \mathsf{P}]$ in $\widehat{\mathcal{S}}$ is isomorphic to $\mathrm{T}_{\mathbf{q}}(\mathcal{S}) \cup \{\hat{1}\}$, via $K \mapsto \mathrm{T}_{\mathbf{q}}(K)$.

Proof. Let $\mathbf{u} \in \mathbb{R}^d$. If $\mathbf{q} + \varepsilon\mathbf{u} \in \mathsf{P}$ for all $\varepsilon > 0$ sufficiently small, then there is a unique cell $\mathsf{G} \in \mathcal{S}$ such that $\mathbf{q} + \varepsilon\mathbf{u} \in \mathsf{G}^\circ$ for all $\varepsilon < \varepsilon_0$. This implies, in particular, that F is a face of G and that $\mathrm{T}_{\mathbf{q}}(\mathsf{G})^\circ \cap \mathrm{T}_{\mathbf{q}}(\mathsf{G}')^\circ = \varnothing$ whenever $\mathsf{G} \neq \mathsf{G}'$. The containment property is a statement about polytopes and hence a consequence of Lemma 3.5.3. This shows that

$$\mathrm{T}_{\mathbf{q}}(\mathcal{S}) \ = \ \{\mathrm{T}_{\mathbf{q}}(\mathsf{G}) \,:\, \mathsf{F} \preceq \mathsf{G} \in \mathcal{S}\} \ \cong \ \{\mathsf{G} \in \mathcal{S} \,:\, \mathsf{F} \preceq \mathsf{G}\}$$

and $|\mathrm{T}_{\mathbf{q}}(\mathcal{S})| = \mathrm{T}_{\mathbf{q}}(\mathsf{P})$. □

We observe that $\mathrm{T}_{\mathbf{q}}(\mathcal{S})$ does not depend on the choice of $\mathbf{q} \in \mathsf{F}^\circ$ and hence we will denote this polyhedral complex by $\mathrm{T}_{\mathsf{F}}(\mathcal{S})$. For the empty face, we define $\mathrm{T}_{\varnothing}(\mathcal{S}) := |\mathcal{S}| = \mathsf{P}$. We are ready to prove Theorem 5.2.1.

Proof of Theorem 5.2.1. We need to consider only the case $\mathsf{G} = \mathsf{P}$. For a cell $\mathsf{F} \in \mathcal{S}$ we compute

$$\begin{aligned} \mu_{\widehat{\mathcal{S}}}(\mathsf{F}, \mathsf{P}) \ &= \ -\sum_{\mathsf{F} \preceq_{\widehat{\mathcal{S}}} \mathsf{G}' \prec_{\widehat{\mathcal{S}}} \mathsf{P}} (-1)^{\dim \mathsf{G}' - \dim \mathsf{F}} \ = \ (-1)^{\dim \mathsf{F}+1} \sum_{\mathsf{C} \in \mathrm{T}_{\mathsf{F}}(\mathcal{S})} (-1)^{\dim \mathsf{C}} \\ &= \ (-1)^{\dim \mathsf{F}+1}\, \chi(\mathrm{T}_{\mathsf{F}}(\mathsf{P}))\,, \end{aligned}$$

where the first equation follows from the defining property of Möbius functions (2.2.1) and $\mu_{\widehat{\mathcal{S}}}(\mathsf{F}, \mathsf{G}') = (-1)^{\dim \mathsf{G}' - \dim \mathsf{F}}$ for all $\mathsf{G}' \prec \mathsf{P}$.

Let $\mathbf{q} \in \mathsf{F}^\circ$. If $\mathsf{F} \in \mathcal{S}$ is contained in the boundary of P, then $\mathbf{q}$ is contained in a proper face of P and, by Lemma 3.5.3, $\mathrm{T}_{\mathsf{F}}(\mathsf{P})$ is a proper polyhedral cone. Hence, our Euler characteristic computations (Corollaries 3.4.8 and 3.4.10) give $\chi(\mathrm{T_F}(\mathsf{P})) = 0$. On the other hand, if $\mathsf{F} \not\subseteq \partial\mathsf{P}$, then $\mathbf{q} \in \mathsf{P}^\circ$. In this case $\mathrm{T_F}(\mathsf{P}) = \mathrm{aff}(\mathsf{P})$ and $\chi(\mathrm{T_F}(\mathsf{P})) = (-1)^{\dim \mathsf{P}}$ finishes the proof. □

With this machinery at hand, we can extend Ehrhart–Macdonald reciprocity from lattice simplices (Theorem 4.6.1) to all lattice polytopes.

Theorem 5.2.3. *Let* $\mathsf{P} \subset \mathbb{R}^d$ *be a lattice polytope and* $\mathrm{ehr}_\mathsf{P}(n)$ *its Ehrhart polynomial. Then for all integers* $n > 0$,

$$(-1)^{\dim \mathsf{P}} \, \mathrm{ehr}_\mathsf{P}(-n) \;=\; \left| n\,\mathsf{P}^\circ \cap \mathbb{Z}^d \right| .$$

In other words, the Ehrhart polynomials of P *and* P° *satisfy*

$$(-1)^{\dim \mathsf{P}} \, \mathrm{ehr}_\mathsf{P}(-n) \;=\; \mathrm{ehr}_{\mathsf{P}^\circ}(n) .$$

Proof. Assume that P is not a lattice simplex and let $\mathcal{S}$ be a lattice triangulation of P. (This exists by Corollary 5.1.6.) Möbius inversion on $\widehat{\mathcal{S}} := \mathcal{S} \cup \{\mathsf{P}\}$ (Corollary 5.1.3) yields

$$\mathrm{ehr}_\mathsf{P}(n) \;=\; -\sum_{\mathsf{F} \in \mathcal{S}} \mu_{\widehat{\mathcal{S}}}(\mathsf{F}, \mathsf{P}) \, \mathrm{ehr}_\mathsf{F}(n) .$$

By Theorem 5.2.1, this simplifies to

$$\mathrm{ehr}_\mathsf{P}(n) \;=\; \sum_{\substack{\mathsf{F} \in \mathcal{S} \\ \mathsf{F} \not\subseteq \partial\mathsf{P}}} (-1)^{\dim \mathsf{P} - \dim \mathsf{F}} \, \mathrm{ehr}_\mathsf{F}(n) .$$

Ehrhart–Macdonald reciprocity for simplices (Theorem 4.6.1) asserts that $\mathrm{ehr}_\mathsf{F}(-n) = (-1)^{\dim \mathsf{F}} |n\,\mathsf{F}^\circ \cap \mathbb{Z}^d|$ for every lattice simplex F. Hence

$$(-1)^{\dim \mathsf{P}} \, \mathrm{ehr}_\mathsf{P}(-n) \;=\; \sum_{\substack{\mathsf{F} \in \mathcal{S} \\ \mathsf{F} \not\subseteq \partial\mathsf{P}}} \mathrm{ehr}_{\mathsf{F}^\circ}(n) , \tag{5.2.1}$$

and the right-hand side counts the number of lattice points in the relative interior of $n\mathsf{P}$, by Lemma 3.3.8 and the definition of a subdivision. □

Our proofs of Theorems 5.1.7 and 5.2.3 can be generalized for a *rational* polytope, and you are invited to prove the following theorem in Exercise 5.14.

Theorem 5.2.4. *If* $\mathsf{P} \subset \mathbb{R}^d$ *is a rational polytope, then for positive integers* n*, the counting function* $\mathrm{ehr}_\mathsf{P}(n)$ *is a quasipolynomial in* n *whose period divides the least common multiple of the denominators of the vertex coordinates of* P*. Furthermore, for all integers* $n > 0$,

$$(-1)^{\dim \mathsf{P}} \, \mathrm{ehr}_\mathsf{P}(-n) \;=\; \left| n\,\mathsf{P}^\circ \cap \mathbb{Z}^d \right| .$$

In other words, the Ehrhart quasipolynomials of P *and* P° *are related as*

$$(-1)^{\dim \mathsf{P}} \, \mathrm{ehr}_\mathsf{P}(-n) \;=\; \mathrm{ehr}_{\mathsf{P}^\circ}(n) .$$

5.3. Beneath, Beyond, and Half-open Decompositions

We promised a second method to obtain nontrivial subdivisions of a given (lattice) polytope. The technique presented in this section is not only algorithmic but it also furnishes a general methodology that avoids (rather: hides) the use of Möbius inversion.

Let $\mathsf{P} \subset \mathbb{R}^d$ be a full-dimensional polytope with vertex set V. The idea is to recursively construct a subdivision of P. So given a vertex $\mathbf{v} \in V$, let $\mathsf{P}' = \operatorname{conv}(V \setminus \{\mathbf{v}\})$. Our next goal is, assuming we have a subdivision $\mathcal{S}'$ of P', to extend $\mathcal{S}'$ to a subdivision of P.

We recall from Section 3.7 that $\mathbf{v}$ is beyond a face F of P' if $\mathbf{v} \notin \mathrm{T}_{\mathsf{F}}(\mathsf{P}')$, and that this is equivalent to all points in F being visible from $\mathbf{v}$. Let $\mathcal{S}'$ be a subdivision of the polyhedron $\mathsf{P}' \subset \mathbb{R}^d$ and $\mathbf{v} \in \mathbb{R}^d$. We define

$$\mathrm{Vis}_{\mathbf{v}}(\mathcal{S}') \;:=\; \left\{\mathsf{F} \in \mathcal{S}' \,:\, \mathbf{v} \notin \mathrm{T}_{\mathsf{F}}(\mathcal{S})\right\},$$

containing the cells of $\mathcal{S}'$ that are visible from $\mathbf{v}$. Note that $\mathrm{T}_{\varnothing}(\mathcal{S}') = \mathsf{P}'$ and, since we assume $\mathbf{v} \notin \mathsf{P}'$, we have $\varnothing \in \mathrm{Vis}_{\mathbf{v}}(\mathcal{S}')$.[3]

Proposition 5.3.1. *Let $\mathcal{S}'$ be a subdivision of the polytope $\mathsf{P}' \subset \mathbb{R}^d$ and let $\mathbf{v} \in \mathbb{R}^d$. Then $\mathrm{Vis}_{\mathbf{v}}(\mathcal{S}')$ is a polyhedral complex.*

Proof. Since $\mathrm{Vis}_{\mathbf{v}}(\mathcal{S}') \subseteq \mathcal{S}'$, the intersection property holds automatically. For the containment property, we note that $\mathrm{T}_{\mathsf{G}}(\mathcal{S}') \subseteq \mathrm{T}_{\mathsf{F}}(\mathcal{S}')$ for faces $\mathsf{G} \preceq \mathsf{F} \in \mathcal{S}'$, and so $\mathbf{v}$ is visible from all points of G if it is visible from all points of F. □

We recall from Section 3.1 that for $\mathbf{v} \notin \operatorname{aff}(\mathsf{F}')$, the polytope

$$\mathbf{v} * \mathsf{F}' \;=\; \operatorname{conv}(\{\mathbf{v}\} \cup \mathsf{F}')$$

is the pyramid over F' with apex $\mathbf{v}$. The next lemma states that under certain conditions, we can take the pyramid over a polyhedral complex.

Lemma 5.3.2. *Let $\mathcal{S}$ be a polyhedral complex and $\mathbf{v} \notin |\mathcal{S}|$. If, for any $\mathbf{p} \in |\mathcal{S}|$, the segment $[\mathbf{v}, \mathbf{p}]$ meets $|\mathcal{S}|$ only in $\mathbf{p}$, then*

$$\mathbf{v} * \mathcal{S} \;:=\; \{\mathbf{v} * \mathsf{F} \,:\, \mathsf{F} \in \mathcal{S}\} \cup \mathcal{S}$$

is a polyhedral complex.

Proof. We first note that $\mathbf{v} \notin \operatorname{aff}(\mathsf{F})$ for every $\mathsf{F} \in \mathcal{S}$—otherwise, $[\mathbf{v}, \mathbf{p}] \cap \mathsf{F}$ would be a segment for any point $\mathbf{p} \in \mathsf{F}^\circ$. Hence, $\mathbf{v} * \mathsf{F}$ is a well-defined pyramid. Moreover, by Exercise 3.40, every face of $\mathbf{v} * \mathsf{F}$ is either a face of F or a pyramid over a face of F. Therefore, $\mathbf{v} * \mathcal{S}$ satisfies the containment property.

[3] Now our notation in (3.7.2) should finally make sense: we studied the support of the polyhedral complex formed by visible faces.

For the intersection property, assume that $\mathsf{F}, \mathsf{G} \in \mathbf{v} * \mathcal{S}$ are cells such that $\mathsf{F}^\circ \cap \mathsf{G}^\circ$ contains a point $\mathbf{p}$. Then F and G cannot both be cells in $\mathcal{S}$, and so we may assume that $\mathsf{F} = \mathbf{v} * \mathsf{F}'$ for some $\mathsf{F}' \in \mathcal{S}$. Now the ray $\{\mathbf{v} + t(\mathbf{p} - \mathbf{v}) : t \geq 0\}$ meets the relative interior of F' in a unique point $\mathbf{p}'$. If $\mathsf{G} \in \mathcal{S}$, then our assumption yields that $\mathsf{F}' = \mathsf{G}$ and so G is a face of F. Otherwise, $\mathsf{G} = \mathbf{v} * \mathsf{G}'$ for some $\mathsf{G}' \in \mathcal{S}$ and again by assumption the segment $[\mathbf{v}, \mathbf{p}']$ meets both F' and G' in their relative interiors which implies that $\mathsf{F}' = \mathsf{G}'$. □

Returning to our initial goal of describing a way of extending a subdivision of P' to P, we can do the following.

Theorem 5.3.3. *Let $\mathcal{S}'$ be a subdivision of the polytope $\mathsf{P}' \subset \mathbb{R}^d$ and let $\mathbf{v} \in \mathbb{R}^d \setminus \mathsf{P}'$. Then*

$$\mathcal{S} \;:=\; \mathcal{S}' \cup (\mathbf{v} * \mathrm{Vis}_{\mathbf{v}}(\mathcal{S}'))$$

is a subdivision of $\mathsf{P} = \mathrm{conv}(\{\mathbf{v}\} \cup \mathsf{P}')$.

Proof. By Lemma 5.3.2, $\mathbf{v} * \mathrm{Vis}_{\mathbf{v}}(\mathcal{S}')$ is a polyhedral complex and

$$(\mathbf{v} * \mathrm{Vis}_{\mathbf{v}}(\mathcal{S}')) \cap \mathcal{S}' \;=\; \mathrm{Vis}_{\mathbf{v}}(\mathcal{S}'),$$

by definition of $\mathrm{Vis}_{\mathbf{v}}(\mathcal{S}')$. Thus, we are left to show that $\mathsf{P} = |\mathcal{S}|$.

Let $\mathbf{p} \in \mathsf{P} \setminus \mathsf{P}'$, and let $\mathbf{r}$ be the first point on the ray $\mathbf{v} + \mathbb{R}_{\geq 0}(\mathbf{p} - \mathbf{v})$ that meets P'. By construction, $\mathbf{r} \in \partial\mathsf{P}'$ is visible from $\mathbf{v}$. Hence, the unique cell $\mathsf{F}' \in \mathcal{S}'$ that contains $\mathbf{r}$ in its relative interior is contained in $\mathrm{Vis}_{\mathbf{v}}(\mathsf{P}')$ and $\mathbf{p}$ lies in $\mathbf{v} * \mathsf{F} \in \mathcal{S}$. □

Theorem 5.3.3 and its proof give a practical algorithm for computing a (lattice) triangulation of a (lattice) polytope $\mathsf{P} = \mathrm{conv}(\mathbf{v}_0, \dots, \mathbf{v}_n)$:

(1) We start with a simplex $\mathsf{P}_0 := \mathrm{conv}(\mathbf{v}_0, \dots, \mathbf{v}_d)$, which is triangulated by its collection of faces $\mathcal{T}_0 := \Phi(\mathsf{P}_0)$.

(2) For every $i = 1, \dots, n - d$, set $\mathsf{P}_i := \mathrm{conv}(\{\mathbf{v}_{d+i}\} \cup \mathsf{P}_{i-1})$. Now Theorem 5.3.3 asserts that

$$\mathcal{T}_i \;:=\; \mathcal{T}_{i-1} \cup \{\mathbf{v}_{d+i} * \mathsf{F}' \;:\; \mathsf{F}' \in \mathrm{Vis}_{\mathbf{v}}(\mathcal{T}_{i-1})\}$$

is a triangulation of P_i.

In particular, $\mathcal{T} := \mathcal{T}_{n-d}$ is a triangulation of $\mathsf{P}_{n-d} = \mathsf{P}$. This triangulation $\mathcal{T}$ is called a **pushing** (or **placing**) **triangulation** and depends only on the labeling of the vertices. Figure 5.9 illustrates the algorithm for a pentagon.

The crucial insight that makes the above algorithm work is that at every iteration of step (2), we know not only the sets of vertices of the cells in $\mathcal{T}_i$ but also a description in terms of facet-defining inequalities. For the initial step (1), this is due to the fact that P_0 is a simplex (see Exercise 5.16); for every subsequent iteration of (2), see Exercises 5.17 and 5.18. In particular, a nice

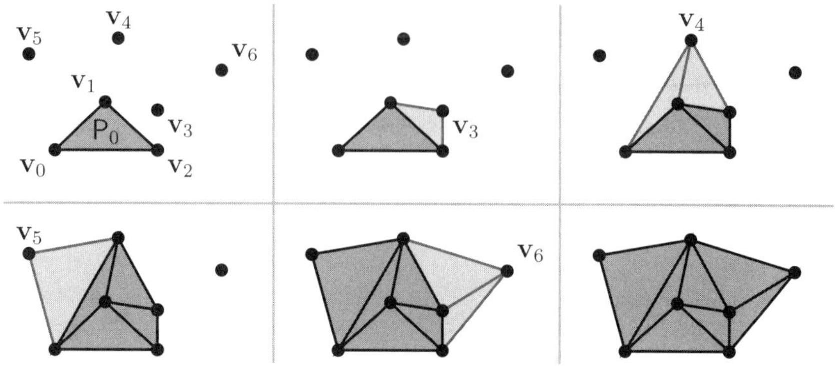

Figure 5.9. A pushing triangulation of a pentagon.

byproduct of our algorithm is that it computes an inequality description for P along the way and hence verifies the first half of the Minkowski–Weyl Theorem 3.2.5 for polytopes; see Exercise 5.20.

Exercise 5.19 reveals that the triangulation constructed by our algorithm is regular, and so in some sense we have really given only *one* method for constructing subdivisions of polytopes. However, the concept of a subdivision or, more generally, a dissection inevitably leads to *overcounting* and Möbius inversion in the setting of, e.g., Ehrhart theory, as in our proof of Theorem 5.2.3. Our use of tangent cones and the notion of points beyond a face in the construction of a pushing triangulation suggests a conceptual perspective that avoids inclusion–exclusion of any sort and that we will elucidate in the following.

Given a full-dimensional polyhedron $\mathsf{P} \subset \mathbb{R}^d$ with facets $\mathsf{F}_1, \mathsf{F}_2, \dots, \mathsf{F}_k$, we call a point $\mathbf{q} \in \mathbb{R}^d$ **generic relative to** P if $\mathbf{q}$ is not contained in any facet-defining hyperplane of P. The key to avoiding Möbius inversion with subdivisions is the following definition: for a full-dimensional polyhedron $\mathsf{P} \subset \mathbb{R}^d$ and $\mathbf{q}$ generic relative to P, let

$$\mathbb{H}_{\mathbf{q}}\mathsf{P} \ := \ \mathsf{P} \setminus |\mathrm{Vis}_{\mathbf{q}}(\mathsf{P})| \, ,$$

which we call a **half-open polyhedron**. By setting $I := \{j \, : \, \mathbf{q} \text{ beyond } \mathsf{F}_j\}$, we may also write

$$\mathbb{H}_{\mathbf{q}}\mathsf{P} \ = \ \mathsf{P} \setminus \bigcup_{j \in I} \mathsf{F}_j \, . \tag{5.3.1}$$

The counterpart to this half-open polyhedron, appealing to our treatment of half-open cones in Section 4.8, is

$$\mathbb{H}^{\mathbf{q}}\mathsf{P} \ := \ \mathsf{P} \setminus \bigcup_{j \notin I} \mathsf{F}_j \, . \tag{5.3.2}$$

Lemma 5.3.4. *Let* $\mathsf{P} \subset \mathbb{R}^d$ *be a full-dimensional polyhedron with dissection* $\mathsf{P} = \mathsf{P}_1 \cup \mathsf{P}_2 \cup \cdots \cup \mathsf{P}_m$. *If* $\mathbf{q} \in \mathbb{R}^d$ *is generic relative to each* P_i, *then*

$$\mathbb{H}_{\mathbf{q}}\mathsf{P} = \mathbb{H}_{\mathbf{q}}\mathsf{P}_1 \uplus \mathbb{H}_{\mathbf{q}}\mathsf{P}_2 \uplus \cdots \uplus \mathbb{H}_{\mathbf{q}}\mathsf{P}_m \quad \textit{and} \tag{5.3.3}$$

$$\mathbb{H}^{\mathbf{q}}\mathsf{P} = \mathbb{H}^{\mathbf{q}}\mathsf{P}_1 \uplus \mathbb{H}^{\mathbf{q}}\mathsf{P}_2 \uplus \cdots \uplus \mathbb{H}^{\mathbf{q}}\mathsf{P}_m \,. \tag{5.3.4}$$

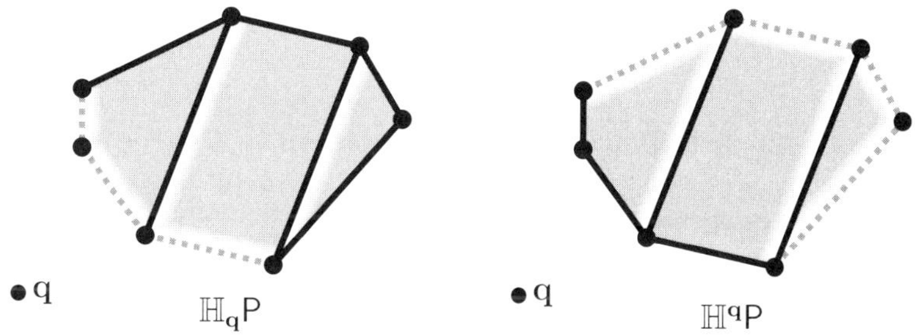

Figure 5.10. Two half-open dissections of a heptagon according to Lemma 5.3.4.

Figure 5.10 illustrates this lemma. If $\mathbf{q} \in P^\circ$, then $\mathbb{H}_{\mathbf{q}}\mathsf{P} = \mathsf{P}$ and $\mathbb{H}^{\mathbf{q}}\mathsf{P} = \mathsf{P}^\circ$.[4] (However, there is no $\mathbf{q}$ such that $\mathbb{H}^{\mathbf{q}}\mathsf{P} = \mathsf{P}$; see Exercise 5.22.) Thus Lemma 5.3.4 immediately implies the following corollary.

Corollary 5.3.5. *Let* $\mathsf{P} \subset \mathbb{R}^d$ *be a full-dimensional polyhedron with dissection* $\mathsf{P} = \mathsf{P}_1 \cup \mathsf{P}_2 \cup \cdots \cup \mathsf{P}_m$. *If* $\mathbf{q} \in \mathsf{P}^\circ$ *is generic relative to all* P_j, *then*

$$\begin{aligned} \mathsf{P} &= \mathbb{H}_{\mathbf{q}}\mathsf{P}_1 \uplus \mathbb{H}_{\mathbf{q}}\mathsf{P}_2 \uplus \cdots \uplus \mathbb{H}_{\mathbf{q}}\mathsf{P}_m \quad \textit{and} \\ \mathsf{P}^\circ &= \mathbb{H}^{\mathbf{q}}\mathsf{P}_1 \uplus \mathbb{H}^{\mathbf{q}}\mathsf{P}_2 \uplus \cdots \uplus \mathbb{H}^{\mathbf{q}}\mathsf{P}_m \,. \end{aligned} \tag{5.3.5}$$

Proof of Lemma 5.3.4. We prove (5.3.3) and leave (5.3.4) to Exercise 5.21.

Since $\mathbb{H}_{\mathbf{q}}\mathsf{P}_j \subseteq \mathbb{H}_{\mathbf{q}}\mathsf{P}$ for $1 \le j \le m$, the right-hand side of (5.3.3) is a subset of $\mathbb{H}_{\mathbf{q}}\mathsf{P}$ and we need to show only the reverse inclusion and that the union is disjoint.

Let $\mathbf{p} \in \mathbb{H}_{\mathbf{q}}\mathsf{P}$. If $\mathbf{p} \in \mathsf{P}_j^\circ$ for some j, then we are done, so suppose $\mathbf{p}$ lies on the boundary of some of the P_js. (There is at least one such P_j because they form a dissection of P.) Since $\mathbf{p}$ is not in a face of P that is visible from $\mathbf{q}$, there is a subinterval

$$[(1-\varepsilon)\,\mathbf{p} + \varepsilon\,\mathbf{q},\, \mathbf{p}] \;\subseteq\; [\mathbf{q}, \mathbf{p}]$$

[4]In particular, a half-open polyhedron might be open or closed (or neither), so our nomenclature should be digested carefully.

for some $\varepsilon > 0$ that is contained in P. By choosing ε smaller, if necessary, we can realize $[(1-\varepsilon)\,\mathbf{p} + \varepsilon\,\mathbf{q},\, \mathbf{p}] \subseteq \mathsf{P}_j$ for some j. (Note that $\mathbf{p} \in \partial\mathsf{P}_j$ and

$$[(1-\varepsilon)\,\mathbf{p} + \varepsilon\,\mathbf{q},\, \mathbf{p}) \ \subseteq \ \mathsf{P}_j^\circ \tag{5.3.6}$$

since $\mathbf{q}$ is generic.) But this means that $\mathbf{p}$ is not on a face of P_j that is visible from $\mathbf{q}$, and so $\mathbf{p} \in \mathbb{H}_\mathbf{q}\mathsf{P}_j$.

If, in addition, $\mathbf{p} \in \partial\mathsf{P}_k$ for some $k \neq j$, then $[\mathbf{q},\, \mathbf{p})$ lies outside of P_k, by convexity, (5.3.6), and the definition of a dissection. This means that $\mathbf{p} \notin \mathbb{H}_\mathbf{q}\mathsf{P}_k$. □

We call (5.3.5) a **half-open decomposition** of P. Half-open decompositions give a cancellation-free way to prove results such as Ehrhart's theorem and Ehrhart–Macdonald reciprocity. Towards the latter, we define the Ehrhart function of the half-open polytope $\mathbb{H}_\mathbf{q}\mathsf{P}$ (naturally) by

$$\operatorname{ehr}_{\mathbb{H}_\mathbf{q}\mathsf{P}}(n) \ := \ \left| n\,\mathbb{H}_\mathbf{q}\mathsf{P} \cap \mathbb{Z}^d \right|$$

for integers $n \geq 1$, with the analogous definition for $\operatorname{ehr}_{\mathbb{H}^\mathbf{q}\mathsf{P}}(n)$. You can already guess the interplay of these counting functions and may try to deduce it via Möbius inversion. Instead, our proof follows the philosophy of Section 4.8.

Proposition 5.3.6. *Let $\triangle \subset \mathbb{R}^d$ be a full-dimensional lattice simplex and $\mathbf{q}$ generic relative to $\triangle$. Then $\operatorname{ehr}_{\mathbb{H}_\mathbf{q}\triangle}(n)$ agrees with a polynomial of degree d, and, for all integers $n \geq 1$,*

$$(-1)^d \operatorname{ehr}_{\mathbb{H}_\mathbf{q}\triangle}(-n) \ = \ \left| n\,\mathbb{H}^\mathbf{q}\triangle \cap \mathbb{Z}^d \right|.$$

In other words, the Ehrhart polynomials of $\mathbb{H}_\mathbf{q}\triangle$ and $\mathbb{H}^\mathbf{q}\triangle$ are related as

$$(-1)^d \operatorname{ehr}_{\mathbb{H}_\mathbf{q}\triangle}(-n) \ = \ \operatorname{ehr}_{\mathbb{H}^\mathbf{q}\triangle}(n)\,.$$

Proof. Let $\triangle \subset \mathbb{R}^d$ be a full-dimensional lattice simplex with vertices $\mathbf{v}_1, \dots, \mathbf{v}_{d+1}$ and facets $\mathsf{F}_1, \dots, \mathsf{F}_{d+1}$. We can assume that the vertices and facets are labeled such that $v_j \notin \mathsf{F}_j$ for all j. For convenience, we can also assume that

$$\{j \in [d+1] \,:\, \mathbf{q} \text{ beyond } \mathsf{F}_j\} \ = \ \{m, m+1, \dots, d+1\}\,.$$

Every point in $\triangle$ has a unique representation as a convex combination of $\mathbf{v}_1, \dots, \mathbf{v}_{d+1}$. Hence, we can give an intrinsic description of the half-open simplex $\mathbb{H}_\mathbf{q}\triangle$ as

$$\mathbb{H}_\mathbf{q}\triangle \ = \ \left\{ \lambda_1\mathbf{v}_1 + \lambda_2\mathbf{v}_2 + \cdots + \lambda_{d+1}\mathbf{v}_{d+1} \,:\, \begin{array}{l} \lambda_1 + \cdots + \lambda_{d+1} = 1 \\ \lambda_1, \dots, \lambda_{m-1} \geq 0 \\ \lambda_m, \dots, \lambda_{d+1} > 0 \end{array} \right\}.$$

The homogenization $\hom(\mathbb{H}_{\mathbf{q}}\triangle)$ is therefore the half-open cone

$$\widehat{\mathsf{C}} := \mathbb{R}_{\geq 0}\begin{pmatrix}\mathbf{v}_1\\1\end{pmatrix} + \cdots + \mathbb{R}_{\geq 0}\begin{pmatrix}\mathbf{v}_{m-1}\\1\end{pmatrix} + \mathbb{R}_{>0}\begin{pmatrix}\mathbf{v}_m\\1\end{pmatrix} + \cdots + \mathbb{R}_{>0}\begin{pmatrix}\mathbf{v}_{d+1}\\1\end{pmatrix}.$$

This gives precisely the setup of Theorem 4.8.1 and our subsequent proof of Theorem 4.6.1. Specializing (4.8.5) to our setting gives

$$\operatorname{Ehr}_{\mathbb{H}_{\mathbf{q}}\triangle}(z) \;=\; \sigma_{\widehat{\mathsf{C}}}(1,\dots,1,z)\,, \tag{5.3.7}$$

where $\operatorname{Ehr}_{\mathbb{H}_{\mathbf{q}}\triangle}(z) := \sum_{n\geq 1}\operatorname{ehr}_{\mathbb{H}_{\mathbf{q}}\triangle}(n)\,z^n$. (Technically, for this definition to be in sync with (5.3.7), we need to assume that $\mathbb{H}_{\mathbf{q}}\triangle \neq \triangle$—which we may, since otherwise Proposition 5.3.6 is simply Theorem 4.6.1, which we have long proved.) By Exercise 5.23, the right-hand side of (5.3.7) is a rational function that yields our first claim.

For the second claim, we refer to Exercise 5.24, which says (in the language of Section 4.8) that $\check{\mathsf{C}} = \hom(\mathbb{H}^{\mathbf{q}}\triangle)$, and so by Theorem 4.8.1,

$$(-1)^{d+1}\operatorname{Ehr}_{\mathbb{H}_{\mathbf{q}}\triangle}(\tfrac{1}{z}) \;=\; \operatorname{Ehr}_{\mathbb{H}^{\mathbf{q}}\triangle}(z)\,,$$

which yields the reciprocity statement, using Theorem 4.1.6. □

As you might have already guessed, the combination of Corollary 5.3.5 and Proposition 5.3.6 yields an alternative proof of Ehrhart–Macdonald reciprocity (Theorem 5.2.3). In fact, we can give a more general reciprocity theorem, namely, the nonsimplicial version of Proposition 5.3.6.

Theorem 5.3.7. *Let $\mathsf{P} \subset \mathbb{R}^d$ be a full-dimensional lattice polytope and let $\mathbf{q} \in \mathbb{R}^d$ be generic relative to P. Then for all integers $n \geq 1$,*

$$(-1)^d\operatorname{ehr}_{\mathbb{H}_{\mathbf{q}}\mathsf{P}}(-n) \;=\; \left|n\,\mathbb{H}^{\mathbf{q}}\mathsf{P} \cap \mathbb{Z}^d\right|.$$

In other words, the Ehrhart polynomials of $\mathbb{H}_{\mathbf{q}}\mathsf{P}$ and $\mathbb{H}^{\mathbf{q}}\mathsf{P}$ are related as

$$(-1)^d\operatorname{ehr}_{\mathbb{H}_{\mathbf{q}}\mathsf{P}}(-n) \;=\; \operatorname{ehr}_{\mathbb{H}^{\mathbf{q}}\mathsf{P}}(n)\,.$$

In particular, the special case $\mathbf{q} \in \mathsf{P}^\circ$ gives an alternative proof of Theorem 5.2.3.

Proof. Let $\mathsf{P} = \triangle_1 \cup \cdots \cup \triangle_m$ be a dissection of P into lattice simplices. If necessary, we can replace $\mathbf{q}$ by some $\mathbf{q}'$ such that $\mathbb{H}_{\mathbf{q}}\mathsf{P} = \mathbb{H}_{\mathbf{q}'}\mathsf{P}$ and $\mathbf{q}'$ is generic relative to each $\triangle_i$. Lemma 5.3.4 implies

$$\begin{aligned}\operatorname{ehr}_{\mathbb{H}_{\mathbf{q}}\mathsf{P}}(n) &= \operatorname{ehr}_{\mathbb{H}_{\mathbf{q}}\triangle_1}(n) + \cdots + \operatorname{ehr}_{\mathbb{H}_{\mathbf{q}}\triangle_m}(n) \quad \text{and}\\ \operatorname{ehr}_{\mathbb{H}^{\mathbf{q}}\mathsf{P}}(n) &= \operatorname{ehr}_{\mathbb{H}^{\mathbf{q}}\triangle_1}(n) + \cdots + \operatorname{ehr}_{\mathbb{H}^{\mathbf{q}}\triangle_m}(n)\,.\end{aligned}$$

Now use Proposition 5.3.6. □

5.4. Stanley Reciprocity

While we have made a number of definitions for general polyhedra, the major constructions in this chapter up to this point have been on *polytopes*. As cones recently made a comeback (in our proof of Proposition 5.3.6), it's time to extend some of our constructs to the conical world.

A subdivision $\mathcal{S}$ of a polyhedral cone C is a **triangulation** if all cells in $\mathcal{S}$ are simplicial cones. If C and the cells in a triangulation $\mathcal{S}$ of C are rational, we call $\mathcal{S}$ **rational**. For arithmetic purposes, we will mostly be interested in the case that C is pointed. A nice side effect of concentrating on pointed cones is that many results for polytopes carry over, such as the following.

Corollary 5.4.1. *Every (rational) pointed cone has a (rational) triangulation.*

Proof. We may assume that $\mathsf{C} \subset \mathbb{R}^d$ is full dimensional. Because C is pointed, there exists a hyperplane H (which we may choose to be rational) such that $\mathsf{C} \cap \mathsf{H} = \{\mathbf{0}\}$. Choose a (rational) point $\mathbf{p} \in \mathsf{C} \setminus \{0\}$; then $\mathbf{p} + \mathsf{H}$ meets every ray of C and

$$\mathsf{P} \ := \ (\mathbf{p} + \mathsf{H}) \cap \mathsf{C}$$

is a polytope. (If C is rational, we can rescale $\mathbf{p}$, if necessary, to make P into a *lattice* polytope.) By Corollary 5.1.6, P admits a (lattice) triangulation $\mathcal{T}$ which, in turn, gives rise to the (rational) triangulation

$$\{\text{cone}(\triangle) \, : \, \triangle \in \mathcal{T}\}$$

of C. □

Half-open decomposition of lattice polytopes naturally led us to Theorem 5.3.7, a generalized version of Ehrhart–Macdonald reciprocity. In the world of cones, we have seen a reciprocity theorem for the integer-point transforms of half-open *simplicial* cones (Theorem 4.8.1). Now we extend it, in similar fashion, to the general case of pointed cones.

Theorem 5.4.2. *Let $\mathsf{C} \subset \mathbb{R}^d$ be a full-dimensional pointed rational cone, and let $\mathbf{q} \in \mathbb{R}^d$ be generic relative to C. Then*

$$\sigma_{\mathbb{H}_{\mathbf{q}}\mathsf{C}}\left(\tfrac{1}{\mathbf{z}}\right) \ = \ (-1)^d \, \sigma_{\mathbb{H}^{\mathbf{q}}\mathsf{C}}(\mathbf{z}) \, .$$

Analogous to Corollary 5.3.5, the special case $\mathbf{q} \in \mathsf{C}^\circ$ gives a result that is worth being mentioned separately.

Corollary 5.4.3. *Let $\mathsf{C} \subset \mathbb{R}^d$ be a full-dimensional pointed rational cone. Then*

$$\sigma_{\mathsf{C}}\left(\tfrac{1}{\mathbf{z}}\right) \ = \ (-1)^d \, \sigma_{\mathsf{C}^\circ}(\mathbf{z}) \, .$$

Theorem 5.4.2 and Corollary 5.4.3 constitute **Stanley reciprocity**; we already saw the simplicial case in Section 4.8.

Proof of Theorem 5.4.2. We repeat the arguments in our proof of Theorem 5.3.7, adjusting them to the integer-point transforms of a pointed rational cone. Let $\mathsf{C} \subset \mathbb{R}^d$ be a rational pointed cone with a rational triangulation $\mathsf{C} = \mathsf{C}_1 \cup \cdots \cup \mathsf{C}_m$. Lemma 5.3.4 implies

$$\begin{aligned} \sigma_{\mathbb{H}_{\mathbf{q}}\mathsf{C}}(\mathbf{z}) &= \sigma_{\mathbb{H}_{\mathbf{q}}\mathsf{C}_1}(\mathbf{z}) + \cdots + \sigma_{\mathbb{H}_{\mathbf{q}}\mathsf{C}_m}(\mathbf{z}) \quad \text{and} \\ \sigma_{\mathbb{H}^{\mathbf{q}}\mathsf{C}}(\mathbf{z}) &= \sigma_{\mathbb{H}^{\mathbf{q}}\mathsf{C}_1}(\mathbf{z}) + \cdots + \sigma_{\mathbb{H}^{\mathbf{q}}\mathsf{C}_m}(\mathbf{z}) . \end{aligned}$$

Now use Theorem 4.8.1 and Exercise 5.25. □

Just as in Section 4.8, Theorem 5.4.2 has an immediate application to Hilbert series.

Corollary 5.4.4. *Let $\mathsf{C} \subset \mathbb{R}^d$ be a full-dimensional rational pointed cone, and fix a grading $\mathbf{a} \in \mathbb{Z}^d$. Then*

$$H_{\mathsf{C}}^{\mathbf{a}}\left(\tfrac{1}{z}\right) = (-1)^d H_{\mathsf{C}^\circ}^{\mathbf{a}}(z) .$$

Again as in Section 4.8, this corollary implies the reciprocity theorem for Ehrhart series, which is equivalent to Ehrhart–Macdonald reciprocity (Theorem 5.2.3). We record it for future reference.

Corollary 5.4.5. *Let $\mathsf{P} \subset \mathbb{R}^d$ be a lattice polytope and $\mathrm{Ehr}_{\mathsf{P}}(n)$ its Ehrhart series. Then*

$$\mathrm{Ehr}_{\mathsf{P}}\left(\tfrac{1}{z}\right) = (-1)^{\dim \mathsf{P}+1}\, \mathrm{Ehr}_{\mathsf{P}^\circ}(z) .$$

We finish this section with an important application of Theorem 3.7.1 (the Brianchon–Gram relation). Applied to integer-point transforms of a rational polytope P, it gives a relation to the integer-point transforms of the tangent cones at the vertices of P, called *Brion's theorem.*

Theorem 5.4.6. *Let $\mathsf{P} \subset \mathbb{R}^d$ be a rational polytope. Then as rational functions,*

$$\sum_{\mathbf{v} \text{ vertex of } \mathsf{P}} \sigma_{\mathrm{T}_{\mathbf{v}}(\mathsf{P})}(\mathbf{z}) = \sigma_{\mathsf{P}}(\mathbf{z}) .$$

Proof. Theorem 3.7.1 gives the identity of indicator functions

$$[P] = \sum_{\varnothing \prec \mathsf{F} \preceq P} (-1)^{\dim \mathsf{F}}\, [\mathrm{T}_{\mathsf{F}}(\mathsf{P})] .$$

Now Exercise 4.34 allows us to sum the above identity over $\mathbb{Z}^d$ in Laurent-series style,

$$\sum_{\mathbf{m} \in \mathbb{Z}^d} [P](\mathbf{m})\, \mathbf{z}^{\mathbf{m}} = \sum_{\varnothing \prec \mathsf{F} \preceq P} (-1)^{\dim \mathsf{F}} \sum_{\mathbf{m} \in \mathbb{Z}^d} [\mathrm{T}_{\mathsf{F}}(\mathsf{P})](\mathbf{m})\, \mathbf{z}^{\mathbf{m}},$$

which is simply

$$\sigma_{\mathsf{P}}(z) = \sum_{\varnothing \prec \mathsf{F} \preceq P} (-1)^{\dim \mathsf{F}} \sigma_{\mathsf{T_F(P)}}(\mathbf{z}).$$

However, $\mathsf{T_F(P)}$ contains a line except when F is a vertex, and so again by Exercise 4.34, the rational generating functions on the right-hand side are zero except for those belonging to vertices of P. □

Half-open decompositions will return to the stage in Section 6.4, where we will directly extract combinatorial information from them.

5.5. h^*-vectors and f-vectors

After having seen two proofs (in Sections 5.1 and 5.3) that the Ehrhart function $\mathrm{ehr}_{\mathsf{P}}(n)$ of a lattice polytope $P \subset \mathbb{R}^d$ is a polynomial, we will now take a look at some fundamental properties of the corresponding Ehrhart series[5]

$$\mathrm{Ehr}_{\mathsf{P}}(z) = \sum_{n \ge 0} \mathrm{ehr}_{\mathsf{P}}(n) z^n.$$

It follows from Proposition 4.1.4 that

$$\mathrm{Ehr}_{\mathsf{P}}(z) = \frac{h_0^* + h_1^* z + \cdots + h_r^* z^r}{(1-z)^{r+1}}$$

for some $h_0^*, \dots, h_r^*$, where $r =: \dim \mathsf{P}$. We call $h_{\mathsf{P}}^*(z) := h_0^* + h_1^* z + \cdots + h_r^* z^r$ the **h^*-polynomial** and $h^*(\mathsf{P}) = (h_0^*, \dots, h_d^*)$ the **h^*-vector** of P, where we set $h_i^* = 0$ for all $i > r$. Note that $h^*(\mathsf{P})$ together with the dimension of P completely determines $\mathrm{Ehr}_{\mathsf{P}}(z)$. The h^*-vector first surfaced in our proof of Theorem 4.6.1, and our goal in this section is to study the salient features of h^*-vectors of lattice polytopes.

If $\triangle \subset \mathbb{R}^d$ is a lattice simplex, then the machinery of Section 4.8 helps us, in a sense, to completely understand $h_\triangle^*(z)$. Consider a half-open lattice simplex $\mathbb{H}_{\mathbf{q}}\triangle \subset \mathbb{R}^d$ of dimension r. Its homogenization

$$\widehat{\mathsf{C}} := \mathrm{hom}(\mathbb{H}_{\mathbf{q}}\triangle) \subset \mathbb{R}^{d+1}$$

is a half-open simplicial cone with integer-point transform

$$\sigma_{\widehat{\mathsf{C}}}(z_1, \dots, z_{d+1}) = \frac{\sigma_{\widehat{\square}}(z_1, \dots, z_{d+1})}{(1 - \mathbf{z}^{\mathbf{v}_1} z_{d+1}) \cdots (1 - \mathbf{z}^{\mathbf{v}_{r+1}} z_{d+1})},$$

where $\mathbf{v}_1, \dots, \mathbf{v}_{r+1}$ are the vertices of $\triangle$, and $\widehat{\square}$ is the fundamental parallelepiped of $\widehat{\mathsf{C}}$, defined in (4.8.3). As in Section 4.6, we recover $h_{\mathbb{H}_{\mathbf{q}}\triangle}^*(z) = \sigma_{\widehat{\square}}(1, \dots, 1, z)$, more explicitly,

$$h_i^*(\mathbb{H}_{\mathbf{q}}\triangle) = \left| \left\{ (\mathbf{x}, x_{d+1}) \in \widehat{\square} \cap \mathbb{Z}^{d+1} : x_{d+1} = i \right\} \right|. \tag{5.5.1}$$

[5]Our original definition (4.6.2) defined this series to have constant term 1; this is consistent with Theorem 5.1.8.

From the definition of the integer-point transform and Theorem 4.8.1, we obtain almost instantly the following nontrivial facts about h^*-vectors and, consequently, about Ehrhart polynomials.

Corollary 5.5.1. *Let $\mathsf{P} \subset \mathbb{R}^d$ be a lattice polytope. Then $h_i^*(\mathsf{P})$ is a nonnegative integer for all $0 \le i \le d$. Moreover, $h_0^*(\mathsf{P}) = 1$ and if we set*

$$m^\circ \; := \; \max\{i \, : \, h_i^*(\mathsf{P}) > 0\}\,,$$

then $n = \dim \mathsf{P} + 1 - m^\circ$ is the smallest dilation factor such that $n\mathsf{P}^\circ$ contains a lattice point.

Proof. Choose a dissection $\mathsf{P} = \triangle_1 \cup \cdots \cup \triangle_k$ into lattice simplices. (This exists by Corollary 5.1.6.) Corollary 5.3.5 gives a half-open decomposition $\mathsf{P} = \mathbb{H}_\mathbf{q}\triangle_1 \uplus \cdots \uplus \mathbb{H}_\mathbf{q}\triangle_k$ for any generic $\mathbf{q} \in \mathsf{P}^\circ$. Since the Ehrhart functions are additive in this situation and all simplices have the same dimension, we conclude that

$$h_i^*(\mathsf{P}) \;=\; h_i^*(\mathbb{H}_\mathbf{q}\triangle_1) + \cdots + h_i^*(\mathbb{H}_\mathbf{q}\triangle_k) \tag{5.5.2}$$

for all $0 \le i \le d$. The first claim now follows from (5.5.1).

For the second part, note that a lattice point $\mathbf{p}$ is in P or P° if and only if it is in $\mathbb{H}_\mathbf{q}\triangle_j$ or $\mathbb{H}^\mathbf{q}\triangle_j$ for some j, respectively. Thus, it suffices to treat half-open simplices and we leave this case to Exercise 5.27. □

Whereas the second part of Corollary 5.5.1 is in some sense a statement about rational generating functions, the nontrivial part is the nonnegativity of the h^*-vector. This, however, was pretty easy with the help of half-open decompositions—we challenge you to try to prove Corollary 5.5.1 without it. In fact, a stronger result holds which we can achieve by pushing the underlying ideas further. The following result subsumes Corollary 5.5.1 by taking $\mathsf{P} = \varnothing$.

Theorem 5.5.2. *Let $\mathsf{P}, \mathsf{Q} \subset \mathbb{R}^d$ be lattice polytopes with $\mathsf{P} \subseteq \mathsf{Q}$. Then*

$$h_i^*(\mathsf{P}) \;\le\; h_i^*(Q)$$

for all $i = 0, \ldots, d$.

Proof. We first assume that $\dim \mathsf{P} = \dim \mathsf{Q} = r$. Repeatedly using Theorem 5.3.3, we can find a dissection into lattice simplices,

$$\mathsf{Q} \;=\; \triangle_1 \cup \triangle_2 \cup \cdots \cup \triangle_n\,,$$

such that for some $m \le n$

$$\mathsf{P} \;=\; \triangle_{m+1} \cup \triangle_2 \cup \cdots \cup \triangle_n$$

is a dissection of P. Pick a point $\mathbf{q} \in \mathsf{P}^\circ$ generic relative to all $\triangle_i$. Then Lemma 5.3.4 yields

$$\mathsf{Q} \setminus \mathsf{P} \;=\; \mathbb{H}_{\mathbf{q}}\triangle_1 \uplus \mathbb{H}_{\mathbf{q}}\triangle_2 \uplus \cdots \uplus \mathbb{H}_{\mathbf{q}}\triangle_m$$

as a disjoint union of half-open lattice simplices of dimension r, and following (5.5.2),

$$h^*(\mathsf{Q}) - h^*(\mathsf{P}) \;=\; \sum_{j=1}^{m} h^*(\mathbb{H}_{\mathbf{q}}\triangle_j)\,.$$

Since we just argued that the h^*-vector is nonnegative for any half-open simplex, we are done for the case $\dim \mathsf{P} = \dim \mathsf{Q}$.

If $\dim \mathsf{Q} - \dim \mathsf{P} =: s > 0$, set $\mathsf{P}^0 := \mathsf{P}$, and for $0 \le i < s$ we recursively define the pyramids

$$\mathsf{P}^{i+1} \;:=\; \mathbf{p}^{i+1} * \mathsf{P}^i \qquad \text{for some} \qquad \mathbf{p}^{i+1} \in \big(\mathsf{Q} \setminus \operatorname{aff}(\mathsf{P}^i)\big) \cap \mathbb{Z}^d.$$

This gives a sequence of nested lattice polytopes

$$\mathsf{P} \;=\; \mathsf{P}^0 \;\subset\; \mathsf{P}^1 \;\subset\; \cdots \;\subset\; \mathsf{P}^s \;\subseteq\; \mathsf{Q}$$

with $\dim \mathsf{P}^{i+1} = \dim \mathsf{P}^i + 1$ for all $i = 0, \dots, s-1$.

By construction, $\dim \mathsf{P}^s = \dim \mathsf{Q}$ (and so the step from P^s to Q is covered by the first half of our proof) and hence it suffices to show the claim for the case that Q is a pyramid over P with apex $\mathbf{v}$. Pick a dissection of $\mathsf{P} = \triangle_1 \cup \cdots \cup \triangle_k$ into lattice simplices which yields, by Exercise 5.30, a dissection of $\mathsf{Q} = \triangle_1' \cup \cdots \cup \triangle_k'$, where $\triangle_i' = \mathbf{v} * \triangle_i$. For a point $\mathbf{q}$ generic relative to $\triangle_1', \dots, \triangle_k'$, we get half-open decompositions of Q as well as of P, and we are left to show that

$$h_i^*(\mathbb{H}_{\mathbf{q}}\triangle_j) \;\le\; h_i^*(\mathbb{H}_{\mathbf{q}}\triangle_j')$$

for all i and j. However, if we denote by $\widehat{\square}_j$ and $\widehat{\square}_j'$ the fundamental parallelepiped of the half-open simplicial cones $\hom(\mathbb{H}_{\mathbf{q}}\triangle_j)$ and $\hom(\mathbb{H}_{\mathbf{q}}\triangle_j')$, respectively, then the last claim follows from (5.5.1) and Exercise 5.31. □

In general, there is no simple interpretation for the numbers $h_i^*(\mathsf{P})$. However, in the rather special situation that the lattice polytope P admits a **unimodular dissection**, that is, a dissection into unimodular simplices, these numbers are quite nice.

Theorem 5.5.3. *Let $\mathsf{P} \subset \mathbb{R}^d$ be a d-dimensional lattice polytope. If P admits a unimodular dissection, then $h_i^*(\mathsf{P})$ is the number of half-open simplices with i facets removed.*

It is wise to pause and let this result sink in. The h^*-vector of P does not depend on any particular dissection—it can be computed just in terms of generating functions. Thus Theorem 5.5.3 implies that if a single unimodular dissection exists, then the number of simplices that will have i facets missing is predetermined for *any* choice of a unimodular dissection and for *any* choice of a generic point. The key lies in (5.5.2), and Theorem 5.5.3 follows directly from the following lemma; see Exercise 5.28.

Lemma 5.5.4. *Let $\triangle \subset \mathbb{R}^d$ be a unimodular d-simplex and $\mathbf{q}$ a point generic relative to $\triangle$. Then*

$$h^*_{\mathbb{H}_{\mathbf{q}}\triangle}(z) \;=\; z^r,$$

where r is the number of facets that are missing in $\mathbb{H}_{\mathbf{q}}\triangle$.

Going a step further, assume that P can be triangulated with unimodular simplices. That is, there is a polyhedral complex $\mathcal{T}$ consisting of unimodular simplices of varying dimensions such that $\mathsf{P} = |\mathcal{T}|$; see Exercise 5.40 for the relation between unimodular dissection and unimodular triangulations. We denote by $f_k(\mathcal{T})$ the number of simplices of dimension k for $-1 \le k \le \dim \mathcal{T}$, and we similarly define $f_k(\mathbb{H}_{\mathbf{q}}\triangle)$ for a half-open simplex. We have to be careful with $f_{-1}(\mathbb{H}_{\mathbf{q}}\triangle)$: we set $f_{-1}(\mathbb{H}_{\mathbf{q}}\triangle) = 0$ if $\mathbb{H}_{\mathbf{q}}\triangle$ is properly half-open and $= 1$ if $\mathbb{H}_{\mathbf{q}}\triangle = \triangle$. Now, if $\triangle_1, \dots, \triangle_m \in \mathcal{T}$ are the inclusion-maximal cells, then

$$\mathsf{P} \;=\; \mathbb{H}_{\mathbf{q}}\triangle_1 \uplus \cdots \uplus \mathbb{H}_{\mathbf{q}}\triangle_m$$

for any generic point $\mathbf{q} \in \mathsf{P}^\circ$. Since each $(k-1)$-dimensional cell in $\mathcal{T}$ is the face of exactly one $\triangle_j$, we infer that

$$f_{k-1}(\mathcal{T}) \;=\; f_{k-1}(\mathbb{H}_{\mathbf{q}}\triangle_1) + \cdots + f_{k-1}(\mathbb{H}_{\mathbf{q}}\triangle_m)\,, \tag{5.5.3}$$

for all $k > 0$. In the next result we use the natural convention that $\binom{n}{-s} = 0$ for $s > 0$.

Proposition 5.5.5. *Let $\mathbb{H}_{\mathbf{q}}\triangle$ be a d-dimensional half-open simplex with $0 \le r \le d$ facets removed. Then*

$$f_{k-1}(\mathbb{H}_{\mathbf{q}}\triangle) \;=\; \binom{d+1-r}{k-r}$$

for all $k \ge 0$.

Proof. Let $\mathsf{F}_1, \dots, \mathsf{F}_r$ be the missing facets and let $\mathbf{v}_i$ be the unique vertex of $\triangle$ with $\mathbf{v}_i \notin \mathsf{F}_i$ for $1 \le i \le r$. A face $\mathsf{F} \preceq \triangle$ is present in $\mathbb{H}_{\mathbf{q}}\triangle = \triangle \setminus \bigcup_i \mathsf{F}_i$ if and only if $\{\mathbf{v}_1, \dots, \mathbf{v}_r\} \subset \mathsf{F}$; see Exercise 5.35. Thus, the $(k-1)$-faces are in bijection with the subsets

$$U \;\subseteq\; \operatorname{vert}(\triangle) \setminus \{\mathbf{v}_1, \dots, \mathbf{v}_r\} \qquad \text{with} \qquad |U| = k - r\,. \qquad \square$$

The following quite amazing result follows now directly from (5.5.3), Theorem 5.5.3, and Proposition 5.5.5.

Theorem 5.5.6. *Let* P *be a* d*-dimensional lattice polytope. If* $\mathcal{T}$ *is a unimodular triangulation of* P*, then*

$$f_{k-1}(\mathcal{T}) \;=\; \sum_{r=0}^{k} \binom{d+1-r}{k-r} h_r^*(\mathsf{P})$$

for all $0 \le k \le d+1$.

Theorem 5.5.6 implies that the h^*-vector is determined by the f-vector of any unimodular triangulation and vice versa. To see this, we introduce the **f-polynomial** of a triangulation $\mathcal{T}$:

$$f_{\mathcal{T}}(z) \;:=\; f_d(\mathcal{T}) + f_{d-1}(\mathcal{T})\, z + \cdots + f_{-1}(\mathcal{T})\, z^{d+1}.$$

It follows from the formal reciprocity of generating functions (Theorem 4.1.6) and Ehrhart–Macdonald reciprocity (Theorem 5.2.3) that

$$h_{\mathsf{P}^\circ}^*(z) \;=\; h_d^*(\mathsf{P})\, z + h_{d-1}^*(\mathsf{P})\, z^2 + \cdots + h_0^*(\mathsf{P})\, z^{d+1}.$$

As in Chapter 4, we emphasize that we view the polynomials $f_{\mathcal{T}}(z)$ and $h_{\mathsf{P}^\circ}^*(z)$ as generating functions: they enumerate certain combinatorial quantities. Theorem 5.5.6 states that these quantities are related and the following result, whose formal verification you are supposed to do in Exercise 5.36, expresses this relation in the most elegant way possible.

Corollary 5.5.7. *Let* P *be a* d*-dimensional lattice polytope and* $\mathcal{T}$ *a unimodular triangulation of* P*. Then, as generating functions,*

$$f_{\mathcal{T}}(z) \;=\; h_{\mathsf{P}^\circ}^*(z+1).$$

In particular, $h_{\mathsf{P}^\circ}^*(z) = f_{\mathcal{T}}(z-1)$ which gives the explicit formula

$$h_r^*(\mathsf{P}) \;=\; \sum_{k=0}^{r} (-1)^{r-k} \binom{d+1-k}{d+1-r} f_{k-1}(\mathcal{T}). \tag{5.5.4}$$

At this point you might (correctly) wonder if there is a similar story involving $h^*(\mathsf{P})$ instead of $h^*(\mathsf{P}^\circ)$. The key to this rests in reciprocity. Let $f_{k-1}^{\mathrm{int}}(\mathcal{T})$ be the number of $(k-1)$-dimensional cells of $\mathcal{T}$ that do not lie in the boundary of P. It follows from Corollary 5.3.5 that for the very same point $\mathbf{q} \in \mathsf{P}^\circ$ used in this corollary that

$$\mathsf{P}^\circ \;=\; \mathbb{H}^{\mathbf{q}}\triangle_1 \uplus \mathbb{H}^{\mathbf{q}}\triangle_2 \uplus \cdots \uplus \mathbb{H}^{\mathbf{q}}\triangle_m.$$

Definitions (5.3.1) and (5.3.2) show that if $\mathbb{H}_{\mathbf{q}}\triangle$ has r missing facets, $\mathbb{H}^{\mathbf{q}}\triangle$ is missing $d+1-r$ facets. The same argument that led to Theorem 5.5.6 gives

$$f_{k-1}^{\mathrm{int}}(\mathcal{T}) \;=\; \sum_{r=1}^{d+1} \binom{d+1-r}{k-r} h_{d+1-r}^*(\mathsf{P}),$$

and thus we conclude:

Proposition 5.5.8. *Let* P *be a d-dimensional lattice polytope and* $\mathcal{T}$ *a unimodular triangulation of* P*. Then, as generating functions,*

$$f_{\mathcal{T}}^{\text{int}}(z) = \sum_{k=0}^{d+1} f_{k-1}^{\text{int}}(\mathcal{T})\, z^{d+1-k} = h_{\mathsf{P}}^*(z+1)\,.$$

As a sneak preview for the next section, we note that Corollary 5.5.7 and Proposition 5.5.8 together show that the number of faces and the number of interior faces of a unimodular triangulation are not independent.

Corollary 5.5.9. *Let* P *be a d-dimensional lattice polytope and* $\mathcal{T}$ *a unimodular triangulation. Then*

$$f_{k-1}^{\text{int}}(\mathcal{T}) = \sum_{l=k}^{d+1} \binom{l}{k} (-1)^{d+1-k} f_{l-1}(\mathcal{T})\,.$$

Proof. Ehrhart–Macdonald reciprocity (Theorem 5.2.3) implies $h_{\mathsf{P}}^*(z) = z^{d+1} h_{\mathsf{P}^\circ}^*(\frac{1}{z})$, and by Corollary 5.5.7

$$h_{\mathsf{P}}^*(z) = z^{d+1} f_{\mathcal{T}}\left(\frac{1-z}{z}\right) = \sum_{k=0}^{d+1} f_{k-1}(\mathcal{T})(1-z)^{d+1-k} z^k.$$

Proposition 5.5.8 now gives

$$f_{\mathcal{T}}^{\text{int}}(z) = h_{\mathsf{P}}^*(z+1) = \sum_{k=0}^{d+1} f_{k-1}(\mathcal{T})(-z)^{d+1-k}(z+1)^k,$$

and the result follows by inspecting coefficients. □

5.6. Self-reciprocal Complexes and Dehn–Sommerville Revisited

Let $\mathcal{K}$ be a complex of lattice polytopes in $\mathbb{R}^d$. We define the Ehrhart function of $\mathcal{K}$ by

$$\operatorname{ehr}_{\mathcal{K}}(n) := \left| n|\mathcal{K}| \cap \mathbb{Z}^d \right|$$

for all integers $n \geq 1$. Corollary 5.1.3 then yields that

$$\operatorname{ehr}_{\mathcal{K}}(n) = \sum_{\mathsf{F} \in \mathcal{K}} -\mu_{\widehat{\mathcal{K}}}(\mathsf{F}, \hat{1}) \operatorname{ehr}_{\mathsf{F}}(n)\,, \tag{5.6.1}$$

where $\widehat{\mathcal{K}}$ is the partially ordered set $\mathcal{K}$ with a maximum $\hat{1}$ adjoined. In fact, if we set $\dim \mathcal{K} := \max\{\dim \mathsf{F} : \mathsf{F} \in \mathcal{K}\}$, then Corollary 5.1.2 immediately implies the following.

Corollary 5.6.1. *If* $\mathcal{K}$ *is a complex of lattice polytopes, then* $\operatorname{ehr}_{\mathcal{K}}(n)$ *agrees with a polynomial of degree* $\dim \mathcal{K}$ *for all positive integers* n.

In this section we want to investigate a special class of polytopal complexes, very much in the spirit of this book: we call a complex $\mathcal{K}$ **self-reciprocal** if for all $n > 0$

$$(5.6.2) \qquad (-1)^{\dim \mathcal{K}} \operatorname{ehr}_{\mathcal{K}}(-n) \;=\; \operatorname{ehr}_{\mathcal{K}}(n)\,.$$

The right-hand side of (5.6.1) also suggests a value for $\operatorname{ehr}_{\mathcal{K}}(n)$ at $n = 0$, namely, the Euler characteristic $\chi(\mathcal{K})$. We will, however, not follow this suggestion and instead decree that

$$\operatorname{ehr}_{\mathcal{K}}(0) \;:=\; 1\,.$$

Unless $\chi(\mathcal{K}) = 1$, this is a rather strange (if not disturbing) convention and in light of Corollary 5.6.1, the function $\operatorname{ehr}_{\mathcal{K}}(n)$ now agrees with a polynomial of degree $\dim \mathcal{K}$ for all n *except for* $n = 0$. (Theorem 5.6.2 below will vindicate our choice.)

We extend the definition of h^*-vectors from Section 5.5 to complexes of lattice polytopes via

$$\operatorname{Ehr}_{\mathcal{K}}(z) \;:=\; 1 + \sum_{n \ge 1} \operatorname{ehr}_{\mathcal{K}}(n) z^n \;=\; \frac{h_0^* + h_1^* z + \cdots + h_{d+1}^* z^{d+1}}{(1-z)^{d+1}}\,,$$

where $d = \dim \mathcal{K}$, and we set $h^*(\mathcal{K}) = (h_0^*, \ldots, h_{d+1}^*)$.

Theorem 5.6.2. *Let $\mathcal{K}$ be a d-dimensional complex of lattice polytopes with $\chi(\mathcal{K}) = 1 - (-1)^{d+1}$. Then $\mathcal{K}$ is self-reciprocal if and only if*

$$(5.6.3) \qquad h_{d+1-i}^*(\mathcal{K}) \;=\; h_i^*(\mathcal{K})$$

for all $0 \le i \le d+1$.

We remark that the condition $\chi(\mathcal{K}) = 1 - (-1)^{d+1}$ is not as esoteric as it might seem; see (5.6.5) and (5.6.6) below. At any rate, in Exercise 5.41 you are asked to verify that $h_{d+1}^*(\mathcal{K}) = (-1)^{d+1}(1 - \chi(\mathcal{K}))$; with the additional property $\chi(\mathcal{K}) = 1 - (-1)^{d+1}$, we obtain $h_{d+1}^*(\mathcal{K}) = 1$, which is certainly necessary for (5.6.3) to hold.

Proof. If (5.6.3) holds, then $(-1)^{d+1} \operatorname{Ehr}_{\mathcal{K}}(\frac{1}{z}) = \operatorname{Ehr}_{\mathcal{K}}(z)$ and Corollary 4.1.8 says that (5.6.2) holds for all $n \ge 1$. Conversely, it follows again from Corollary 4.1.8 that

$$(-1)^{d+1} \operatorname{Ehr}_{\mathcal{K}}(\tfrac{1}{z}) \;=\; (-1)^{d+1}(1 - \chi(\mathcal{K})) + (-1)^d \sum_{n \ge 1} \operatorname{ehr}_{\mathcal{K}}(-n) z^n \;=\; \operatorname{Ehr}_{\mathcal{K}}(z)\,,$$

where we used (5.6.2) and that $\chi(\mathcal{K}) = 1 - (-1)^{d+1}$. Multiplying both sides by $(1-z)^{d+1}$ and comparing coefficients proves the claim. $\square$

If $\mathcal{K}$ is the subdivision of a lattice polytope P then, since $\chi(P) = 1$, we have $\mathrm{ehr}_{\mathcal{K}}(n) = \mathrm{ehr}_{\mathsf{P}}(n)$ for all $n \geq 0$ and we are consistent with the definition of Ehrhart functions for polytopes. However, subdivisions of polytopes can never be self-reciprocal: if $\mathcal{K}$ is the subdivision of a lattice polytope P, then, by Ehrhart–Macdonald reciprocity (Theorem 5.2.3),

$$(-1)^{\dim \mathcal{K}} \mathrm{ehr}_{\mathcal{K}}(-n) \;=\; \mathrm{ehr}_{\mathsf{P}^\circ}(n) \;<\; \mathrm{ehr}_{\mathsf{P}}(n) \qquad \text{for all } n > 0\,.$$

On the other hand, the boundaries of polytopes are good examples of self-reciprocal complexes. The **boundary complex** $\Phi(\partial\mathsf{P})$ of a polytope P is the collection of all proper faces of P.

Proposition 5.6.3. *The boundary complex of any lattice polytope is self-reciprocal.*

Proof. Let P be a d-dimensional lattice polytope. Then

$$(5.6.4)\quad \mathrm{ehr}_{\Phi(\partial\mathsf{P})}(n) \;=\; \mathrm{ehr}_{\mathsf{P}}(n) - \mathrm{ehr}_{\mathsf{P}^\circ}(n) \;=\; \mathrm{ehr}_{\mathsf{P}}(n) - (-1)^d\, \mathrm{ehr}_{\mathsf{P}}(-n)\,,$$

by Ehrhart–Macdonald reciprocity (Theorem 5.2.3). The boundary complex is of dimension $d-1$, and for $n > 0$ we compute

$$\begin{aligned}(-1)^{d-1} \mathrm{ehr}_{\Phi(\partial\mathsf{P})}(-n) \;&=\; (-1)^{d-1} \mathrm{ehr}_{\mathsf{P}}(-n) - (-1)^{2d-1} \mathrm{ehr}_{\mathsf{P}}(n) \\ &=\; \mathrm{ehr}_{\mathsf{P}}(n) - (-1)^d \mathrm{ehr}_{\mathsf{P}}(-n) \\ &=\; \mathrm{ehr}_{\Phi(\partial\mathsf{P})}(n)\,. \end{aligned}$$

□

From $\mathsf{P} = \partial\mathsf{P} \uplus \mathsf{P}^\circ$, we compute with the help of Theorem 3.4.1 and Corollary 3.4.6

$$(5.6.5)\qquad \chi(\partial\mathsf{P}) \;=\; \chi(\mathsf{P}) - \chi(\mathsf{P}^\circ) \;=\; 1 - (-1)^{\dim \mathsf{P}},$$

and Theorem 5.6.2 implies that $h_i^*(\Phi(\partial\mathsf{P})) = h_{d-i}^*(\Phi(\partial\mathsf{P}))$ for all $0 \leq i \leq d$. In this particular case, the symmetry of the h^*-vector can also be seen quite directly:

$$\begin{aligned}\mathrm{Ehr}_{\Phi(\partial\mathsf{P})}(z) \;&=\; \sum_{n\geq 0} \mathrm{ehr}_{\Phi(\partial\mathsf{P})}(n) z^n \;=\; \mathrm{Ehr}_{\mathsf{P}}(z) - \mathrm{Ehr}_{\mathsf{P}^\circ}(z) \\ &=\; \mathrm{Ehr}_{\mathsf{P}}(z) - (-1)^{d+1}\, \mathrm{Ehr}_{\mathsf{P}}(\tfrac{1}{z}) \;=\; \frac{h_{\mathsf{P}}^*(z) - z^{d+1} h_{\mathsf{P}}^*(\frac{1}{z})}{(1-z)^{d+1}}\,,\end{aligned}$$

where the third equation follows from Corollary 5.4.5. We note that

$$\tilde{h}(z) \;:=\; h_{\mathsf{P}}^*(z) - z^{d+1} h_{\mathsf{P}}^*(\tfrac{1}{z})$$

has a root at $z = 1$ and hence $\tilde{h}(z) = (1-z)\, h_{\Phi(\partial\mathsf{P})}^*(z)$. Moreover, $z^{d+1}\tilde{h}(\frac{1}{z}) = -\tilde{h}(z)$ which implies $z^d\, h_{\Phi(\partial\mathsf{P})}^*(\frac{1}{z}) = h_{\Phi(\partial\mathsf{P})}^*(z)$.

There are much more general classes of self-reciprocal complexes and (at least) one particular class that has a strong combinatorial flavor. We recall from Chapter 2 that a graded poset Π with minimum and maximum is Eulerian if $\mu_\Pi(a,b) = (-1)^{l(a,b)}$ for any $a \preceq b$, where $l(a,b)$ is the length of a maximal chain in the interval $[a,b]_\Pi$. Viewed as a poset, a polyhedral complex $\mathcal{K}$ is graded if it is **pure**, that is, if every inclusion-maximal cell is of the same dimension $\dim \mathcal{K}$. We will say that $\mathcal{K}$ is an **Eulerian complex** if $\widehat{\mathcal{K}} := \mathcal{K} \cup \{\hat{1}\}$ is an Eulerian poset. For example $\Phi(\partial \mathsf{P})$ is an Eulerian complex, for any polytope P.

Theorem 5.6.4. *If $\mathcal{K}$ is an Eulerian complex of lattice polytopes, then $\mathcal{K}$ is self-reciprocal.*

Proof. The length of a maximal chain starting in $\mathsf{F} \in \mathcal{K}$ and ending in $\hat{1}$ is $\dim \mathcal{K} + 1 - \dim \mathsf{F}$. Hence, if $\widehat{\mathcal{K}}$ is Eulerian, then

$$\mathrm{ehr}_{\mathcal{K}}(n) \;=\; \sum_{\mathsf{F}\in\mathcal{K}} -\mu_{\widehat{\mathcal{K}}}(\mathsf{F},\hat{1})\,\mathrm{ehr}_{\mathsf{F}}(n) \;=\; \sum_{\mathsf{F}\in\mathcal{K}} (-1)^{\dim\mathcal{K}-\dim\mathsf{F}}\,\mathrm{ehr}_{\mathsf{F}}(n)\,,$$

and for $n > 0$ we thus compute with Ehrhart–Macdonald reciprocity (Theorem 5.2.3)

$$\begin{aligned}(-1)^{\dim\mathcal{K}}\,\mathrm{ehr}_{\mathcal{K}}(-n) \;&=\; \sum_{\mathsf{F}\in\mathcal{K}} (-1)^{\dim\mathsf{F}}\,\mathrm{ehr}_{\mathsf{F}}(-n) \;=\; \sum_{\mathsf{F}\in\mathcal{K}} \mathrm{ehr}_{\mathsf{F}^\circ}(n)\\ &=\; \mathrm{ehr}_{\mathcal{K}}(n)\,.\end{aligned}$$ □

Not all self-reciprocal complexes are necessarily Eulerian. For example, the disjoint union of two self-reciprocal complexes of the same dimension is self-reciprocal but never Eulerian; see Exercise 5.42. However, if $\mathcal{K}$ is a d-dimensional Eulerian complex, then Theorem 2.4.6 implies that

$$(5.6.6)\qquad (-1)^{d+1} \;=\; \mu_{\widehat{\mathcal{K}}}(\hat{0},\hat{1}) \;=\; 1 - \sum_{F\in\mathcal{K}} \mu_{\widehat{\mathcal{K}}}(\hat{0},F) \;=\; 1 - \chi(\mathcal{K})$$

and so Theorem 5.6.2 applies. If $\mathcal{K}$ is a polyhedral complex of unimodular simplices, then, thinking back to Corollary 5.5.9, we could hope that the symmetry of the h^*-vector implies conditions on the face numbers $f_i(\mathcal{K})$. It turns out that we can do this in a more general way.

Let V be a finite set. We recall from Section 4.9 that an abstract simplicial complex Γ is a nonempty collection of finite subsets of V with the property

$$(5.6.7)\qquad \sigma \in \Gamma,\ \sigma' \subseteq \sigma \quad\Longrightarrow\quad \sigma' \in \Gamma.$$

Back then we agreed that if P is a simplicial polytope, then the vertex sets of proper faces of P form an abstract simplicial complex Γ_{P}. More generally, if $\mathcal{K}$ is a polyhedral complex consisting of simplices—that is, in the language of Section 5.1, $\mathcal{K}$ is a geometric simplicial complex—,then

$$\Gamma_{\mathcal{K}} \;:=\; \{\mathrm{vert}(\mathsf{F}) \,:\, \mathsf{F}\in\mathcal{K}\}$$

is an abstract simplicial complex. The geometric information is lost in the passage from $\mathcal{K}$ to $\Gamma_{\mathcal{K}}$ but all *combinatorial* information is retained; in particular, $\mathcal{K}$ and $\Gamma_{\mathcal{K}}$ are isomorphic as posets. Every k-simplex has $k+1$ vertices and $\dim \sigma = |\sigma| - 1$ for $\sigma \in \Gamma$ as well as $\dim \Gamma = \max\{\dim \sigma : \sigma \in \Gamma\}$ is consistent with the geometry.

Conversely, for an abstract simplicial complex Γ on the ground set V, we can construct the following realization as a geometric simplicial complex. Let $\{\mathbf{e}_v : v \in V\}$ be the standard basis of the vector space $\mathbb{R}^V$ and consider the unimodular $(|V| - 1)$-dimensional simplex

$$\triangle \ := \ \operatorname{conv}\{\mathbf{e}_v : v \in V\} \ = \ \left\{\mathbf{p} \in \mathbb{R}^V_{\geq 0} : \sum_{v \in V} p_v = 1\right\}.$$

The faces of $\triangle$ are given by $\triangle[A] := \operatorname{conv}\{\mathbf{e}_v : v \in A\}$ for all $\varnothing \neq A \subseteq V$. Thus to Γ we can associate the geometric simplicial complex

$$\mathcal{R}[\Gamma] \ := \ \{\triangle[\sigma] : \sigma \in \Gamma\} \ \subseteq \ \Phi(\triangle),$$

the **canonical realization** of Γ. Here are two simple but pivotal facts about $\mathcal{R}[\Gamma]$.

Proposition 5.6.5. *Let Γ be an abstract simplicial complex. The canonical realization $\mathcal{R}[\Gamma]$ is a complex of unimodular simplices and $\Gamma \cong \mathcal{R}[\Gamma]$ as posets.*

For $0 \leq i \leq \dim \Gamma$, set $f_i(\Gamma)$ to be the number of faces $\sigma \in \Gamma$ with $\dim \sigma = |\sigma| - 1 = i$. The empty set $\varnothing$ is always a face of Γ of dimension -1 which we record by $f_{-1}(\Gamma) = f_{-1}(\mathcal{R}[\Gamma]) = 1$. Since $\mathcal{R}[\Gamma]$ is a complex of unimodular simplices, we can compute its Ehrhart series as

$$\begin{aligned}\operatorname{Ehr}_{\mathcal{R}[\Gamma]}(z) \ &= \ \sum_{S \in \Gamma} \operatorname{Ehr}_{S^\circ}(z) \ = \ \sum_{S \in \Gamma} \frac{z^{|S|}}{(1-z)^{|S|}} \\ &= \ \frac{\sum_{i=0}^{d+1} f_{i-1}(\Gamma)\, z^i\, (1-z)^{d+1-i}}{(1-z)^{d+1}}. \qquad (5.6.8)\end{aligned}$$

(This mirrors some of our computations in Section 5.5.) Thus the h^*-vector of $\mathcal{R}[\Gamma]$ depends only on the f-vector of Γ, and we define the **h-vector** $h(\Gamma) = (h_0, \ldots, h_{d+1})$ of a d-dimensional abstract simplicial complex Γ through $h_i(\Gamma) := h_i^*(\mathcal{R}[\Gamma])$ for $0 \leq i \leq d+1$. So with (5.6.8),

$$\sum_{i=0}^{d+1} f_{i-1}(\Gamma)\, z^i\, (1-z)^{d+1-i} \ = \ \sum_{i=0}^{d+1} h_i(\Gamma)\, z^i. \qquad (5.6.9)$$

At this point, we take a short break to repeat our computation in (5.6.8) for the (geometric) simplicial complex given by a unimodular triangulation of a lattice polytope P. The computation is literally the same, except that on the left-hand side of (5.6.8), $\mathrm{Ehr}_{\mathcal{R}[\Gamma]}(z)$ needs to be replaced by $\mathrm{Ehr}_{\mathsf{P}}(z)$. The numerator polynomial now becomes the h^*-vector of P from Section 5.5, and (5.6.9) implies:

Corollary 5.6.6. *Let* P *be a* d*-dimensional lattice polytope with a unimodular triangulation* $\mathcal{T}$. *Then*

$$\mathrm{Ehr}_{\mathsf{P}}(z) \;=\; \frac{\sum_{i=0}^{d} h_i(\mathcal{T})\, z^i}{(1-z)^{d+1}}\,.$$

In particular, the h^**-vector of* P *depends only on the combinatorics of* $\mathcal{T}$.

The h-vector is *like* the h^*-vector but it is not exactly the same thing. For starters—unlike in the geometric scenario of a unimodular triangulation—it is not true that $h_i(\Gamma) \geq 0$ for all simplicial complexes Γ; check out Exercise 5.38. In particular, $h^*(\mathcal{K})$ and $h(\Gamma_{\mathcal{K}})$ are typically different.

Next we reap some fruit from Theorem 5.6.2.

Corollary 5.6.7. *Let* Γ *be a* d*-dimensional Eulerian simplicial complex. Then, for* $0 \leq i \leq d+1$,

$$h_i(\Gamma) \;=\; h_{d+1-i}(\Gamma)\,.$$

In particular, for $0 \leq j \leq d+1$,

$$f_{j-1}(\Gamma) \;=\; \sum_{k=j}^{d+1} \binom{k}{j} (-1)^{d+1-k} f_{k-1}(\Gamma)\,.$$

Proof. The first claim is exactly Theorem 5.6.2 applied to the canonical realization $\mathcal{R}[\Gamma]$. For the second claim, we replace z by $\frac{1}{z}$ in (5.6.9) and multiply the resulting equation by z^{d+1}. This gives

$$\sum_{i=0}^{d+1} f_{i-1}(\Gamma)(z-1)^{d+1-i} \;=\; \sum_{i=0}^{d+1} h_{d+1-i}(\Gamma)\, z^i.$$

But $h_i(\Gamma) = h_{d+1-i}(\Gamma)$, and so with (5.6.9) we conclude

$$\sum_{i=0}^{d+1} f_{i-1}(\Gamma)(z-1)^{d+1-i} \;=\; \sum_{i=0}^{d+1} f_{i-1}(\Gamma) z^i (1-z)^{d+1-i}.$$

Now substituting $z+1$ for z and comparing coefficients yields the second claim. □

Applying Corollary 5.6.7 to the case that Γ is the $(d-1)$-dimensional boundary complex of a d-dimensional simplicial polytope P yields exactly the Dehn–Sommerville relations for $f(\mathsf{P})$ obtained in Theorem 3.5.5. We could now employ the same reasoning as in Section 3.5 to the zeta polynomials of Eulerian simplicial complexes; however, we will forgo this in favor of a more geometric perspective on zeta polynomials of abstract simplicial complexes. For a finite set V, let

$$\square_V \ := \ \left\{ \mathbf{q} \in \mathbb{R}^V \ : 0 \le q_v < 1 \text{ for all } v \in V \right\},$$

the standard half-open unit cube in $\mathbb{R}^V$. For any $A \subseteq V$, we can identify $\square_A$ as a (half-open) face of $\square_V$, and for $B \subseteq V$ we have $\square_A \cap \square_B = \square_{A \cap B}$. It follows that for an abstract simplicial complex Γ on vertices V, the collection of half-open parallelepipeds

$$\square_\Gamma \ := \ \{\square_S \ : \ S \in \Gamma\}$$

satisfies the intersection property of Section 5.1. As before, we consider the Ehrhart function $\operatorname{ehr}_{\square_\Gamma}(n)$ of the support $|\square_\Gamma| := \bigcup_{S \in \Gamma} \square_S$. In sync with Lemma 3.3.8,

$$\square_A \ = \ \biguplus_{B \subseteq A} \square_B^\circ \,,$$

and so

$$\operatorname{ehr}_{\square_\Gamma}(n) \ = \ \sum_{A \in \Gamma} \operatorname{ehr}_{\square_A^\circ}(n) \ = \ \sum_{A \in \Gamma} (n-1)^{|A|} \ = \ \sum_{i=0}^{\dim \Gamma + 1} f_{i-1}(\Gamma)(n-1)^i \,.$$

A do-it-yourself example is given in Exercise 5.39.

Now, if Γ is an Eulerian simplicial complex of dimension d, then by reasoning similarly to that in our proof of Theorem 5.6.4,

$$\begin{aligned} \operatorname{ehr}_{\square_\Gamma}(n) \ &= \ \sum_{A \in \Gamma} -\mu_{\hat{\Gamma}}(A, \hat{1}) \operatorname{ehr}_{\square_A}(n) \ = \ \sum_{A \in \Gamma} (-1)^{d - \dim A} \, n^{|A|} \\ &= \ \sum_{i=0}^{d+1} f_{i-1}(\Gamma)(-1)^{d+1-i} \, n^i. \end{aligned}$$

Equating these two expressions for $\operatorname{ehr}_{\square_\Gamma}(n)$ and comparing coefficients again yields the linear relations of Corollary 5.6.7.

We close this section with yet another connection. We recall from Section 4.9 that for a graded poset Π with $\hat{0}$ and $\hat{1}$ and an order preserving and ranked $\phi : \Pi \to \mathbb{Z}_{\ge 0}$, a (Π, ϕ)-chain partition of $n > 0$ stems from a multichain $\hat{0} \prec c_1 \preceq c_2 \preceq \cdots \preceq c_m \prec \hat{1}$ such that

$$n \ = \ \operatorname{rk}(c_1) + \operatorname{rk}(c_2) + \cdots + \operatorname{rk}(c_m) \,,$$

and the number of chain partitions of n is denoted by $cp_{\Pi,\phi}(n)$.

Proposition 5.6.8. *Let Γ be a pure simplicial complex on the ground set V with rank function $\operatorname{rk} = \operatorname{rk}_\Gamma$. Then*

$$\operatorname{ehr}_{\mathcal{R}[\Gamma]}(n) \ = \ cp_{\Gamma \cup \{\hat{1}\}, \operatorname{rk}}(n)$$

for all $n > 0$.

Proof. A point $\mathbf{p} \in \mathbb{Z}^V$ is contained in $n|\mathcal{R}[\Gamma]|$ if and only if $\mathbf{p} \geq 0$, $|\mathbf{p}| = n$, and $\operatorname{supp}(\mathbf{p}) \in \Gamma$. The claim now follows from Lemma 4.9.5. $\square$

Thus for an Eulerian complex Γ, we discovered Corollary 5.6.7 already in (4.9.5) in Section 4.9.

5.7. A Combinatorial Triangulation

We finish our study of subdivisions of polytopes with a construction that brings us back to purely combinatorial considerations. Fix a polytope $\mathsf{P} \subset \mathbb{R}^d$ and let $\Phi(\mathsf{P})$ be its face lattice. For a vertex $\mathbf{v} \in \mathsf{P}$, we define the **antistar** of $\mathbf{v}$ as the collection of faces of P not containing $\mathbf{v}$:

$$\operatorname{Ast}_{\mathbf{v}}(\mathsf{P}) := \{\mathsf{F} \in \Phi(\mathsf{P}) : \mathbf{v} \notin \mathsf{F}\}.$$

Like the collection of faces visible from a given point, the antistar is a polyhedral complex. Indeed, since $\operatorname{Ast}_{\mathbf{v}}(\mathsf{P}) \subseteq \Phi(\mathsf{P})$, it inherits the intersection property from the face lattice. Moreover, if F is a face that does not contain $\mathbf{v}$, so does every face $\mathsf{G} \prec \mathsf{F}$. We note the following simple but useful fact about antistars (see Exercise 5.43).

Lemma 5.7.1. *Let P be a polytope and $\mathbf{v}$ a vertex. Then $\operatorname{Ast}_{\mathbf{v}}(\mathsf{G}) \subseteq \operatorname{Ast}_{\mathbf{v}}(\mathsf{P})$ is a subcomplex for any face $\mathsf{G} \preceq \mathsf{P}$ with $\mathbf{v} \in \mathsf{G}$.*

The connection between antistars and subdivisions is the following.

Proposition 5.7.2. *For a polytope P with vertex $\mathbf{v}$,*

$$\mathbf{v} * \operatorname{Ast}_{\mathbf{v}}(\mathsf{P}) = \{\mathbf{v} * \mathsf{F} : \mathsf{F} \in \operatorname{Ast}_{\mathbf{v}}(\mathsf{P})\} \cup \operatorname{Ast}_{\mathbf{v}}(\mathsf{P})$$

is a subdivision of P.

Proof. Let $\mathbf{p} \in \mathsf{P}$ be an arbitrary point that is different from $\mathbf{v}$. The ray $\mathbf{v} + \mathbb{R}_{\geq 0}(\mathbf{p} - \mathbf{v})$ intersects P in a segment with endpoints $\mathbf{v}$ and $\mathbf{r} \in \partial \mathsf{P}$. Let F be the unique face that contains $\mathbf{r}$ in its relative interior. We claim that $\mathsf{F} \in \operatorname{Ast}_{\mathbf{v}}(\mathsf{P})$. Otherwise, $\mathbf{r} + \varepsilon(\mathbf{v} - \mathbf{r}) \in \mathsf{F} \subseteq \mathsf{P}$ for some $\varepsilon > 0$, which would contradict the choice of $\mathbf{r}$ as an endpoint of our segment.

By invoking Lemma 5.3.2, this shows that $\mathbf{v} * \operatorname{Ast}_{\mathbf{v}}(\mathsf{P})$ is a polyhedral complex and, at the same time, that $\mathsf{P} = |\mathbf{v} * \operatorname{Ast}_{\mathbf{v}}(\mathsf{P})|$, which finishes the proof. $\square$

Note that $\mathbf{v} * \operatorname{Ast}_{\mathbf{v}}(\mathsf{P}) = \Phi(\mathsf{P})$ if P is a simplex. In all other cases, $\mathbf{v} * \operatorname{Ast}_{\mathbf{v}}(\mathsf{P})$ is a proper subdivision of P, and it is a triangulation if and only if all faces in $\operatorname{Ast}_{\mathbf{v}}(\mathsf{P})$ are simplices. This happens, for example, when P is simplicial. Figure 5.11 illustrates Proposition 5.7.2 for a hexagon and Figure 5.12 shows the resulting subdivision (into pyramids over squares) for a cube.

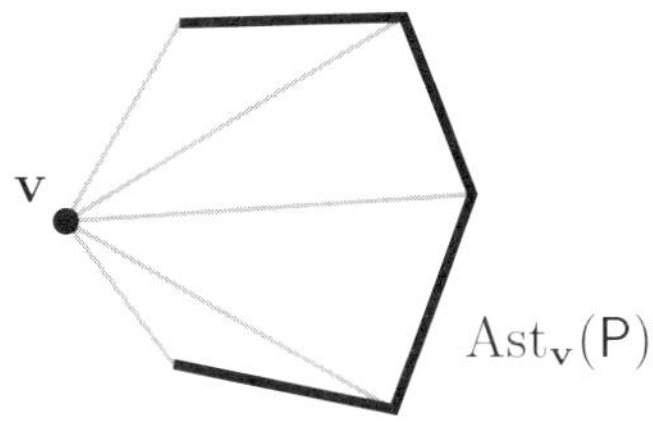

Figure 5.11. Pulling a vertex of a hexagon.

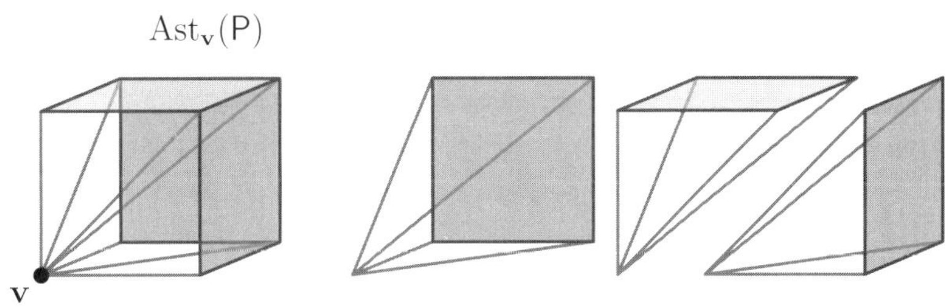

Figure 5.12. The antistar of a vertex in a 3-cube and the resulting subdivision into pyramids over squares.

To obtain a triangulation of a general (nonsimplicial) polytope, the idea is to use $\mathbf{v} * \mathcal{T}$, where $\mathcal{T}$ is a triangulation of $\mathrm{Ast}_{\mathbf{v}}(\mathsf{P})$ and we can get $\mathcal{T}$ by recursively applying Proposition 5.7.2 to elements in $\mathrm{Ast}_{\mathbf{v}}(\mathsf{P})$.

For a fixed ordering $\mathbf{v}_1, \dots, \mathbf{v}_n$ of the vertices of P, we denote by $\mathbf{v}_\mathsf{F}$ the smallest vertex of a face $\mathsf{F} \preceq \mathsf{P}$ in this ordering.

Theorem 5.7.3. *Let* P *be a polytope and order its vertices* $\mathbf{v}_1, \dots, \mathbf{v}_n$. *If* P *is simplicial, set* $\mathrm{Pull}(\mathsf{P}) := \mathbf{v}_\mathsf{P} * \mathrm{Ast}_{\mathbf{v}_\mathsf{P}}(\mathsf{P})$. *Otherwise,*

$$\mathrm{Pull}(\mathsf{P}) \; := \bigcup_{\mathsf{F} \in \mathrm{Ast}_{\mathbf{v}_\mathsf{P}}(\mathsf{P})} \mathbf{v}_\mathsf{P} * \mathrm{Pull}(\mathsf{F})\,. \tag{5.7.1}$$

Then $\mathrm{Pull}(\mathsf{P})$ *is a triangulation of* P.

The triangulation obtained in Theorem 5.7.3 is called the **pulling triangulation** of P with respect to the chosen order on the vertices.

Proof. Since every polytope of dimension at most 1 is a simplex and every 2-dimensional polytope is simplicial, the claim is true in dimensions $d \le 2$ and we can proceed by induction on $d = \dim \mathsf{P}$.

Let P be a polytope of dimension $d > 2$. For any $\mathsf{F} \in \mathrm{Ast}_{\mathbf{v}_\mathsf{P}}(\mathsf{P})$, $\mathrm{Pull}(\mathsf{F})$ is a triangulation of F by induction. What we need to check is that the collection of polyhedral complexes

$$\{\mathrm{Pull}(\mathsf{F}) \,:\, \mathsf{F} \in \mathrm{Ast}_{\mathbf{v}_\mathsf{P}}(\mathsf{P})\}$$

all fit together to give a triangulation of $\mathrm{Ast}_{\mathbf{v}_\mathsf{P}}(\mathsf{P})$. Let $\mathsf{F}, \mathsf{F}' \in \mathrm{Ast}_{\mathbf{v}_\mathsf{P}}(\mathsf{P})$ such that $\mathsf{G} = \mathsf{F} \cap \mathsf{F}'$ is not empty. If G is a face in $\mathrm{Ast}_{\mathbf{v}_\mathsf{F}}(\mathsf{F})$ or $\mathrm{Ast}_{\mathbf{v}_{\mathsf{F}'}}(\mathsf{F}')$, then by construction the restriction of $\mathrm{Pull}(\mathsf{F})$ and $\mathrm{Pull}(\mathsf{F}')$ to G is exactly $\mathrm{Pull}(\mathsf{G})$. If G is neither in $\mathrm{Ast}_{\mathbf{v}_\mathsf{F}}(\mathsf{F})$ nor in $\mathrm{Ast}_{\mathbf{v}_{\mathsf{F}'}}(\mathsf{F}')$, then $\mathbf{v}_\mathsf{F} = \mathbf{v}_{\mathsf{F}'}$ and the claim follows from Lemma 5.7.1. Thus

$$\mathcal{T} \;:=\; \bigcup_{\mathsf{F} \in \mathrm{Ast}_{\mathbf{v}_\mathsf{P}}(\mathsf{P})} \mathrm{Pull}(\mathsf{F}) \tag{5.7.2}$$

is a triangulation of $|\mathrm{Ast}_{\mathbf{v}_\mathsf{P}}(\mathsf{P})|$ and $\mathrm{Pull}(\mathsf{P}) = \mathbf{v}_\mathsf{P} * \mathcal{T}$ finishes the proof. □

Note that the pulling triangulation of Theorem 5.7.3 makes no reference to the geometry of P and can be constructed solely from the knowledge of $\Phi(\mathsf{P})$. For this reason pulling triangulations are a favorite tool to obtain combinatorial results about polytopes. To get a better feel for that, consider the d-dimensional cube $C_d = [0,1]^d$. We can identify its vertices $\mathrm{vert}(C_d) = \{0,1\}^d$ with subsets of $[d]$ under the correspondence that takes $A \subseteq [d]$ to the point $\mathbf{e}_A \in \{0,1\}^d$ with

$$(\mathbf{e}_A)_i \;=\; \begin{cases} 1 & \text{if } i \in A, \\ 0 & \text{otherwise.} \end{cases}$$

Choose any total ordering of the vertices of C_d such that $\mathbf{e}_A$ comes before $\mathbf{e}_B$ whenever $A \subseteq B$; see Exercise 5.37. In any such ordering $\mathbf{v}_{C_d} = \mathbf{e}_\varnothing$.

Proposition 5.7.4. *Let $\mathcal{T} = \mathrm{Pull}(C_d)$ be the pulling triangulation of the d-cube $C_d = [0,1]^d$ with respect to an ordering of its vertices that refines the inclusion-order on subsets of $[d]$. Then*

$$\mathrm{conv}(\mathbf{e}_{A_1}, \mathbf{e}_{A_2}, \dots, \mathbf{e}_{A_k}) \;\in\; \mathcal{T}$$

for distinct $A_1, \dots, A_k \subseteq [d]$ if and only if $A_i \subset A_j$ or $A_j \subset A_i$ for all $1 \le i < j \le k$.

Proof. We argue by induction on the dimension d. For $d = 1$, C_1 is a simplex with vertices $0 = \mathbf{e}_\varnothing$ and $1 = \mathbf{e}_{\{1\}}$.

For $d > 1$, we observe that the facets not containing $\mathbf{v}_{C_d} = \mathbf{e}_\varnothing$ are

$$\mathsf{F}_i \;:=\; \left\{\mathbf{p} \in [0,1]^d \,:\, p_i = 1\right\},$$

for $i = 1, \dots, d$. So F_i is the cube $[0,1]^{d-1}$ embedded in the hyperplane $\{\mathbf{x} \in \mathbb{R}^d \,:\, x_i = 1\}$ with minimal vertex $\mathbf{v}_{\mathsf{F}_i} = \mathbf{e}_i$. By induction the simplices of $\mathrm{Pull}(\mathsf{F}_i)$ correspond exactly to the chains

$$A_1 \;\subset\; A_2 \;\subset\; \cdots \;\subset\; A_k \;\subseteq\; [d] \setminus \{i\}\,.$$

Thus, the simplices of $\operatorname{Pull}(C_d)$ correspond exactly to all chains of $[d]$, which proves the claim. □

We have seen this triangulation before—this is exactly the triangulation given in Proposition 5.1.9: for a maximal chain $\varnothing = A_0 \subset A_1 \subset \cdots \subset A_d$, we have $A_j \setminus A_{j-1} = \{a_{d-j+1}\}$ for some $a_{d-j+1} \in [d]$. Since each a_j can appear only once, the map $\sigma : [d] \to [d]$ given by $\sigma(j) := a_j$ is a permutation of $[d]$. The point corresponding to A_i is then

$$\mathbf{e}_{A_i} \; = \; \sum_{j=1}^{i} \mathbf{e}_{\sigma(1)} + \mathbf{e}_{\sigma(2)} + \cdots + \mathbf{e}_{\sigma(j)} \; = \; \mathbf{u}^{\sigma}.$$

As a byproduct, we obtain the following from Proposition 5.1.9.

Corollary 5.7.5. *Every pulling triangulation of $[0,1]^d$ is unimodular.*

We now illustrate the power of pulling triangulations on a different class of polytopes. For integers $0 < k < d$, we define the (d,k)**-hypersimplex** as the polytope

$$\triangle(d,k) \; := \; \operatorname{conv}\{\mathbf{e}_A \, : \, A \subseteq [d], \; |A| = k\}.$$

For example, for $k = 1$, $\triangle(d,1) = \operatorname{conv}(\mathbf{e}_1, \dots, \mathbf{e}_d)$ is a unimodular simplex, and $\triangle(d,k) \cong \triangle(d, d-k)$ under the linear transformation $\mathbf{x} \mapsto \mathbf{e}_{[d]} - \mathbf{x}$. In Exercise 5.44 you will show that

$$\triangle(d,k) \; = \; \left\{\mathbf{x} \in [0,1]^d \, : \, x_1 + x_2 + \cdots + x_d = k\right\}. \tag{5.7.3}$$

In particular, for $i = 1, \dots, d$,

$$\begin{aligned} \mathsf{F}_i^0 \; &:= \; \triangle(d,k) \cap \left\{\mathbf{x} \in \mathbb{R}^d \, : \, x_i = 0\right\} \; \cong \; \triangle(d-1,k) \quad \text{and} \\ \mathsf{F}_i^1 \; &:= \; \triangle(d,k) \cap \left\{\mathbf{x} \in \mathbb{R}^d \, : \, x_i = 1\right\} \; \cong \; \triangle(d-1,k-1), \end{aligned} \tag{5.7.4}$$

where in both cases the isomorphism is with respect to the projection $\mathbb{R}^d \to \mathbb{R}^{d-1}$ that deletes the i-th coordinate.

To get an impression what a hypersimplex looks like, we note that $\triangle(d,1)$ and $\triangle(d,d-1)$ are just unimodular simplices. The first nontrivial case is $\Delta(4,2)$. This is a 3-dimensional polytope with $\binom{4}{2} = 6$ vertices and $2 \cdot 4 = 8$ facets, each of which is a unimodular triangle. It is not difficult to verify that $\Delta(4,2)$ is, in fact, affinely isomorphic (in the sense of Exercise 5.45) to an octahedron. For the case $\triangle(4,2)$, the following result can be verified quite directly.

Proposition 5.7.6. *Every pulling triangulation of a hypersimplex is unimodular.*

Proof. We proceed once more by induction on d. For $d = 2$ and hence $k = 1$, $\triangle(d,k)$ is the unimodular simplex with vertices $(1,0)$ and $(0,1)$.

For $d > 2$, let $\mathbf{v}$ be the first vertex in an arbitrary but fixed ordering of the vertices of $\triangle(d,k)$. Since each facet of $\triangle(d,k)$ is again a hypersimplex, we obtain by induction that (5.7.2) is a unimodular triangulation $\mathcal{T}$ of the antistar of $\mathbf{v}$. Any inclusion-maximal $\mathsf{F} \in \mathcal{T}$ is contained in the hyperplane $\{\mathbf{x} \in \mathbb{R}^d : x_i = 1 - v_i\}$ for some $i = 1, \dots, d$. You should check (Exercise 5.46) that if F has a unimodular triangulation, then $\mathbf{v} * \mathsf{F}$ has a unimodular triangulation. This completes the proof. □

Proposition 5.7.6 gives the quite remarkable (and rare) property that every pulling triangulation is unimodular. Such lattice polytopes are called **compressed** and the unimodular simplices, the cube, and the hypersimplices are examples; see Exercise 5.47 for more.

We can use (5.7.4) to determine the number of simplices in any such triangulation of a hypersimplex. For a permutation $\pi \in \mathfrak{S}_d$, we call $1 \le i < d-1$ an **ascent** if $\pi(i) < \pi(i+1)$ and a **descent** if $\pi(i) > \pi(i+1)$.

Theorem 5.7.7. *Let $0 \le k \le d-1$ and let $\mathcal{T}$ be a unimodular triangulation of the d-dimensional hypersimplex $\triangle(d+1,k+1)$. Then the number of full-dimensional simplices in $\mathcal{T}$ is the number $s(d,k)$ of permutations in $\mathfrak{S}_d$ with exactly k descents.*

For example, for $(d,k) = (3,1)$, the hypersimplex $\triangle(4,2)$ is triangulated by four unimodular simplices. At the same time, there are $d! = 6$ permutations of which the following 4 have exactly one descent: [132], [312], [231], and [213]. (Here we use the *one-line notation* for a permutation $\pi \in \mathfrak{S}_d$ and write it as $[\pi(1)\,\pi(2)\,\cdots\,\pi(d)]$.)

The numbers $s(d,k)$ are called the **Eulerian numbers**. We will see them again in Chapter 6.

Proof. It follows from Proposition 5.7.6 and Corollary 5.5.7 that it suffices to prove the claim for a pulling triangulation $\mathcal{T} = \mathrm{Pull}(\triangle(d+1,k+1))$. Let's write $s(d,k)$ for the number of d-simplices of $\mathcal{T}$. If $k = 0$ or $k = d-1$, then $\triangle(d+1,k+1)$ is already a unimodular simplex and therefore $s(d,0) = s(d,d-1) = 1$. We claim that

$$s(d,k) \;=\; (d-k)\,s(d-1,k) + (k+1)\,s(d-1,k-1)\,. \tag{5.7.5}$$

Indeed, the number of d-simplices of $\mathcal{T}$ is equal to the number of $(d-1)$-simplices in the restriction of $\mathcal{T}$ to $\mathrm{Ast}_{\mathbf{v}}(\triangle(d+1,k+1))$ for the vertex $\mathbf{v} = \mathbf{v}_{\triangle(d+1,k+1)}$ that was first in the pulling order. From (5.7.4), we know that the antistar is composed of $d+1$ facets of $\triangle(d+1,k)$, namely, $d-k$ facets $\mathsf{F}_i^0 \cong \triangle(d,k+1)$ corresponding to those positions i with $v_i = 0$ and

$k+1$ facets $\mathsf{F}_i^1 \cong \triangle(d,k)$ for $v_i = 1$. Each of these facets is triangulated by $s(d-1,k)$ and $s(d-1,k-1)$ simplices, respectively.

The permutations $[12\cdots d]$ and $[d(d-1)\cdots 1]$ are the only permutations in $\mathfrak{S}_d$ with 0 and $d-1$ descents, respectively. Hence, to complete the proof, it suffices to show that the number of permutations $\tau \in \mathfrak{S}_d$ with k descents satisfies the recurrence (5.7.5). This is done in Exercise 5.49. □

Corollary 5.5.7 says that the number $f_r(\mathcal{T})$ of r-dimensional simplices for $0 \le r \le d-1$ in any unimodular triangulation $\mathcal{T}$ of $\triangle(d,k)$ has to be the same. However, it seems to be a quite challenging problem to find a meaningful interpretation for this number. We will revisit this problem in Section 7.4.

Notes

Subdivisions and triangulations of polyhedra and, more generally, manifolds go back at least to the beginning of the 20th century. In particular geometric simplicial complexes are a common means to build complex objects from simple ones. The study of simplicial complexes independent of a geometric embedding (i.e., abstract simplicial complexes) belongs to the field of (combinatorial) topology; see, for example, [**88**, **126**].

It seems to be hard to credit a single mathematician for inventing (or discovering) regular subdivisions. Certainly Hermann Minkowski [**123**] knew that the projection of a full-dimensional polytope onto a hyperplane H yields two subdivisions of the projection. Boris Delaunay [**53**] described an important class of subdivisions (see Exercise 5.5) that is still today of great importance in discretizing (and solving) differential equations. The name *regular* or *coherent* goes back to Israel Gelfand, Mikhail Kapranov, and Andrei Zelevinsky [**70**] who initiated much of modern research into subdivisions, in particular the poset of subdivisions of a polytope $\mathsf{P} = \operatorname{conv}(V)$ with vertices in V. The partial order relation is given by refinement. A sneak peak: for fixed V there is a polytope, the *secondary polytope* of V, whose face lattice is isomorphic to the poset of regular subdivisions. The key to the construction of that polytope is Proposition 5.1.10. This result is also the reason why these subdivisions are sometimes called *convex*. Exercise 5.2 shows that there are nonregular subdivisions and the whole story becomes more subtle. A comprehensive treatise is given in [**51**].

The triangulation of the cube given in Proposition 5.1.9 was described by Hans Freudenthal [**66**].

Theorem 5.2.1 holds for all polyhedral complexes $\mathcal{S}$ whose support $|\mathcal{S}|$ is homeomorphic to a ball but a different proof is needed with techniques from topology; see, for example, [**33**, Part II].

Eugène Ehrhart laid the foundation for lattice-point enumeration in rational polyhedra, starting with Theorems 4.6.1, 4.7.2, and 5.1.7 in 1962 [**57**] as a teacher at a *lycée* in Strasbourg, France. (Ehrhart received his doctorate later, at age 60 on the urging of some colleagues.) As already mentioned in the Notes for Chapter 4, our approach follows Ehrhart's original lines of thought; an alternative proof from first combinatorial principles can be found in [**147**]. The reciprocity theorem for Ehrhart polynomials (Theorem 5.2.3) was conjectured (and proved in several special cases) by Ehrhart and proved by I. G. Macdonald [**115**]. Theorem 5.2.3 is a particular instance of a reciprocity relation for simple lattice-invariant valuations due to Peter McMullen [**120**], who also proved a parallel extension of Ehrhart–Macdonald reciprocity to general lattice-invariant valuations.

Ehrhart–Macdonald reciprocity takes on a special form for reflexive polytopes which we define and study in Exercise 5.13. The term *reflexive polytope* was coined by Victor Batyrev, who motivated these polytopes by applications of mirror symmetry in string theory [**18**]. That the Ehrhart series of a reflexive polytope exhibits an unexpected symmetry (Exercise 5.13) was discovered by Takayuki Hibi [**84**]. The number of reflexive polytopes in dimension d is known only for $d \leq 4$ [**105, 106**]; see also [**1**, Sequence A090045].

The placing triangulation was described by Branko Grünbaum [**78**, Section 5.2]. The algorithm described underneath Theorem 5.3.3 is called the *beneath-beyond method* and, in more sophisticated versions, it is widely used in practice; see [**51**].

The half-open decompositions in Lemma 5.3.4 (and beyond) first surfaced in the computational approach to Ehrhart quasipolynomials by Matthias Köppe and Sven Verdoolaege [**103**]. The fact that a half-open decomposition is a disjoint union of half-open polytopes makes computations much simpler, both in theory and in practice; see, for example, [**96, 97**].

Richard Stanley developed much of the theory of Ehrhart (quasi)polynomials, initially from a commutative-algebra point of view. Theorem 5.4.2 appeared in [**162**], the paper that coined the term *combinatorial reciprocity theorem*, Corollary 5.5.1 (the simplicial case of which we mentioned in our proof of Theorem 4.6.1) in [**163**], and Theorem 5.5.2 in [**168**]. The nonnegativity constraints in Corollary 5.5.1 serve as the starting point when trying to classify Ehrhart polynomials, though a complete classification is known only in two dimensions [**21**]. Theorem 5.5.2 was proved in [**168**], our proof is taken from [**97**]. *Reciprocal domains* were studied by Ehrhart [**59, 60**] and in [**162**]. For (much) more about Ehrhart (quasi)polynomials, see [**16, 24, 83, 170**].

Theorem 5.4.6 is due to Michel Brion [**40**]; his proof was quite a bit more involved than the one we give here. Alternative proofs can be found in

[**22**, **92**]; see also [**110**, **111**, **180**] for more decomposition theorems with the same philosophy. Theorem 5.4.6 motivated Alexander Barvinok to devise an efficient algorithm for Ehrhart quasipolynomials [**14**]. Barvinok's algorithm, which is described in detail in [**17**], has been implemented in the software packages `barvinok` [**181**] and `LattE` [**49**, **50**, **102**].

Theorem 5.5.6 (and the equivalent Corollary 5.5.7) is due to Richard Stanley [**163**]. As we mentioned already, lattice polytopes that admit unimodular triangulations are quite special. There are other, less restrictive, classes of lattice polytopes which come with numerous applications (e.g., to semigroup algebras) and open problems [**42**]. A far-reaching extension of Theorem 5.5.6 to general lattice polytopes, still relating its h^*-polynomial with the f-polynomial of a fixed triangulation, is due to Ulrich Betke and Peter McMullen [**28**]. Their theorem becomes particularly powerful when the polytope has an interior lattice point; this consequence was fully realized only by Alan Stapledon [**173**] who extended the Betke–McMullen theorem further; see also [**172**, **174**] which give the current state of the art regarding inequalities among Ehrhart coefficients. Related work includes [**12**, **43**, **133**, **141**].

Corollary 5.5.9 is due to Victor Klee [**100**]. Other Dehn–Sommerville-type relations include [**2**, **127**].

The pulling triangulation was described by John F. P. Hudson [**88**, Lemma 1.4] as a technique to refine polyhedral complexes. The consequence of Theorem 5.7.7 was already known to Pierre-Simon Laplace: the volume of the hypersimplex $\Delta(d+1, k+1)$ times $d!$ is the number of permutations $\sigma \in \mathfrak{S}_d$ with k descents; see [**65**], which also contains a short after-thought by Richard Stanley in which he constructs a piecewise-linear map from the Freudenthal triangulation to the collection of all hypersimplices that maps Δ_σ with $\sigma \in \mathfrak{S}_d$ to $\triangle(d+1, k+1)$ precisely when σ has k descents. We will revisit this in Section 7.4.

Exercises

5.1 ☖ Let $\mathcal{S}$ be a subdivision of a polytope P. Show that if $\mathsf{F} \in \mathcal{S}$ with $\dim \mathsf{F} < \dim \mathsf{P}$, then there is some $\mathsf{G} \in \mathcal{S}$ with $\mathsf{F} \subset \mathsf{G}$. In particular, $\mathcal{S}$ is a graded poset.

5.2 ☖ Prove that the subdivision of the triangle in Figure 5.13 (with three additional vertices) is not regular. Give an example of a nonregular subdivision in every dimension ≥ 3.

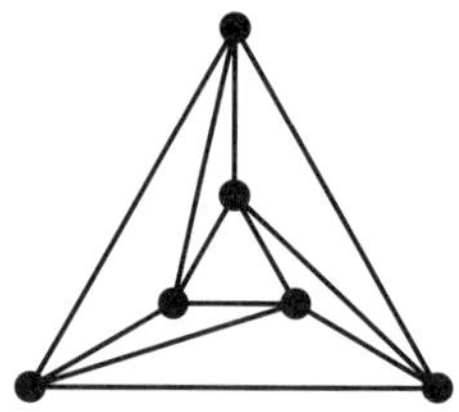

Figure 5.13. A nonregular triangulation.

5.3 Let $\mathsf{P} \subset \mathbb{R}^d$ be a full-dimensional polytope. For every nonempty face $\mathsf{F} \subset \mathsf{P}$, let $\mathbf{p}_\mathsf{F}$ be a point in F°.

(a) Let $F = \{\mathsf{F}_0 \subset \mathsf{F}_1 \subset \cdots \subset \mathsf{F}_k\}$ be a chain of nonempty faces. Show that

$$\mathsf{T}(F) \ := \ \operatorname{conv}\{\mathbf{p}_{\mathsf{F}_i} : i = 0, 1, \dots, k\}$$

is a k-dimensional simplex.

(b) Show that

$$\mathcal{B}(\mathsf{P}) \ := \ \{\mathsf{T}(F) : F \text{ chain of nonempty faces of } \mathsf{P}\}$$

is a triangulation of P, called a **barycentric subdivision**.

(c) Show that, as posets, $\mathcal{B}(\mathsf{P})$ is isomorphic to the order complex of $\Phi(\mathsf{P})$.

5.4 ⬠ Let $V := \{\mathbf{v}_1, \mathbf{v}_2, \dots, \mathbf{v}_n\} \subset \mathbb{R}^d$ be a configuration of $n \geq d+2$ points such that $\mathsf{P} := \operatorname{conv}(V)$ is full dimensional. A stronger condition than all bounded faces of $\mathsf{E}^\omega(V)$ being simplices (as in Corollary 5.1.6) is that no hyperplane in $\mathbb{R}^{d+1}$ contains more than $d+1$ points of V^ω. Indeed, each supporting hyperplane of $\mathsf{E}^\omega(V)$ will then contain $k \leq d+1$ points which are thus the vertices of a $(k-1)$-simplex. In the following you will show that there is a (sufficiently large) $h \in \mathbb{R}$ such that for $\omega(\mathbf{v}_j) := h^j$, $\mathsf{E}^\omega(V)$ satisfies this stronger condition.

(a) Consider the case $n = d+2$. Then V^ω is contained in a hyperplane if and only if the points V^ω are affinely dependent, that is, the $(d+2) \times (d+2)$-matrix

$$\begin{pmatrix} \mathbf{v}_1 & \mathbf{v}_2 & \cdots & \mathbf{v}_{d+2} \\ h & h^2 & \cdots & h^{d+2} \\ 1 & 1 & \cdots & 1 \end{pmatrix}$$

has determinant 0. Use Laplace expansion to show that this can happen only for finitely many values of h.

(b) Argue that the same holds true for $n > d+2$ by considering all $(d+2)$-subsets of V.

5.5 ⬠ Let $V \subset \mathbb{R}^d$ be a finite set such that $\mathsf{P} = \operatorname{conv}(V)$ is full dimensional and define $\omega : V \to \mathbb{R}$ by

$$\omega(\mathbf{p}) := \sum_{i=1}^{d} p_i^2 .$$

Show that $\mathcal{S}^\omega(V)$ has the following interesting property: the vertices of every full-dimensional cell F lie on the boundary of a unique ball and this ball does not contain any element of $V \setminus \mathsf{F}$.

5.6 ⬠ Show that the vectors $\mathbf{u}_0^\sigma, \mathbf{u}_1^\sigma, \ldots, \mathbf{u}_{d-1}^\sigma$ as defined in (5.1.7) form a lattice basis of $\mathbb{Z}^d$ for any $\sigma \in \mathfrak{S}_d$. In particular, $\triangle_\sigma$ is a unimodular simplex.

5.7 Let $\mathcal{S} = \mathcal{S}^\omega(V)$ be a regular subdivision of the polytope $\mathsf{P} = \operatorname{conv}(V)$. Use Theorem 3.5.1 and the fact that

$$\widehat{\mathcal{S}} \cong \Phi^{\mathrm{bnd}}(\mathsf{E}^\omega(V)) \cup \{\mathsf{E}^\omega(V)\} \subset \Phi(\mathsf{E}^\omega(V))$$

to give an independent proof of Theorem 5.2.1 for *regular* subdivisions. (*Hint:* The tricky case is $\mathsf{G} = \mathsf{P}$ and $\mathsf{F} \subseteq \partial\mathsf{P}$; here first show that there is a bijection between the bounded and unbounded faces of $\mathsf{E}^\omega(V)$ that contain the face of $\mathsf{E}^\omega(V)$ that projects to F.)

5.8 Compute the Ehrhart polynomial of the pyramid with vertices $(0,0,0)$, $(2,0,0)$, $(0,2,0)$, $(2,2,0)$, and $(1,1,1)$.

5.9 Compute the Ehrhart polynomial of the **octahedron** with vertices $(\pm1,0,0)$, $(0,\pm1,0)$, and $(0,0,\pm1)$. Generalize to the d-dimensional cross polytope, the convex hull of the unit vectors in $\mathbb{R}^d$ and their negatives, which made its debut in (3.1.13).

5.10 ⬠ Show that if P is a d-dimensional lattice polytope in $\mathbb{R}^d$, then the degree of its Ehrhart polynomial $\operatorname{ehr}_{\mathsf{P}}(n)$ is d and the leading coefficient is the volume of P. What can you say if P is not full dimensional?

5.11 Find and prove an interpretation of the second leading coefficient of $\operatorname{ehr}_{\mathsf{P}}(n)$ for a full-dimensional lattice polytope P. (*Hint:* Start by computing the Ehrhart polynomial of the boundary of P.)

5.12 For a lattice d-polytope $\mathsf{P} \subset \mathbb{R}^d$, consider its Ehrhart series

$$\operatorname{Ehr}_{\mathsf{P}}(z) = \sum_{n \ge 0} \operatorname{ehr}_{\mathsf{P}}(n)\, z^n .$$

By Proposition 4.5.1 and Theorem 5.1.7, we can write

$$\operatorname{Ehr}_{\mathsf{P}}(z) = \frac{h_d^*\, z^d + h_{d-1}^*\, z^{d-1} + \cdots + h_0^*}{(1-z)^{d+1}}$$

for some $h_0^*, h_1^*, \ldots, h_d^*$. Prove:

(a) $h_0^* = 1$.

(b) $h_1^* = \left|\mathsf{P} \cap \mathbb{Z}^d\right| - d - 1$.
(c) $h_d^* = \left|\mathsf{P}^\circ \cap \mathbb{Z}^d\right|$.
(d) $h_0^* + h_1^* + \cdots + h_d^* = d! \operatorname{vol}(\mathsf{P})$.
(This extends Exercise 4.42 from lattice simplices to arbitrary lattice polytopes.)

5.13 A **reflexive polytope** is a lattice polytope P such that the origin is the unique interior lattice point of P and[6]

$$\operatorname{ehr}_{\mathsf{P}^\circ}(n) \quad = \quad \operatorname{ehr}_{\mathsf{P}}(n-1) \qquad \text{for all } n \in \mathbb{Z}_{>0}\,. \tag{5.7.6}$$

Prove that if P is a lattice d-polytope that contains the origin in its interior and that has the Ehrhart series

$$\operatorname{Ehr}_{\mathsf{P}}(z) \quad = \quad \frac{h_d^*\, z^d + h_{d-1}^*\, z^{d-1} + \cdots + h_1^*\, z + h_0^*}{(1-z)^{d+1}}\,,$$

then P is reflexive if and only if $h_k^* = h_{d-k}^*$ for all $0 \le k \le \frac{d}{2}$.

5.14 ✩ Prove Theorem 5.2.4: If P is a rational polytope in $\mathbb{R}^d$, then for positive integers n, the counting function $\operatorname{ehr}_{\mathsf{P}}(n)$ is a quasipolynomial in n whose period divides the least common multiple of the denominators of the vertex coordinates of P. Furthermore, for all integers $n > 0$,

$$(-1)^{\dim \mathsf{P}} \operatorname{ehr}_{\mathsf{P}}(-n) \quad = \quad \left|n\, \mathsf{P}^\circ \cap \mathbb{Z}^d\right|.$$

5.15 ✩ Generalize Theorem 5.1.8 to rational polytopes, as follows: Let P be a rational polytope and $\operatorname{ehr}_{\mathsf{P}}(n)$ its Ehrhart quasipolynomial. Then $\operatorname{ehr}_{\mathsf{P}}(0) = 1$. (*Hint:* This is the constant term for one of the constituents of $\operatorname{ehr}_{\mathsf{P}}(n)$.)

5.16 ✩ Let $\mathbf{v}_0, \ldots, \mathbf{v}_d \in \mathbb{R}^d$ be affinely independent, and so $\operatorname{conv}(\mathbf{v}_0, \ldots, \mathbf{v}_d)$ is a d-simplex. Describe how to find the facet-defining hyperplane H_i for the facet $\mathsf{F}_i := \operatorname{conv}(\mathbf{v}_0, \ldots, \mathbf{v}_{i-1}, \mathbf{v}_{i+1}, \ldots, \mathbf{v}_d)$ for $i = 0, \ldots, d$.

5.17 ✩ In the setting of Theorem 5.3.3, prove that $\mathsf{F} = \mathbf{v} * \mathsf{F}'$ is a cell of $\mathcal{S}$ that was not present in $\mathcal{S}'$ if and only if F' is the facet of some cell in $\mathcal{S}'$ such that $\mathbf{v}$ is beyond F'. (*Hint:* Exercise 3.69.)

5.18 ✩ Consider a d-simplex with facet-defining hyperplanes

$$\mathsf{H}_i \quad = \quad \left\{\mathbf{x} \in \mathbb{R}^d \,:\, \langle \mathbf{a}_i, \mathbf{x}\rangle = b_i\right\}$$

for $i = 0, \ldots, d$, and let $\mathbf{v}$ be a point that is beyond F_0, the facet corresponding to H_0. Show that $\mathbf{v} * \mathsf{F}_0$ is given by all $\mathbf{x} \in \mathbb{R}^d$ such that

$$\begin{aligned} \langle -\mathbf{a}_0, \mathbf{x}\rangle &\le b_0\,, \\ \langle \mathbf{a}_i + \delta_i\, \mathbf{a}_0, \mathbf{x}\rangle &\le b_i + \delta_i\, b_0\,, \end{aligned}$$

[6] More generally, if the 1 on the right-hand side of (5.7.6) is replaced by an arbitrary fixed positive integer, we call P **Gorenstein**. You may think about how Exercise 5.13 can be extended to Gorenstein polytopes.

where $\delta_i := \frac{b_i - \langle \mathbf{a}_i, \mathbf{v} \rangle}{\langle \mathbf{a}_0, \mathbf{v} \rangle - b_0}$ for $i = 1, \ldots, d$.

5.19 ⬠ Let P be a d-polytope with ordered vertices $\mathbf{v}_1, \ldots, \mathbf{v}_n$. Show that for $h > 0$ sufficiently large, the subdivision $\mathcal{S}^\omega(\mathbf{v}_1, \ldots, \mathbf{v}_n)$ with $\omega(\mathbf{v}_i) = h^i$ is exactly the pushing triangulation of P with the given order.

5.20 Let P be a polytope with dissection $\mathsf{P} = \mathsf{P}_1 \cup \mathsf{P}_2 \cup \cdots \cup \mathsf{P}_m$, and let H be a hyperplane. Show that H is facet-defining for P if and only if all P_i are contained in the same halfspace of H, and H is facet-defining for at least one P_j.

5.21 ⬠ Prove (5.3.4): Let $\mathsf{P} \subset \mathbb{R}^d$ be a full-dimensional polyhedron with dissection $\mathsf{P} = \mathsf{P}_1 \cup \mathsf{P}_2 \cup \cdots \cup \mathsf{P}_m$. If $\mathbf{q} \in \mathbb{R}^d$ is generic relative to all P_i, then

$$\mathbb{H}^{\mathbf{q}}\mathsf{P} \;=\; \mathbb{H}^{\mathbf{q}}\mathsf{P}_1 \uplus \mathbb{H}^{\mathbf{q}}\mathsf{P}_2 \uplus \cdots \uplus \mathbb{H}^{\mathbf{q}}\mathsf{P}_m \,.$$

5.22 ⬠ Let $\mathsf{P} \subset \mathbb{R}^d$ be a full-dimensional polytope. Show that there is no $\mathbf{q} \in \mathbb{R}^d$ such that $\mathbb{H}^{\mathbf{q}}\mathsf{P} = \mathsf{P}$.

5.23 ⬠ Finish the first half of our proof of Proposition 5.3.6 by showing (using the notation from our proof) that

$$\sigma_{\widehat{\mathsf{C}}}(0, \ldots, 0, z) \;=\; \frac{h(z)}{(1-z)^{d+1}}$$

for some polynomial $h(z)$ of degree $\le d+1$. Deduce from (5.3.7) and Proposition 4.1.4 that $\operatorname{ehr}_{\mathbb{H}_{\mathbf{q}}\triangle}(n)$ is a polynomial in n of degree d. (*Hint:* You might have to use long division to write $\sigma_{\widehat{\mathsf{C}}}(0, \ldots, 0, z)$ as a constant plus a proper rational function.)

5.24 ⬠ Using the notation of Section 4.8 and our proof of Proposition 5.3.6, show that $\check{\mathsf{C}} = \hom(\mathbb{H}^{\mathbf{q}}\triangle)$ and (using Theorem 4.8.1)

$$(-1)^{d+1} \operatorname{Ehr}_{\mathbb{H}_{\mathbf{q}}\triangle}\left(\tfrac{1}{z}\right) \;=\; \operatorname{Ehr}_{\mathbb{H}^{\mathbf{q}}\triangle}(z) \,.$$

5.25 ⬠ Fix linearly independent vectors $\mathbf{v}_1, \mathbf{v}_2, \ldots, \mathbf{v}_d \in \mathbb{Z}^d$ and consider the simplicial cone

$$\mathsf{C} \;:=\; \mathbb{R}_{\ge 0}\mathbf{v}_1 + \mathbb{R}_{\ge 0}\mathbf{v}_2 + \cdots + \mathbb{R}_{\ge 0}\mathbf{v}_d \,.$$

Prove that, for

$$\widehat{\mathsf{C}} \;:=\; \mathbb{R}_{\ge 0}\mathbf{v}_1 + \cdots + \mathbb{R}_{\ge 0}\mathbf{v}_{m-1} + \mathbb{R}_{>0}\mathbf{v}_m + \cdots + \mathbb{R}_{>0}\mathbf{v}_d \,,$$

there exists $\mathbf{q} \in \mathbb{R}^d$ (generic relative to C) such that

$$\widehat{\mathsf{C}} \;=\; \mathbb{H}_{\mathbf{q}}\mathsf{C} \,.$$

Conversely, show that, for every generic $\mathbf{q} \in \mathbb{R}^d$ relative to C, the half-open cone $\mathbb{H}_{\mathbf{q}}\mathsf{C}$ is of the form $\widehat{\mathsf{C}}$ for some reordering of the $\mathbf{v}_j$s and some m.

5.26 Let $\mathsf{P} \subset \mathbb{R}^d$ be a full-dimensional polytope and let $\mathbf{q} \in \mathbb{R}^d$ be generic relative to P. Prove that $\chi(\mathbb{H}_{\mathbf{q}}\mathsf{P}) = 0$ unless $\mathbb{H}_{\mathbf{q}}\mathsf{P} = \mathsf{P}$.

5.27 ⬠ Prove the following generalization of Corollary 5.5.1: For a half-open simplex $\mathbb{H}_{\mathbf{q}}\triangle$ of dimension r, let

$$m := \min\{i : h_i^*(\mathbb{H}_{\mathbf{q}}\triangle) \neq 0\}$$

and

$$m^\circ := \max\{i : h_i^*(\mathbb{H}_{\mathbf{q}}\triangle) \neq 0\}.$$

Then m and m° are the smallest dilation factors such that $\mathrm{ehr}_{\mathbb{H}_{\mathbf{q}}\triangle}(m) > 0$ and $\mathrm{ehr}_{\mathbb{H}^{\mathbf{q}}\triangle}(r+1-m^\circ) > 0$, respectively. Note that if $\mathbb{H}_{\mathbf{q}}\triangle = \triangle$, then $m = 0$.

5.28 ⬠ Give a proof of Lemma 5.5.4: Let $\triangle \subset \mathbb{R}^d$ be a unimodular d-simplex and $\mathbf{q}$ a point generic relative to $\triangle$. Then

$$h^*_{\mathbb{H}_{\mathbf{q}}\triangle}(z) = z^r,$$

where r is the number of facets that are missing in $\mathbb{H}_{\mathbf{q}}\triangle$.

5.29 A *generalized* half-open decomposition of a polytope P is a collection of polytopes $\mathsf{P}_1, \dots, \mathsf{P}_m \subset \mathsf{P}$ and points $\mathbf{q}_1, \dots, \mathbf{q}_m \in \mathsf{P}^\circ$ such that

$$\mathsf{P} = \mathbb{H}_{\mathbf{q}_1}\mathsf{P}_1 \uplus \mathbb{H}_{\mathbf{q}_2}\mathsf{P}_2 \uplus \cdots \uplus \mathbb{H}_{\mathbf{q}_m}\mathsf{P}_m.$$

The *ordinary* half-open decompositions in Section 5.3 are thus the special case $\mathbf{q}_i = \mathbf{q}$ for all i. Find a generalized half-open decomposition of a polytope that is not ordinary.

5.30 ⬠ Let Q be a pyramid over P with apex $\mathbf{v}$. Show that a dissection of $\mathsf{P} = \triangle_1 \cup \cdots \cup \triangle_k$ into simplices induces a dissection of $\mathsf{Q} = \Delta_1' \cup \cdots \cup \Delta_k'$, where $\Delta_i' = \mathbf{v} * \triangle_i$.

5.31 ⬠ Following the setup of the last part of our proof of Theorem 5.5.2, let $\mathsf{P} \subset \mathbb{R}^d$ be a lattice polytope and $\mathbf{v} \in \mathbb{Z}^d \setminus \mathrm{aff}(\mathsf{P})$. Given a dissection of $\mathsf{P} = \triangle_1 \cup \cdots \cup \triangle_k$ into lattice simplices, let $\triangle_i' := \mathbf{v} * \triangle_i$. For a point $\mathbf{q}$ generic relative to $\triangle_1', \dots, \triangle_k'$, let $\widehat{\Pi}_j$ and $\widehat{\Pi}_j'$ be the fundamental parallelepiped of the half-open simplicial cones $\mathrm{hom}(\mathbb{H}_{\mathbf{q}}\triangle_j)$ and $\mathrm{hom}(\mathbb{H}_{\mathbf{q}}\triangle_j')$, respectively. Show that $\widehat{\Pi}_j \subseteq \widehat{\Pi}_j'$ for all j.

5.32 Come up with and prove a generalization of Theorem 5.5.2 for *rational* polytopes.

5.33 Let $\triangle$ be the convex hull of $j+1$ unit vectors in $\mathbb{R}^d$, and consider $\mathrm{cone}(\triangle)$, the conical hull of the same $j+1$ unit vectors.
(a) Show that $\mathrm{ehr}_\triangle(n) = \binom{n+j}{j}$ and $\mathrm{ehr}_{\triangle^\circ}(n) = \binom{n-1}{j}$.
(b) Show that

$$\mathrm{Ehr}_\triangle(z) = \sigma_{\mathrm{cone}(\triangle)}(z, z, \dots, z) = \left(\frac{1}{1-z}\right)^{j+1}$$

and

$$\mathrm{Ehr}_{\triangle^\circ}(z) \;=\; \sigma_{\mathrm{cone}(\triangle)^\circ}(z, z, \dots, z) \;=\; \left(\frac{z}{1-z}\right)^{j+1}.$$

5.34 Find a lattice polytope P that has a unimodular dissection but no unimodular triangulation.

5.35 ⬠ Let $\triangle$ be a d-simplex, $\mathsf{F} \prec \triangle$ a facet, and $\mathbf{v}$ the unique vertex not contained in F. Show that for any face $\mathsf{G} \preceq \triangle$, we have $\mathsf{G}^\circ \not\subseteq \mathsf{F}$ if and only if $\mathbf{v} \in \mathsf{G}$.

5.36 ⬠ Use Theorem 5.5.6 to prove Corollary 5.5.7 and (5.5.4).

5.37 ⬠ Define a partial order on the collection of subsets of $[d]$ by setting $A \preceq B$ if and only if $\min(A \setminus B) > \min(B \setminus A)$. (Set $\min(\varnothing) := \infty$.) Show that this is a total order that satisfies $A \subseteq B$ implies $A \preceq B$.

5.38 Every simple graph $G = (V, E)$ can be thought of as a 1-dimensional simplicial complex $\Gamma = \{\varnothing\} \cup V \cup E$. Characterize which graphs have nonnegative h-vector.

5.39 Let Γ be the $(d-1)$-dimensional abstract simplex. That is, Γ is the complex of *all* subsets of $[d]$. Show that

$$\mathrm{ehr}_{\square_\Gamma}(n) \;=\; n^d .$$

5.40 ⬠ Find a 3-dimensional lattice polytope that has a unimodular dissection that is not a unimodular triangulation. Can you find a 3-dimensional lattice polytope that has a unimodular dissection but no unimodular triangulation? (*Hint:* A suitable construction to do this can be found in [**148**].)

5.41 ⬠ Verify that for a d-dimensional complex $\mathcal{K}$ of lattice polytopes,

$$h^*_{d+1}(\mathcal{K}) \;=\; (-1)^{d+1}(1 - \chi(\mathcal{K})) .$$

5.42 ⬠ Show that a 0-dimensional complex is Eulerian if and only if it has at most two vertices.

5.43 ⬠ Carry out a proof for Lemma 5.7.1: Let P be a polytope and $\mathbf{v}$ a vertex. Then $\mathrm{Ast}_{\mathbf{v}}(\mathsf{G}) \subseteq \mathrm{Ast}_{\mathbf{v}}(\mathsf{P})$ is a subcomplex for any face $\mathsf{G} \preceq \mathsf{P}$ with $\mathbf{v} \in \mathsf{G}$.

5.44 ⬠ Prove (5.7.3):

$$\triangle(d, k) \;=\; \left\{ \mathbf{x} \in [0,1]^d \,:\, x_1 + x_2 + \cdots + x_d = k \right\}.$$

5.45 ⬠ Let P be the octahedron from Exercise 5.9, with vertices $(\pm 1, 0, 0)$, $(0, \pm 1, 0)$, and $(0, 0, \pm 1)$. Show that there is an injective affine map $L : \mathbb{R}^3 \to \mathbb{R}^4$ such that $L(\mathsf{P}) = \Delta(4, 2)$.

5.46 ⬠ Verify the claim made in the proof of Proposition 5.7.6 that, if F has a unimodular triangulation, then $\mathbf{v} * \mathsf{F}$ has a unimodular triangulation.

5.47 Let $\mathsf{P} \subset \mathbb{R}^d$ be a full-dimensional lattice polytope. Suppose that for any facet-defining hyperplane $\mathsf{H} = \{\mathbf{x} \in \mathbb{R}^d : \langle \mathbf{a}, \mathbf{x} \rangle = b\}$ there is some $r \in \mathbb{R}$ such that

$$\langle \mathbf{a}, \mathbf{p} \rangle \in \{b, b + r\} \qquad \text{for all} \qquad \mathbf{p} \in \mathsf{P} \cap \mathbb{Z}^d.$$

(a) Show that every facet of P, when considered as a lattice polytope, has the same property.

(b) Show that any such P is a compressed polytope.

(c) Show the converse: If P is compressed, then it satisfies the above property.

5.48 Let $\mathsf{P} \subset \mathbb{R}^d$ be a full-dimensional polytope, not necessarily a lattice polytope. Assume that P satisfies the condition of Exercise 5.47 for every vertex $\mathbf{p} \in \operatorname{vert}(\mathsf{P})$; such a polytope is called **2-level**.

(a) Show that there is a linear transformation $T : \mathbb{R}^d \to \mathbb{R}^d$ such that $\operatorname{vert}(T(\mathsf{P})) \subseteq \{0, 1\}^d$. Such a polytope is called a **0/1-polytope**. (*Hint:* Pick d facets that intersect in a vertex and whose normals are linearly independent.)

(b) Conclude that up to linear transformation there are only finitely many 2-level (and hence compressed) polytopes.

(c) Can you count them in dimensions $1, 2, 3, 4, \dots$?

5.49 ☖ This exercise completes our proof of Theorem 5.7.7. We first observe that every position i except for $i = d$ is either a descent or an ascent. Now come up, and prove, a recurrence relation for $s(d, k)$; it will involve the quantities $s(d-1, k)$ and $s(d-1, k-1)$. (*Hint:* Write a permutation in $\mathfrak{S}_{d-1}$ in one-line notation; this way it is easier to go from $\mathfrak{S}_{d-1}$ to $\mathfrak{S}_d$ by inserting a number at some position.)

5.50 Let $\triangle = \operatorname{conv}(\mathbf{u}_0, \dots, \mathbf{u}_d) \subset \mathbb{R}^d$, $\triangle' = \operatorname{conv}(\mathbf{v}_0, \dots, \mathbf{v}_e) \subset \mathbb{R}^e$ be two unimodular simplices and let $\mathsf{P} = \triangle \times \triangle'$ be their Cartesian product.

(a) Show that any $(d + e)$-simplex spanned by the vertices of P is unimodular.

(b) We can identify the vertices of P with the nodes of the square grid $\{0, \dots, d\} \times \{0, \dots, e\}$. A **lattice path** from $(0, 0)$ to (d, e) is a path on the grid that uses only unit steps $\to$ and $\uparrow$. Show that any such path encodes a unique $(d + e)$-simplex of P.

(c) Show that the collection of all such simplices yields a triangulation of P.

(d) Compute the h^*-vector of P.

Chapter 6

Partially Ordered Sets, Geometrically

Order is not sufficient. What is required, is something much more complex. It is order entering upon novelty; so that the massiveness of order does not degenerate into mere repetition; and so that the novelty is always reflected upon a background of system.
Alfred North Whitehead

Now we return to the combinatorics of posets of Chapter 2. We will employ the machinery of the previous chapters to study posets and their combinatorics from a geometric point of view. To do so, we will follow a sound approach of modern mathematics: objects can be understood by studying their relations to other objects through structure-preserving maps. The objects that we want to study in this chapter are posets and, as we pinpointed in Section 1.3, the structure preserving maps are the (strictly) order-preserving maps. We will consider the collection of *all* order-preserving maps from Π to a particular poset which, at the same time, gives Π a geometric incarnation. The **order cone** of a finite poset Π is the set

$$\mathsf{K}_\Pi \;:=\; \{\phi : \Pi \to \mathbb{R}_{\geq 0} \;:\; \phi \text{ order preserving}\}\,.$$

The order cone lives in the finite-dimensional vector space $\mathbb{R}^\Pi \cong \mathbb{R}^{|\Pi|}$ and, as the name suggests, is a polyhedral cone:

$$\text{(6.0.1)} \qquad \mathsf{K}_\Pi \;=\; \left\{ \phi \in \mathbb{R}^\Pi \;:\; \begin{array}{rl} 0 \leq \phi(m) & \text{for all } m \in \Pi \\ \phi(a) \leq \phi(b) & \text{for all } a \preceq_\Pi b \end{array} \right\}.$$

Our investigations of its geometric aspects will shed a different light on several results from Chapters 2 and 4.

Since we will exclusively be dealing with *finite* posets $(\Pi, \preceq)$, we decree in this chapter that $\Pi = [d]$. That is, we will view $\preceq$ as a partial order on the elements $1, 2, \ldots, d$; in particular, we might write $10 \prec 3$ even though this looks unnatural. We call Π **naturally labelled** if $i \preceq j$ implies $i \le j$ for all $i, j \in [d]$. As we will see, it will be at times convenient to assume that Π is naturally labelled.

With the convention $\Pi = [d]$, we have a canonical isomorphism $\mathbb{R}^\Pi \cong \mathbb{R}^d$ and we will interchangeably write ϕ_i and $\phi(i)$ when considering $\phi \in \mathbb{R}^\Pi$.

6.1. The Geometry of Order Cones

As we can see already from their definition, order cones inevitably live in high dimensions. It is thus all the more important to have some natural families and examples at hand to develop our geometric intuition.

We recall that the d-antichain A_d is the poset with elements $1, 2, \ldots, d$ and the trivial order relation $\preceq$, i.e., $i \preceq j$ implies $i = j$ for all $i, j \in \mathrm{A}_d$. Thus, in (6.0.1) there are no linear inequalities other than nonnegativity, and we conclude

$$\mathsf{K}_{\mathrm{A}_d} \;=\; \mathbb{R}^d_{\ge 0}\,.$$

Diametrically to antichains are the chains. It turns out that here a slight variation is helpful. For a permutation $\tau \in \mathfrak{S}_d$, we define a partial order $\preceq_\tau$ on $[d]$ through

$$i \preceq_\tau j \qquad :\Longleftrightarrow \qquad \tau^{-1}(i) \;\le\; \tau^{-1}(j)$$

for all $i, j \in [d]$. This is a total order on $[d]$ with minimal element $\tau(1)$ and maximal element $\tau(d)$, and a little bit of thinking reveals that the order cone of $([d], \preceq_\tau)$ is given by

$$\mathsf{K}_\tau \;:=\; \left\{ \phi \in \mathbb{R}^d \,:\, 0 \le \phi_{\tau(1)} \le \phi_{\tau(2)} \le \cdots \le \phi_{\tau(d)} \right\}. \tag{6.1.1}$$

Notice that this representation uses far less, namely d, of the $\binom{d+1}{2}$ inequalities of (6.0.1). A similar reduction can be observed for general order cones. We recall that $a \prec b$ is a **cover relation** in Π if there are no elements "between" a and b, i.e., there is no $p \in \Pi$ with $a \prec p \prec b$. In this case we use the notation $a \prec\!\!\cdot\; b$.

Proposition 6.1.1. *An irredundant representation of* K_Π *is given by*

$$\mathsf{K}_\Pi \;=\; \left\{ \phi \in \mathbb{R}^\Pi \,:\, \begin{array}{rl} 0 \le \phi(m) & \text{if } m \in \Pi \text{ is a minimum} \\ \phi(a) \le \phi(b) & \text{if } a \prec\!\!\cdot\; b \end{array} \right\}.$$

Proof. We show that the stated inequalities imply the ones in (6.0.1). For an order relation $a \preceq b$, there are elements $a_0, a_1, \ldots, a_k \in \Pi$ such that $a = a_0 \prec\!\cdot\, a_1 \prec\!\cdot\, \cdots \prec\!\cdot\, a_k = b$. Hence

$$\phi(b) - \phi(a) \;=\; (\phi(a_k) - \phi(a_{k-1})) + \cdots + (\phi(a_1) - \phi(a_0)) \;\geq\; 0\,.$$

Moreover, for any $a \in \Pi$ there is a minimum $m \in \Pi$ such that $m \preceq a$ and repeating our argument yields $0 \leq \phi(a)$. □

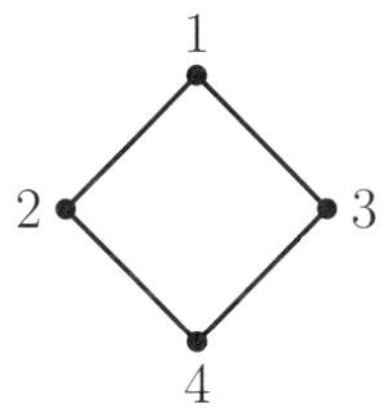

Figure 6.1. The poset $\diamond$.

Our next sample is the diamond poset $\diamond$ pictured in Figure 6.1, whose order cone

$$\mathsf{K}_\diamond \;=\; \left\{ \phi \in \mathbb{R}^{\diamond}_{\geq 0} \;:\; \begin{array}{ccc} \phi(1) & \geq & \phi(2) \\ \text{\rotatebox{90}{$\leq$}} & & \text{\rotatebox{90}{$\leq$}} \\ \phi(3) & \geq & \phi(4) \end{array} \right\}$$

made an appearance in Section 4.7, where we painted a geometric picture of plane partitions. Whereas the order cones $\mathsf{K}_{\mathrm{A}_d}$ and $\mathsf{K}_{\mathcal{T}}$ are simplicial (even unimodular), this is not true for $\mathsf{K}_\diamond$.

The dimension of $\mathsf{K}_\diamond$ is 4: the strictly order-preserving map

$$(\mathfrak{l}(4), \mathfrak{l}(3), \mathfrak{l}(2), \mathfrak{l}(1)) \;=\; (1, 2, 3, 4)$$

satisfies all defining inequalities strictly and hence lies in the interior of $\mathsf{K}_\diamond$. We can expand this argument to prove:

Proposition 6.1.2. *For a finite poset* Π, *the order cone* K_Π *is of dimension* $|\Pi|$.

Proof. Guided by the diamond-poset example, we will construct a strictly order-preserving map $\mathfrak{l} : \Pi \to [d]$ that strictly satisfies all defining inequalities of K_Π and hence lies in the interior of K_Π. This shows that $\mathsf{K}_\Pi \subset \mathbb{R}^\Pi$ is full dimensional.

We construct ϕ by induction on $d = |\Pi|$. For the base case $d = 1$, say, $\Pi = \{1\}$, we simply set $\mathfrak{l}(1) := 1$. For $d > 1$, let $M \in \Pi$ be a maximum. By induction, there is a strictly order preserving map $\mathfrak{l} : \Pi \setminus \{M\} \to [d-1]$. We can complete $\mathfrak{l}$ to a map $\Pi \to [d]$ by setting $\mathfrak{l}(M) = d$. □

Working from the bottom up, we could have also employed the strategy of picking the low-hanging fruits: setting $\mathfrak{l}(a_1) := 1$ for a minimum $a_1 \in \Pi$, then $\mathfrak{l}(a_2) := 2$ for a minimum $a_2 \in \Pi \setminus \{a_1\}$, and so on, we can construct a point in the interior of K_Π.

Any map $\mathfrak{l}$ constructed in this way is a **linear extension**, that is, a strictly order-preserving bijection

$$\mathfrak{l} : (\Pi, \preceq) \ \to \ ([d], \leq) .$$

Since we assume $\Pi = \{1, \dots, d\}$, we may view $\mathfrak{l}$ as a permutation of $[d]$. We denote the set of linear extensions of Π by $\operatorname{Lin}(\Pi)$. That is, $\operatorname{Lin}(\Pi)$ is the collection of all permutations $\mathfrak{l} \in \mathfrak{S}_d$ for which

$$(6.1.2) \qquad i \prec_\Pi j \quad \Longrightarrow \quad \mathfrak{l}(i) < \mathfrak{l}(j)$$

for all $i, j \in [d]$.

A poset Π is **connected** if $\Pi \neq \Pi_1 \uplus \Pi_2$ for some posets Π_1, Π_2, where $\uplus$ means that there are no relations between an element of Π_1 and an element of Π_2. That is, Π is connected if its Hasse diagram is connected as a graph. The following lemma implies that we lose nothing by restricting our study to connected posets. You are asked to supply a proof in Exercise 6.1.

Lemma 6.1.3. *Let* Π_1, Π_2 *be posets. Then*

$$\mathsf{K}_{\Pi_1 \uplus \Pi_2} \ = \ \mathsf{K}_{\Pi_1} \times \mathsf{K}_{\Pi_2} .$$

To study the facial structure of K_Π, we start by forcing equality in one of the irredundant inequalities of Proposition 6.1.1. If $m \in \Pi$ is a minimum, then

$$\mathsf{K}_\Pi \cap \{\phi(m) = 0\} \ = \ \mathsf{K}_{\Pi \setminus \{m\}} .$$

Figure 6.2 shows an example for the diamond poset.

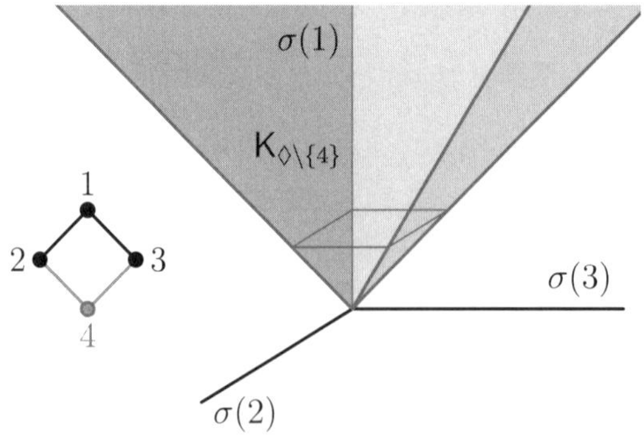

Figure 6.2. A facet of $\mathsf{K}_\diamond$ stemming from removing a minimum from $\diamond$.

The facet stemming from a cover relation $a \prec\!\cdot\, b$ is slightly more subtle to describe:

$$\mathsf{K}_\Pi \cap \{\phi(a) = \phi(b)\} \;=\; \mathsf{K}_{\Pi'}\,,$$

where Π' is the poset obtained from Π by identifying the elements a and b. Hence, the Hasse diagram of Π' is the result of contracting the edge corresponding to the cover relation $a \prec\!\cdot\, b$. This is plausible: every order-preserving map $\phi : \Pi \to \mathbb{R}_{\geq 0}$ that assigns a and b the same value is genuinely an element of $\mathsf{K}_{\Pi'}$. As we hinted at already, there is a subtlety here, and so for a poset Π, we define $\check{\Pi}$ as the poset obtained from Π by adding an element $\hat{0}$ that is smaller than every element in Π.

Theorem 6.1.4. *Let Π be a connected poset. For every surjective order-preserving map $\Psi : \Pi \to \check{\Pi}'$ there is a face of K_Π isomorphic to $\mathsf{K}_{\Pi'}$. Conversely, for every face $\mathsf{F} \preceq \mathsf{K}_\Pi$ there is a poset Π' and a surjective order-preserving map $\Psi : \Pi \to \check{\Pi}'$ such that $\mathsf{F} \cong \mathsf{K}_{\Pi'}$.*

Proof. Let $\Psi : \Pi \to \check{\Pi}'$ be a surjective order-preserving map. Every order-preserving map $\phi : \Pi' \to \mathbb{R}_{\geq 0}$ naturally extends to $\check{\Pi}'$ by setting $\phi(\hat{0}) := 0$. Hence we obtain a linear map $T : \mathsf{K}_{\Pi'} \to \mathsf{K}_\Pi$ that maps ϕ to $\phi \circ \Psi$, and $\mathsf{K}_{\Pi'}$ is isomorphic to $T(\mathsf{K}_{\Pi'})$. An order-preserving map $\phi : \Pi \to \mathbb{R}_{\geq 0}$ is in the image of T if and only if

$$\phi(a) = \phi(b) \;\text{ if }\; \Psi(a) = \Psi(b) \qquad \text{and} \qquad \phi(a) = 0 \;\text{ if }\; \Psi(a) = \hat{0}\,.$$

It follows from (6.0.1) that these conditions define a nonempty face F.

For the converse statement, for a given face $\mathsf{F} \preceq \mathsf{K}_\Pi$ we define an equivalence relation $\sim$ on $\check{\Pi}$ by setting $a \sim b$ if $\phi(a) = \phi(b)$ for all $\phi \in \mathsf{F}$. We use the usual notation $\overline{a}$ for the equivalence class of a and $\check{\Pi}' := \Pi/\!\sim$ for the set of all equivalence classes. In Exercise 6.2 you are asked to show that $\check{\Pi}'$ is a poset with partial order $\overline{a} \preceq' \overline{b}$ if there are elements $a_0, a_1, \dots, a_k \in \check{\Pi}$ such that $a \sim a_0 \preceq a_1 \sim a_2 \preceq \cdots \preceq a_k \sim b$. The map $\check{\Pi} \to \check{\Pi}'$ that takes a to $\overline{a}$ is surjective and order preserving and thus finishes the proof. $\square$

Proposition 6.1.2 now immediately implies:

Corollary 6.1.5. *Let F be a face of K_Π with corresponding surjective order preserving map $\Pi \to \check{\Pi}'$. Then $\dim \mathsf{F} = |\Pi'|$.*

Theorem 6.1.4 and Corollary 6.1.5 tell us the generators of K_Π. Namely, if $\mathrm{cone}(\mathbf{u})$ is a ray of K_Π with corresponding surjective order-preserving map $\Pi \to \check{\Pi}'$, then $\check{\Pi}'$ is a chain on two elements.

We recall from Chapter 2 that a subset $F \subseteq \Pi$ is a **filter** if $a \in F$ and $b \succeq a$ imply $b \in F$. For instance, Figure 6.3 shows the five nonempty filters of $\Diamond$. We also recall that $F \subseteq \Pi$ is a filter if and only if $\Pi \setminus F$ is an order

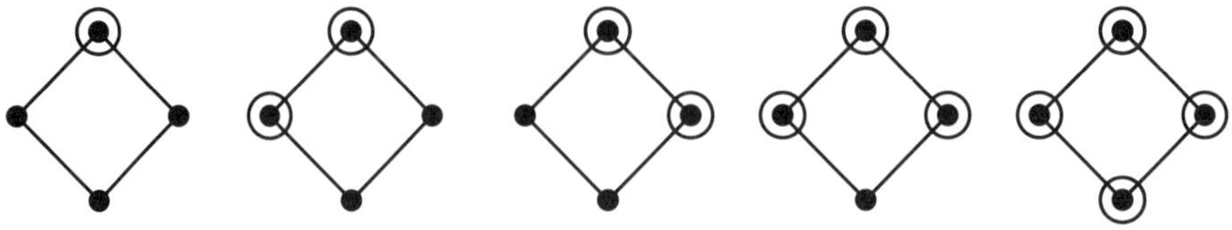

Figure 6.3. The five nonempty filters of $\diamond$.

ideal, and so filters and order ideals of Π are in one-to-one correspondence. Thus filters are in bijection with order-preserving maps $\phi : \Pi \to [2]$, and so filters get us on the trail started in the last paragraph. For a filter $F \subseteq \Pi$ we write $\mathbf{e}_F : \Pi \to \{0, 1\}$ for the order-preserving map

$$\mathbf{e}_F(a) \;=\; \begin{cases} 1 & \text{if } a \in F, \\ 0 & \text{otherwise.} \end{cases}$$

Every subset of the antichain A_d is a filter. But not every $\mathbf{e}_F$ spans a generator of the corresponding order cone $\mathsf{K}_{\mathrm{A}_d} \cong \mathbb{R}^d$, which has exactly d generators; so some care is needed.

Theorem 6.1.6. *Let Π be a poset. A minimal set of generators for K_Π is given by $\{\mathbf{e}_F : \varnothing \neq F \subseteq \Pi$ connected filter$\}$.*

For example, $\mathsf{K}_\diamond$ is a 4-dimensional polyhedral cone with five facets and five generators given by the filters corresponding to

$$\{1\}, \quad \{1, 2\}, \quad \{1, 3\}, \quad \{1, 2, 3\}, \quad \{1, 2, 3, 4\}.$$

Since $\mathsf{K}_\diamond$ is a pointed cone, we may intersect it with a hyperplane meeting all its generators to get a 3-dimensional picture of it; see Figure 6.4.

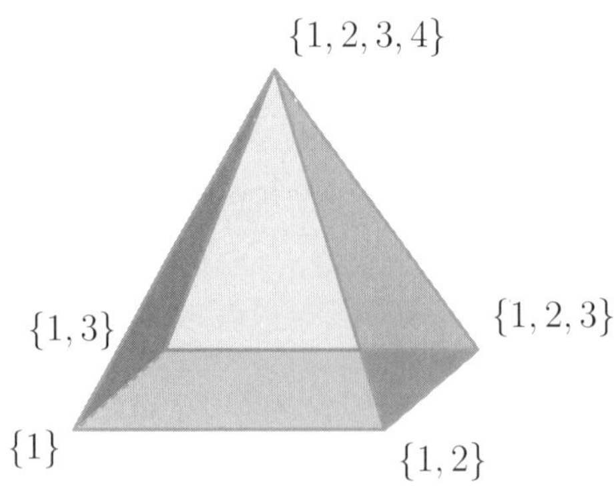

Figure 6.4. A 3-dimensional representation of $\mathsf{K}_\diamond$.

Proof. We first show that the $\mathbf{e}_F$ for nonempty connected filters F generate K_Π. In fact, it suffices to show that the $\mathbf{e}_F$ for *all* filters—connected or not—generate K_Π: if $F = F_1 \uplus F_2$ is a disconnected filter, then $\mathbf{e}_F = \mathbf{e}_{F_1} + \mathbf{e}_{F_2}$.

Let $\phi \in \mathsf{K}_\Pi$. We will prove that ϕ is a nonnegative linear combination of $\mathbf{e}_F$ for some nonempty filters F by induction on

$$s(\phi) \;:=\; |\{a \in \Pi : \phi(a) > 0\}|\,,$$

the size of the support of ϕ. If $s(\phi) = 0$, then $\phi(a) = 0$ for all $a \in \Pi$ and there is nothing to prove. For $s(\phi) > 0$, the key observation is that

$$F \;:=\; \{a \in \Pi : \phi(a) > 0\}$$

is a filter: if $b \succeq a$ and $a \in F$, then $\phi(b) \geq \phi(a) > 0$ and $b \in F$. Let $\mu := \min\{\phi(a) : a \in F\}$ and define $\phi' := \phi - \mu\,\mathbf{e}_F$. By construction,

$$\phi(a) \geq \mu \;\text{ if }\; a \in F \qquad \text{and} \qquad \phi(a) = 0 \;\text{ otherwise.}$$

Consequently, $\phi'(a) \geq 0$ for all $a \in \Pi$. Moreover, for $b \succeq a$, we have $\phi'(b) - \phi'(a) = \phi(b) - \phi(a) \geq 0$ whenever $a \in F$ or $b \notin F$. If $a \notin F$ but $b \in F$, then $\phi(a) = 0$ and

$$\phi'(b) - \phi'(a) \;=\; \phi(b) - \mu \;\geq\; 0\,.$$

This shows that $\phi' \in \mathsf{K}_\Pi$. There is at least one element $a \in F$ with $\phi(a) = \mu$. This yields $s(\phi') < s(\phi)$ and so we are done by induction and the fact that $\phi = \phi' + \mu\,\mathbf{e}_F$.

To show that we cannot forgo any single connected filter $F \neq \varnothing$, suppose that

$$\mathbf{e}_F \;=\; \lambda_1\,\mathbf{e}_{F_1} + \lambda_2\,\mathbf{e}_{F_2} + \cdots + \lambda_r\,\mathbf{e}_{F_r}\,,$$

where $F_1, \dots, F_r$ are nonempty connected filters and $\lambda_1, \dots, \lambda_r > 0$. We may further assume that r is minimal. If $a \in F_i$ for some $1 \leq i \leq r$, then $\mathbf{e}_F(a) \geq \lambda_i\,\mathbf{e}_{F_i}(a) > 0$ and hence $F_i \subseteq F$. From the minimality of r it follows that there is a minimal element $m \in F_1$ that is not contained in any F_j with $j > 1$. From $1 = \mathbf{e}_F(m) = \lambda_1\,\mathbf{e}_{F_1}(m)$ we deduce that $\lambda_1 = 1$. Since we assume that F is connected, there is an element $a \in F_1$ such that $a \in F_j$ for some $1 < j \leq r$—otherwise we would have $F = F_1 \uplus F'$ with $F' = F_2 \cup \cdots \cup F_r$. But then

$$1 \;=\; \mathbf{e}_F(a) \;\geq\; \lambda_1\,\mathbf{e}_{F_1}(a) + \lambda_j\,\mathbf{e}_{F_j}(a) \;=\; 1 + \lambda_j\,,$$

which implies $\lambda_j = 0$. Since this contradicts our assumption that r is minimal, it can only be that $r = 1$. □

6.2. Subdivisions, Linear Extensions, and Permutations

We now turn to the question of how to subdivide an order cone. We have already seen a subdivision (in fact, a triangulation) of $\mathsf{K}_\diamond$ in (4.7.1):

$$\mathsf{K}_\diamond \;=\; \left\{\mathbf{x} \in \mathbb{R}^4_{\geq 0} \,:\, x_4 \leq x_3 \leq x_2 \leq x_1\right\} \cup \left\{\mathbf{x} \in \mathbb{R}^4_{\geq 0} \,:\, x_4 \leq x_2 \leq x_3 \leq x_1\right\}.$$

The two cones on the right-hand side are order cones themselves, namely, those for the permutations $\tau, \tau' \in \mathfrak{S}_4$ given by

$$\begin{aligned} \tau\,(1) &= 4\,, & \tau\,(2) &= 3\,, & \tau\,(3) &= 2\,, & \tau\,(4) &= 1\,, \\ \tau'(1) &= 4\,, & \tau'(2) &= 2\,, & \tau'(3) &= 3\,, & \tau'(4) &= 1\,. \end{aligned}$$

In this example we can witness that the original order cone is subdivided into order cones. This suggests studying the relationship between pairs of posets and their respective order cones. For two partial orders $\preceq_1, \preceq_2$ on the same ground set Π we say that $\preceq_2$ **refines** $\preceq_1$ if for all $a, b \in \Pi$

$$a \;\preceq_1\; b \qquad \Longrightarrow \qquad a \;\preceq_2\; b\,. \tag{6.2.1}$$

Informally speaking, $\preceq_2$ possibly puts more relations on Π than $\preceq_1$. We can say the following on the geometric side; since the ground set Π is fixed for now, we simply write $\mathsf{K}_\preceq$ for the order cone of the poset $(\Pi, \preceq)$.

Proposition 6.2.1. *Let $\preceq_1$ and $\preceq_2$ be partial orders on Π. Then $\preceq_2$ refines $\preceq_1$ if and only if $\mathsf{K}_{\preceq_2} \subseteq \mathsf{K}_{\preceq_1}$.*

Proof. For $i = 1, 2$, let

$$R_i \;:=\; \left\{(a,b) \in \Pi \times \Pi \,:\, a \preceq_i b\right\}.$$

Then $\preceq_1$ is refined by $\preceq_2$ if and only if $R_1 \subseteq R_2$. By construction (6.0.1), $(a, b) \in R_i$ if and only if $\mathsf{K}_{\preceq_i}$ is contained in the halfspace

$$\left\{\phi \in \mathbb{R}^\Pi \,:\, \phi(a) \leq \phi(b)\right\}.$$

Hence, $R_1 \subseteq R_2$ if and only if $\mathsf{K}_{\preceq_2} \subseteq \mathsf{K}_{\preceq_1}$. □

From the definition (or, geometrically, from Proposition 6.2.1) it becomes apparent that *refinement* is actually a partial order on posets with the same ground set. To make use of this, we define for a given poset $(\Pi, \preceq)$, the partially ordered set

$$\mathcal{N}(\Pi, \preceq) \;:=\; \left\{\preceq' \,:\, \preceq' \text{ refines } \preceq\right\}.$$

By construction, $\preceq$ is the maximum in $\mathcal{N}(\Pi, \preceq)$. What are the minimal elements? If $\preceq_2$ strictly refines $\preceq_1$, then there are elements $a, b \in \Pi$ that are **incomparable**, i.e., $a \not\preceq_1 b$ and $b \not\preceq_1 a$, but comparable with respect to $\preceq_2$. Thus, $\preceq'$ cannot be refined if and only if for any $a, b \in \Pi$, either $a \preceq' b$ or $b \preceq' a$. That is, $\preceq'$ is a minimal element of $\mathcal{N}(\Pi, \preceq)$ if and only if $(\Pi, \preceq')$ is a linear or total order.

Since we assume that $\Pi = \{1, 2, \ldots, d\}$, any total order is of the form $\preceq_\tau$ for some $\tau \in \mathfrak{S}_d$. Now, if $\preceq_\tau$ refines $\preceq$, then (6.2.1) together with the definition of $\preceq_\tau$ yields

$$i \prec_\Pi j \implies \tau^{-1}(i) < \tau^{-1}(j)$$

for all $i, j \in [d]$. That is, $\preceq_\tau$ refines $\preceq$ if and only if $\tau^{-1} \in \mathrm{Lin}(\Pi)$. We define the **Jordan–Hölder set** of Π as

$$\begin{aligned} \mathrm{JH}(\Pi) &:= \{\tau \in \mathfrak{S}_d : \tau^{-1} \in \mathrm{Lin}(\Pi)\} \\ &= \{\tau \in \mathfrak{S}_d : \preceq_\tau \text{ refines } \preceq\}. \end{aligned}$$

A linear extension $\mathfrak{l} \in \mathrm{Lin}(\Pi)$ tells us the position $\mathfrak{l}(a)$ of an element $a \in \Pi$ in a linear order that respects the partial order $\preceq$. Dually, each $\tau \in \mathrm{JH}(\Pi)$ gives the elements of Π labels: the element $\tau(1) \in \Pi$ is the minimal element in the refined order $\preceq_\tau$.

The Jordan–Hölder set $\mathrm{JH}(\Pi)$ is exactly the set of minimal elements of $\mathcal{N}(\Pi, \preceq)$. In Exercise 6.3 you will explore some further properties of $\mathcal{N}(\Pi, \preceq)$. Now, for each $\preceq' \in \mathcal{N}(\Pi, \preceq)$, we have $\mathsf{K}_{\preceq'} \subseteq \mathsf{K}_\Pi$ by Proposition 6.2.1. This gives plenty of supply of nice cones to subdivide K_Π.

A **crosscut** in a poset $\mathcal{N}$ is an antichain $\{c_1, \ldots, c_s\} \subseteq \mathcal{N} \setminus \{\hat{0}\}$ such that for every maximal chain $C \subseteq \mathcal{N}$, there is a unique $c_i \in C$.

Theorem 6.2.2. *Let $(\Pi, \preceq)$ be a poset and $\mathcal{N} = \mathcal{N}(\Pi, \preceq)$ its poset of refinements. Let $\preceq_1, \preceq_2, \ldots, \preceq_s \in \mathcal{N}$ be a collection of refinements of Π. Then*

$$\mathsf{K}_\Pi = \mathsf{K}_{\preceq_1} \cup \mathsf{K}_{\preceq_2} \cup \cdots \cup \mathsf{K}_{\preceq_s}$$

is a dissection of K_Π if and only if $\preceq_1, \preceq_2, \ldots, \preceq_s$ is a crosscut in $\mathcal{N}$ such that every minimal element is covered uniquely.

Proof. To ease notation, we set $\mathsf{K}_i = \mathsf{K}_{\preceq_i}$ for $i = 1, \ldots, s$. We first assume that $\preceq_1, \preceq_2, \ldots, \preceq_s$ is a crosscut. Due to Proposition 6.2.1, we only need to show that $\mathsf{K}_\Pi \subseteq \mathsf{K}_1 \cup \cdots \cup \mathsf{K}_s$ and that $\dim(\mathsf{K}_i \cap \mathsf{K}_j) < |\Pi|$ for all $i \neq j$.

We may assume that Π is naturally labelled, i.e., $i \prec_\Pi j$ implies $i < j$ for all $i, j \in \Pi = [d]$. For any $\phi \in \mathsf{K}_\Pi$, we define a refinement $\preceq_\phi$ of $\preceq$ as follows. For $i, j \in \Pi$, we set

$$i \preceq_\phi j \quad \text{if} \quad \phi_i < \phi_j \quad \text{or} \quad (\phi_i = \phi_j \text{ and } i \le j).$$

You are asked to check in Exercise 6.4 that $\preceq_\phi$ is a partial order relation on Π and, in fact, a linear order. Moreover, if $i \prec j$, then $\phi_i \le \phi_j$ by assumption and $i < j$ by the fact that our labeling is natural; hence $i \prec_\phi j$. Consequently, $\preceq_\phi$ is a minimal element in $\mathcal{N}$ and $\phi \in \mathsf{K}_{\preceq_\phi}$. By definition of a crosscut, any maximal chain in $\mathcal{N}$ that starts in $\preceq_\phi$ contains $\preceq_i$ for some unique $1 \le i \le s$. That is, $\preceq_i$ is refined by $\preceq_\phi$ and therefore $\phi \in \mathsf{K}_{\preceq_\phi} \subseteq \mathsf{K}_{\preceq_i}$ using Proposition 6.2.1. Hence $\mathsf{K}_\Pi \subseteq \mathsf{K}_1 \cup \cdots \cup \mathsf{K}_s$.

If $\dim \mathsf{K}_i \cap \mathsf{K}_j = |\Pi|$, then pick a generic ϕ in $\mathsf{K}_i \cap \mathsf{K}_j$. Genericity implies that ϕ is in the interior of $\mathsf{K}_{\preceq_\phi}$ and hence $\mathsf{K}_{\preceq_\phi} \subseteq \mathsf{K}_i \cap \mathsf{K}_j$. Proposition 6.2.1 implies that $\preceq_\phi$ refines both $\preceq_i$ and $\preceq_j$ which contradicts the assumption that every minimal element is uniquely covered.

For sufficiency, we can play the arguments backwards: if $\preceq_i$ refines $\preceq_j$, then $\mathsf{K}_i \subseteq \mathsf{K}_j$, which would contradict the assumption that $\mathsf{K}_1, \dots, \mathsf{K}_s$ form a dissection. Every maximal chain C in $\mathcal{N}$ starts in a minimal element $\preceq_\tau$ with $\tau \in \mathrm{JH}(\Pi)$. For any ϕ in the relative interior of K_τ, there is a unique i with $\phi \in \mathsf{K}_i$ and from Proposition 6.2.1, we deduce that $\preceq_i \in C$. Thus, $\preceq_1, \dots, \preceq_s$ form a crosscut such that every minimal element is uniquely covered. □

A canonical crosscut of a poset that uniquely covers the minimal elements is the set of minimal elements itself. Hence the elements in $\mathrm{JH}(\Pi)$ yield a subdivision of K_Π. Moreover, the cones K_τ for $\tau \in \mathrm{JH}(\Pi)$ are simplicial and even unimodular. We record the result of applying Theorem 6.2.2 in the following important corollary.

Corollary 6.2.3. *Let* Π *be a finite poset. Then*

$$\mathsf{K}_\Pi \; = \bigcup_{\tau \in \mathrm{JH}(\Pi)} \mathsf{K}_\tau$$

is a dissection into unimodular cones.

In Exercise 6.5, you are asked to verify that this is actually a triangulation of K_Π, i.e., that the cones meet in faces.

Figure 6.5 illustrates the two linear extensions of $\Diamond$. Thinking back to (4.7.1) we now realize that the two maximal simplicial cones in the triangulation of $\mathsf{K}_\Diamond$ in Section 4.7 are naturally indexed by the two linear extensions of $\Diamond$.

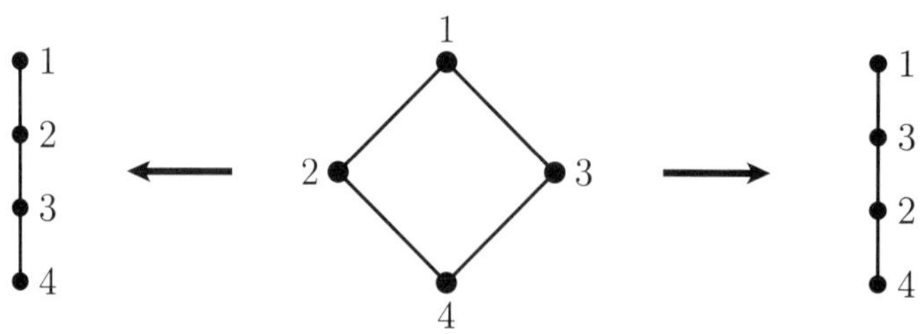

Figure 6.5. The two linear extensions of $\Diamond$.

For the antichain A_d, *every* permutation $\tau \in \mathfrak{S}_d$ represents a linear extension, and $\mathrm{JH}(A_d) = \mathfrak{S}_d$. Corollary 6.2.3 yields

$$\mathbb{R}^d_{\geq 0} \;=\; \mathsf{K}_{\mathrm{A}_d} \;=\; \bigcup_{\tau \in \mathfrak{S}_d} \mathsf{K}_{\Gamma_\tau}\,. \tag{6.2.2}$$

Figure 6.6 shows a picture for $d = 3$.

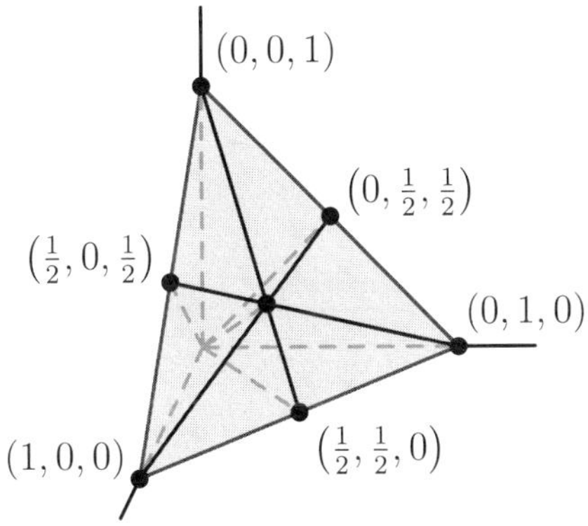

Figure 6.6. The triangulation (6.2.2) for $d = 3$.

Our next goal is to understand *all* the half-open decompositions of the triangulation of K_Π furnished by Corollary 6.2.3. We recall that for $\tau \in \mathfrak{S}_d$, every point $\phi \in \mathsf{K}^\circ_\tau$ satisfies

$$0 \;<\; \phi_{\tau(1)} \;<\; \phi_{\tau(2)} \;<\; \cdots \;<\; \phi_{\tau(d)}\,.$$

In particular, any such ϕ is generic relative to K_σ, for all $\sigma \in \mathfrak{S}_d$. The key insight, formalized below, is that $\mathbb{H}_\phi \mathsf{K}_\sigma$ will depend only on the two permutations τ and σ.

We recall that for a permutation $\rho \in \mathfrak{S}_d$, an index $1 \leq i \leq d-1$ is a descent if $\rho(i) > \rho(i+1)$. We collect the descents of ρ in the set

$$\mathrm{Des}(\rho) \;:=\; \{i \in [d-1] \,:\, \rho(i) > \rho(i+1)\}\,.$$

Lemma 6.2.4. *Let $\tau, \sigma \in \mathfrak{S}_d$ and let $\phi \in \mathsf{K}^\circ_\tau$. Then*

$$\mathbb{H}_\phi \mathsf{K}_\sigma \;=\; \left\{ \mathbf{x} \in \mathbb{R}^d \,:\, \begin{array}{l} 0 \leq x_{\sigma(1)} \leq x_{\sigma(2)} \leq \cdots \leq x_{\sigma(d)} \\ x_{\sigma(i)} < x_{\sigma(i+1)} \text{ if } i \in \mathrm{Des}\left(\tau^{-1} \circ \sigma\right) \end{array} \right\}.$$

Proof. We need to determine the indices i for which $\phi_{\sigma(i)} > \phi_{\sigma(i+1)}$. By construction,

$$\phi_s \;>\; \phi_t \quad \text{if and only if} \quad \tau^{-1}(s) \;>\; \tau^{-1}(t) \tag{6.2.3}$$

for $1 \leq s, t \leq d$. Hence $\phi_{\sigma(i)} > \phi_{\sigma(i+1)}$ if and only if

$$\tau^{-1}(\sigma(i)) \;>\; \tau^{-1}(\sigma(i+1))\,,$$

that is, if and only if $i \in \mathrm{Des}(\tau^{-1} \circ \sigma)$. □

If i is not a descent of ρ, then it is an ascent: $\rho(i) < \rho(i+1)$. We likewise record the ascents of ρ in the set $\mathrm{Asc}(\rho)$. The half-open cone reciprocal to $\mathbb{H}_\phi \mathsf{K}_\sigma$ excludes the complementary facets, that is, those corresponding to ascents of $\tau^{-1} \circ \sigma$ and, additionally, the facet $x_{\sigma(1)} = 0$:

$$\mathbb{H}^\phi \mathsf{K}_\sigma \;=\; \left\{ \mathbf{x} \in \mathbb{R}^d \;:\; \begin{array}{l} 0 < x_{\sigma(1)} \le x_{\sigma(2)} \le \cdots \le x_{\sigma(d)} \\ x_{\sigma(i)} < x_{\sigma(i+1)} \text{ if } i \in \mathrm{Asc}(\tau^{-1} \circ \sigma) \end{array} \right\}.$$

If Π is naturally labelled, then the identity permutation $\tau(i) = i$ for $i \in [d]$ is in $\mathrm{JH}(\Pi)$ and we can conveniently construct a half-open decomposition with respect to $\phi(i) := i$ for all $i \in \Pi$. In anticipation of the following sections, we exercise this on the antichain A_d. Corollary 5.3.5 gives in this case

$$\mathbb{R}^d_{\ge 0} \;=\; \biguplus_{\sigma \in \mathfrak{S}_d} \left\{ \mathbf{x} \in \mathbb{R}^d \;:\; \begin{array}{l} 0 \le x_{\sigma(1)} \le x_{\sigma(2)} \le \cdots \le x_{\sigma(d)} \\ x_{\sigma(i)} < x_{\sigma(i+1)} \text{ if } i \in \mathrm{Des}(\sigma) \end{array} \right\}, \tag{6.2.4}$$

$$\mathbb{R}^d_{> 0} \;=\; \biguplus_{\sigma \in \mathfrak{S}_d} \left\{ \mathbf{x} \in \mathbb{R}^d \;:\; \begin{array}{l} 0 < x_{\sigma(1)} \le x_{\sigma(2)} \le \cdots \le x_{\sigma(d)} \\ x_{\sigma(i)} < x_{\sigma(i+1)} \text{ if } i \in \mathrm{Asc}(\sigma) \end{array} \right\}. \tag{6.2.5}$$

6.3. Order Polytopes and Order Polynomials

For a given poset $(\Pi, \preceq)$, let

$$\widehat{\Pi} \;:=\; \Pi \cup \{\hat{1}\},$$

the poset obtained by adding an element $\hat{1}$ to Π that is larger than any member of Π and keeping the remaining relations. By construction, every order-preserving map $\phi \in \mathsf{K}_{\widehat{\Pi}}$ satisfies $\phi(a) \le \phi(\hat{1})$ for all $a \in \Pi$. In particular, all nonempty filters of $\widehat{\Pi}$ contain $\hat{1}$ and, using Theorem 6.1.6, we find that every ray of $\mathsf{K}_{\widehat{\Pi}}$ inevitably meets the hyperplane $\mathsf{H} = \{\phi : \phi(\hat{1}) = 1\}$. The polytope $\mathsf{H} \cap \mathsf{K}_{\widehat{\Pi}}$ is called the **order polytope** of Π and is given by

$$\mathsf{O}_\Pi \;:=\; \left\{ \phi \in \mathbb{R}^\Pi \;:\; \begin{array}{ll} 0 \le \phi(p) \le 1 & \text{for all } p \in \Pi \\ \phi(a) \le \phi(b) & \text{for all } a \preceq b \end{array} \right\}. \tag{6.3.1}$$

Our two polyhedral constructions for posets Π are intimately connected: the homogenization $\mathrm{hom}(\mathsf{O}_\Pi)$ is canonically isomorphic to $\mathsf{K}_{\widehat{\Pi}}$; see Exercise 6.7. The correspondence between polytopes and their homogenizations allows us to deduce several properties of O_Π from those of $\mathsf{K}_{\widehat{\Pi}}$. To start with a simple implication,

$$\dim \mathsf{O}_\Pi \;=\; \dim \mathsf{K}_{\widehat{\Pi}} - 1 \;=\; |\widehat{\Pi}| - 1 \;=\; |\Pi|\,.$$

More importantly, all nonempty filters of $\widehat{\Pi}$ contain $\hat{1}$ and hence are connected. Theorem 6.1.6 thus shows the following.

Corollary 6.3.1. *Let Π be a finite poset. The vertices of O_Π are exactly $\mathbf{e}_F$, where $F \subseteq \Pi$ ranges over all filters of Π.*

Proof. The vertices of O_Π correspond to the rays of $\mathsf{K}_{\widehat{\Pi}}$. Theorem 6.1.6 says that the rays are spanned by the nonempty connected filters of $\widehat{\Pi}$. All nonempty filters of $\widehat{\Pi}$ are connected and the map $F \mapsto F \cup \{\hat{1}\}$ gives a bijection between the filters of Π and the nonempty filters of $\widehat{\Pi}$. □

We can verify this corollary for the sample posets we have seen so far: if $\Pi = \mathrm{A}_d$ is an antichain, every subset is a filter, and so the vertices of O_Π are all possible vectors with 0/1 entries. This confirms that $\mathsf{O}_\Pi = [0,1]^d$.

If $\Pi = [d]$ is a (naturally labelled) chain, the filters are of the form $\{j, j+1, \dots, d\}$ for any $0 \le j \le d$, and so O_Π is a (unimodular) simplex with vertices $\mathbf{0}$, $\mathbf{e}_d$, $\mathbf{e}_{d-1} + \mathbf{e}_d$, $\dots$, $\mathbf{e}_1 + \mathbf{e}_2 + \cdots + \mathbf{e}_d$. (Similar simplices surfaced in Section 5.1.)

If $\Pi = \diamond$, the filters are the ones shown in Figure 6.3 plus the empty filter, and so O_Π has vertices $\mathbf{0}$, $\mathbf{e}_1$, $\mathbf{e}_1 + \mathbf{e}_2$, $\mathbf{e}_1 + \mathbf{e}_3$, $\mathbf{e}_1 + \mathbf{e}_2 + \mathbf{e}_3$, and $\mathbf{e}_1 + \mathbf{e}_2 + \mathbf{e}_3 + \mathbf{e}_4$.

Corollary 6.3.1 says, in particular, that any order polytope is a lattice polytope. Stronger even, its vertices have coordinates in $\{0,1\}$ and so an order polytope is a 0/1-polytope (like those appearing in Exercise 5.48). By Ehrhart's Theorem (Corollary 5.1.2), $\mathrm{ehr}_{\mathsf{O}_\Pi}(n)$ is a polynomial in n. In fact, we know this polynomial quite well.

Proposition 6.3.2. *Let Π be a finite poset. Then*

$$\Omega_\Pi(n) \quad = \quad \mathrm{ehr}_{\mathsf{O}_\Pi}(n-1)\,.$$

Proof. Let $f \in (n-1)\mathsf{O}_\Pi \cap \mathbb{Z}^\Pi$. Thus, by definition, f is an order-preserving map with range $f(\Pi) \subseteq \{0, 1, \dots, n-1\}$. So, if we define the map ϕ by $\phi(p) := f(p) + 1$ for $p \in \Pi$, then ϕ is order preserving with $\phi(\Pi) \subseteq [n]$. This argument works also in the other direction and proves that the lattice points in $(n-1)\mathsf{O}_\Pi$ are in bijection with order-preserving maps into the n-chain. □

Proposition 6.3.2, with assistance from Corollaries 5.1.2 and 6.3.1, gives an alternative, geometric proof that $\Omega_\Pi(n)$ agrees with a polynomial of degree $|\Pi|$ (Proposition 1.3.1). This geometric point of view also yields an alternative proof for the reciprocity theorem for order polynomials.

Second proof of Theorem 1.3.2. Essentially by the same argument as in our proof of Proposition 6.3.2, we realize that the lattice points in $(n+1)\mathsf{O}_\Pi^\circ$ are in bijection with strictly order-preserving maps into the n-chain and hence

$$\Omega_\Pi^\circ(n) \;=\; \mathrm{ehr}_{\mathsf{O}_\Pi^\circ}(n+1)\,. \tag{6.3.2}$$

Thus, by Ehrhart–Macdonald reciprocity (Theorem 5.2.3),

$$\Omega_\Pi^\circ(-n) \;=\; \mathrm{ehr}_{\mathsf{O}_\Pi^\circ}(-n+1) \;=\; (-1)^{|\Pi|}\,\mathrm{ehr}_{\mathsf{O}_\Pi}(n-1) \;=\; (-1)^{|\Pi|}\Omega_\Pi(n)\,. \ \square$$

Much more can be said about the facial structure of O_Π, and we defer this to Exercises 6.8 and 6.12. In the remainder of this section, we will focus on a canonical subdivision of O_Π. Of course, Theorem 6.2.2 and, in particular, Corollary 6.2.3 apply to O_Π. We choose, however, to use the combinatorial triangulation of Section 5.7. To start, we need to fix an ordering on the vertices of O_Π, that is, the filters of Π. We already used a *partial* order on filters: we recall from Section 2.1 that the collection of filters $\mathcal{J}(\Pi)$ of Π ordered by inclusion is the Birkhoff lattice. Now choose any refinement to a total order, i.e., pick any linear extension of $\mathcal{J}(\Pi)$. In Exercise 6.13, you will show that this choice will not affect the resulting triangulation.

Following the procedure outlined in (5.7.1), we pick the first vertex in our chosen order, which is the unique minimum $F = \varnothing$ of $\mathcal{J}(\Pi)$. Next, we need to determine the faces of O_Π that do not contain $\mathbf{e}_F = \mathbf{0}$. Actually, we need to know only the facets of O_Π, and from Proposition 6.1.1 applied to $\mathsf{K}_{\hat{\Pi}}$, we infer that these are exactly

$$\mathsf{F}_M \;=\; \mathsf{O}_\Pi \cap \left\{\phi \in \mathbb{R}^\Pi \,:\, \phi(M) = 1\right\}$$

for a maximum $M \in \Pi$.

Theorem 6.3.3. *Let Π be a finite poset. The simplices of a pulling triangulation of O_Π for an ordering that refines that of $\mathcal{J}(\Pi)$ are given by*

$$\mathsf{Z}(\mathcal{F}) \;:=\; \mathrm{conv}\left(\mathbf{e}_{F_0}, \mathbf{e}_{F_1}, \dots, \mathbf{e}_{F_k}\right),$$

where $\mathcal{F} = \{F_0 \subset F_1 \subset \cdots \subset F_k\}$ is a chain of filters in Π. In particular, as a partially ordered set, $\mathrm{Pull}(\mathsf{O}_\Pi)$ *is isomorphic to the order complex* $\Delta(\mathcal{J}(\Pi))$.

Proof. We prove the claim by induction on $d = |\Pi|$. For $d = 1$ and $\Pi = \{a\}$, O_Π is already a simplex with vertices $\mathbf{e}_\varnothing = 0$ and $\mathbf{e}_\Pi = 1$.

For $d \geq 2$, according to (5.7.1), the pulling triangulation of O_Π is obtained by adding the vertex $\mathbf{e}_\varnothing$ to all simplices of $\mathrm{Pull}(\mathsf{F}_M)$ for a maximum $M \in \Pi$. Now, F_M is linearly isomorphic to $\mathsf{O}_{\Pi \setminus \{M\}}$ by projecting onto $\mathbb{R}^{\Pi \setminus \{M\}}$. The vertex to pull in F_M is $\mathbf{e}_{\{M\}}$ which, under the linear isomorphism, corresponds to the empty filter in $\Pi \setminus \{M\}$. Hence, by induction, the simplices in $\mathrm{Pull}(\mathsf{F}_M)$ are given by chains of filters of the form

$$\{M\} \;\subseteq\; F_1 \;\subset\; F_2 \;\subset\; \cdots \;\subset\; F_k\,,$$

and adding $F_0 = \varnothing$ to these chains completes the proof. □

We will denote the triangulation of Theorem 6.3.3 by $\mathcal{T}_\Pi$ and refer to it as the **canonical triangulation** of O_Π. To be honest, the pulling triangulation of Theorem 6.3.3 is actually the triangulation of $\mathsf{K}_{\widehat{\Pi}}$ given in Corollary 6.2.3 restricted to the hyperplane

$$\left\{\phi \in \mathbb{R}^{\widehat{\Pi}} \,:\, \phi(\hat{1}) = 1\right\}.$$

To make this more concrete, we need to relate linear extensions of Π to saturated chains in $\mathcal{J}(\Pi)$. We saw this connection already a couple of times. If $\mathfrak{l} \in \mathrm{Lin}(\Pi)$ is a linear extension of Π, then

$$F_i \;:=\; \{a \in \Pi \,:\, \mathfrak{l}(a) > |\Pi| - i\}\,, \tag{6.3.3}$$

for $i = 0, \ldots, |\Pi|$, yields a maximal chain of filters. Conversely, if $F_0 \subset F_1 \subset \cdots \subset F_k$ is a saturated chain of filters, then $F_0 = \varnothing$, $F_k = \Pi$, and $|F_i \setminus F_{i-1}| = 1$ for all $1 \le i \le k$; see Exercise 6.14. In particular, $k = |\Pi|$. If we define a map $\mathfrak{l} : \Pi \to [k]$ by

$$\mathfrak{l}(a) \;:=\; \min\,(i : a \in F_i) \tag{6.3.4}$$

for $a \in \Pi$, then you may verify that $\mathfrak{l}$ is a linear extension. This shows the following.

Lemma 6.3.4. *Saturated chains in $\mathcal{J}(\Pi)$ are in bijection with linear extensions of Π.*

Every linear extension of $\widehat{\Pi}$ yields a linear extension of Π and vice versa.

Proposition 6.3.5. *Let $\mathcal{F}$ be a saturated chain of filters in Π and let $\mathfrak{l}$ be the corresponding linear extension. The simplex $\mathsf{Z}(\mathcal{F})$ of $\mathcal{T}_\Pi$ satisfies*

$$\mathsf{Z}(\mathcal{F}) \;=\; \left\{\phi \,:\, 0 \le \phi\left(\mathfrak{l}^{-1}(1)\right) \le \phi\left(\mathfrak{l}^{-1}(2)\right) \le \cdots \le \phi\left(\mathfrak{l}^{-1}(d)\right) \le 1\right\}.$$

Proof. Let $a_1, \ldots, a_d$ be the elements of Π such that $F_i \setminus F_{i-1} = \{a_{d+1-i}\}$ and $\mathfrak{l}(a_i) = i$ for $i = 1, \ldots, d$. In particular, $F_j = \{a_{d+1-j}, \ldots, a_d\}$ for $j = 0, \ldots, d$. Therefore, $\mathbf{e}_{F_j}(a_1) \ge 0$ for all j and $= 0$ only if $j = d$. Likewise, $\mathbf{e}_{F_j}(a_d) \le 1$ for all j and < 1 only if $j = 0$. By the same token, $\mathbf{e}_{F_{j-1}}(a_i) \le \mathbf{e}_{F_j}(a_i)$ for all j, and the inequality is strict only if $j + i = d + 1$. Thus, the $d+1$ linear inequalities given in the proposition are facet defining for $\mathsf{Z}(\mathcal{F})$. □

If we assume that $\Pi = \{1, 2, \ldots, d\}$ carries a natural labeling, then the Jordan–Hölder set $\mathrm{JH}(\Pi)$ is comprised of all permutations $\tau \in \mathfrak{S}_d$ such that τ^{-1} is a linear extension of Π. The inequality description featured in Proposition 6.3.5 then says that the maximal cells in the canonical triangulation are exactly the simplices

$$\triangle_\tau = \left\{ \mathbf{x} \in \mathbb{R}^d : 0 \le x_{\tau(1)} \le x_{\tau(2)} \le \cdots \le x_{\tau(d)} \le 1 \right\}$$

from Section 5.1. By Exercise 5.6, these are unimodular simplices.

Corollary 6.3.6. *The canonical triangulation $\mathcal{T}_\Pi$ of O_Π is regular and unimodular.*

In Exercise 6.15 you will show that all pulling triangulations of O_P are unimodular. In particular, if $\Pi = \mathrm{A}_d$, then $\mathsf{O}_\Pi = [0,1]^d$ and the canonical triangulation is exactly the one from Proposition 5.1.9.

Next, we zoom in on the faces triangulating the interior of O_Π.

Proposition 6.3.7. *Let $\mathcal{F} = \{F_0 \subset \cdots \subset F_k\}$ be a chain of filters of Π. The corresponding simplex $\mathsf{Z}(\mathcal{F})$ in $\mathcal{T}_\Pi$ is contained in the boundary of O_Π if and only if $F_0 \ne \varnothing$, $F_k \ne \Pi$, or there are $a, b \in F_j \setminus F_{j-1}$ for some $1 \le j \le k$ such that $a \prec_\Pi b$.*

Proof. If $F_0 \ne \varnothing$, then there is some $p \in \bigcap_{i=0}^k F_i$. Consequently, $\mathbf{e}_{F_i}(p) = 1$ for all i and by (6.3.1), $\mathsf{Z}(\mathcal{F})$ lies in the supporting hyperplane $\{\phi : \phi(p) = 1\}$. Similarly, if there is some $p \in \Pi \setminus F_k$, then $\mathsf{Z}(\mathcal{F}) \subset \{\phi : \phi(p) = 0\}$. The last condition implies that there is no i with $b \in F_i$ and $a \notin F_i$. Thus, the defining property of filters forces $\mathbf{e}_{F_i}(a) = \mathbf{e}_{F_i}(b)$ for all i, and so $\mathsf{Z}(\mathcal{F})$ lies in the boundary of O_Π. Conversely, if $\mathsf{Z}(\mathcal{F})$ lies in the boundary of O_Π, then all points in $\mathsf{Z}(\mathcal{F})$ satisfy one of the defining inequalities of O_Π, and reversing the arguments yields the claim. □

Since the triangulation $\mathcal{T}_\Pi$ is unimodular, we know from Corollary 5.6.6 that the Ehrhart polynomial of O_Π depends only on the combinatorics of $\Delta(\mathcal{J}(\Pi))$. We write $f_k^{\text{int}} = f_k^{\text{int}}(\mathcal{T}_\Pi)$ for the number of k-dimensional faces that are contained in the interior of O_Π. Recalling that $\binom{n-1}{k}$ is the Ehrhart polynomial for the interior of a unimodular k-simplex, we deduce that

$$\Omega_\Pi^\circ(n) = \operatorname{ehr}_{\mathsf{O}_\Pi^\circ}(n+1) = f_d^{\text{int}}\binom{n}{d} + f_{d-1}^{\text{int}}\binom{n}{d-1} + \cdots + f_1^{\text{int}}\binom{n}{1}. \tag{6.3.5}$$

From the proof of Proposition 1.3.1, we can interpret the numbers f_k^{int}, and Exercise 6.16 asks for bijective proof.

Corollary 6.3.8. *Let Π be a finite poset. The number of interior k-faces in the canonical triangulation $\mathcal{T}_\Pi$ of O_Π equals the number of surjective strictly order-preserving maps from Π to the k-chain.*

Corollary 6.3.8 also gives a clear geometric reason for Corollary 2.2.3. All full-dimensional cells in $\mathcal{T}_\Pi$ are contained in the interior. This yields the following beautiful fact that we will revisit in the next chapter.

Corollary 6.3.9. *Let Π be a poset on d elements. The leading coefficient of $d! \cdot \Omega_\Pi(n)$ equals the number of linear extensions of Π.*

Proof. The leading coefficient of $\Omega_\Pi(n) = (-1)^d \, \Omega_\Pi^\circ(-n)$ is $\frac{1}{d!} f_d^{\text{int}}(\mathcal{T}_\Pi)$. The claim then follows with help from Lemma 6.3.4 or Exercise 6.16. □

The existence of a unimodular triangulation of O_Π brings forward the question of what the h^*-vector of $\operatorname{Ehr}_{\mathsf{O}_\Pi}(z)$ might tell us about Π. This is what we will look into next. As a first step, we observe that the simplices triangulating O_Π are restrictions of the simplicial cones of Section 6.2. By a slight abuse of notation, we set $\tau(d+1) = d+1$ for any permutation $\tau \in \mathfrak{S}_d$.

Corollary 6.3.10. *The canonical triangulation $\mathcal{T}_\Pi$ is the restriction of the triangulation of $\mathsf{K}_{\hat{\Pi}}$ given in Corollary 6.2.3 by*

$$\triangle_\tau \;=\; \mathsf{K}_\tau \cap \{\phi \,:\, \phi_{d+1} = 1\}$$

for $\tau \in \mathrm{JH}(\Pi)$.

Theorem 5.5.3 states that $h_i^*(\mathsf{O}_\Pi)$ equals the number of simplices in a half-open decomposition of the canonical triangulation $\mathcal{T}_\Pi$, from which i facets were removed. Since we assume that $\Pi = \{1, \dots, d\}$ is naturally labelled, we can use $\phi \in \mathbb{R}^d$ with $\phi(i) = \frac{i}{d+1}$ for $i \in \Pi$ as the point from which we construct a half-open decomposition. Lemma 6.2.4 together with Corollary 6.3.10 then yields

$$\mathbb{H}_\phi \triangle_\tau \;=\; \left\{ \mathbf{x} \in \mathbb{R}^d \,:\, \begin{array}{c} 0 \le x_{\tau(1)} \le x_{\tau(2)} \le \cdots \le x_{\tau(d)} \\ x_{\tau(i)} < x_{\tau(i+1)} \text{ if } i \in \operatorname{Des}(\tau) \end{array} \right\}. \tag{6.3.6}$$

We write $\operatorname{des}(\tau) := |\operatorname{Des}(\tau)|$ for the number of descents of $\tau \in \mathfrak{S}_d$.

Theorem 6.3.11. *Let $\Pi = \{1, \dots, d\}$ be a naturally labelled poset with Jordan–Hölder set $\mathrm{JH}(\Pi) \subseteq \mathfrak{S}_d$. Then*

$$h_{\mathsf{O}_\Pi}^*(z) \;=\; \sum_{\tau \in \mathrm{JH}(\Pi)} z^{\operatorname{des}(\tau)} .$$

That is, $h_i^*(\mathsf{O}_\Pi)$ equals the number of permutations $\tau \in \mathrm{JH}(\Pi)$ with i descents.

Proof. Equation (6.3.6) together with Lemma 5.5.4 gives

$$h_{\mathbb{H}_\phi \triangle_\tau}^*(z) \;=\; z^{\operatorname{des}(\tau)}$$

for $\tau \in \mathrm{JH}(\Pi)$. By the additivity (5.5.2) of h^*-vectors of half-open polytopes in a half-open decomposition, we then conclude

$$h_{\mathsf{O}_\Pi}^*(z) \;=\; \sum_{\tau \in \mathrm{JH}(\Pi)} h_{\mathbb{H}_\phi \triangle_\tau}^*(z) \;=\; \sum_{\tau \in \mathrm{JH}(\Pi)} z^{\operatorname{des}(\tau)} . \qquad \square$$

With Proposition 6.3.2, we can rephrase Theorem 6.3.11 in terms of order polynomials.

Corollary 6.3.12. *Let* $\Pi = \{1, \dots, d\}$ *be a naturally labelled poset with Jordan–Hölder set* $\mathrm{JH}(\Pi) \subseteq \mathfrak{S}_d$. *Then*

$$\frac{1}{z} \sum_{n \ge 1} \Omega_\Pi(n)\, z^n \;=\; \frac{\sum_{\tau \in \mathrm{JH}(\Pi)} z^{\mathrm{des}(\tau)}}{(1-z)^{d+1}} \,.$$

To see Theorem 6.3.11 in action, we consider our favorite two families of posets. First, if Π is a (naturally labelled) chain, we already know that O_Π is unimodular simplex, and consequently $h^*_{\mathsf{O}_\Pi}(z) = 1$. This is confirmed by the fact that, in this case, $\mathrm{JH}(\Pi)$ consists only of the identity permutation.

For the special case that $\Pi = \mathrm{A}_d$ is the antichain on d elements and hence $\mathsf{O}_\Pi = [0,1]^d$ is the unit cube, Theorem 6.3.11 recovers the famous **Eulerian polynomial**

$$s_d(z) \;:=\; s(d,0) + s(d,1)\, z + \cdots + s(d,d-1)\, z^{d-1} \,.$$

The Eulerian numbers $s(d,k)$ count permutations $\tau \in \mathfrak{S}_d$ with k descents and were introduced in Theorem 5.7.7.

Corollary 6.3.13. *For* $d \ge 1$,

$$h^*_{[0,1]^d}(z) \;=\; h^*_{\mathsf{O}_{\mathrm{A}_d}}(z) \;=\; \sum_{\tau \in \mathfrak{S}_d} z^{\mathrm{des}(\tau)} \;=\; s_d(z) \,.$$

6.4. The Arithmetic of Order Cones and P-Partitions

We now bring in the machinery of Chapter 4. We start by translating (6.2.4) into an identity of rational generating functions (and leave the analogous integer-point transform version of (6.2.5) to Exercise 6.19). The integer-point transform of the left-hand side of (6.2.4) is easy:

$$\sigma_{\mathbb{R}^d_{\ge 0}}(\mathbf{z}) \;=\; \frac{1}{(1-z_1)(1-z_2)\cdots(1-z_d)} \,.$$

And from our experience gained in Section 4.8, we effortlessly compute for the half-open cones of Lemma 6.2.4 with $\phi = (1, 2, \dots, d)$

$$\mathbb{H}_\phi \mathsf{K}_\sigma \;=\; \sum_{j \in \mathrm{Des}(\sigma)} \mathbb{R}_{>0}\, \mathbf{u}_j^\sigma \;+\; \sum_{j \in \{0,1,\dots,d-1\} \setminus \mathrm{Des}(\sigma)} \mathbb{R}_{\ge 0}\, \mathbf{u}_j^\sigma \,, \tag{6.4.1}$$

which comes with the integer-point transform

$$\sigma_{\mathbb{H}_\phi \mathsf{K}_\sigma}(\mathbf{z}) \;=\; \frac{\prod_{j \in \mathrm{Des}(\sigma)} \mathbf{z}^{\mathbf{u}_j^\sigma}}{\prod_{j=0}^{d-1} \left(1 - \mathbf{z}^{\mathbf{u}_j^\sigma}\right)} \;=\; \frac{\prod_{j \in \mathrm{Des}(\sigma)} z_{\sigma(j+1)} \cdots z_{\sigma(d)}}{\prod_{j=0}^{d-1} \left(1 - z_{\sigma(j+1)} \cdots z_{\sigma(d)}\right)} \tag{6.4.2}$$

(Exercise 6.6). So the integer-point transform version of (6.2.4) is as follows.

Theorem 6.4.1.

$$\frac{1}{(1-z_1)(1-z_2)\cdots(1-z_d)} = \sum_{\sigma\in\mathfrak{S}_d} \frac{\prod_{j\in\mathrm{Des}(\sigma)} \mathbf{z}^{\mathbf{u}_j^\sigma}}{\prod_{j=0}^{d-1}\left(1-\mathbf{z}^{\mathbf{u}_j^\sigma}\right)}.$$

Thinking back to Section 4.7, at this point we cannot help specializing the identity in Theorem 6.4.1 to $z_1 = z_2 = \cdots = z_d = q$, which gives

$$\frac{1}{(1-q)^d} = \frac{\sum_{\sigma\in\mathfrak{S}_d}\prod_{j\in\mathrm{Des}(\sigma)} q^{d-j}}{(1-q)(1-q^2)\cdots(1-q^d)} = \frac{\sum_{\sigma\in\mathfrak{S}_d}\prod_{j\in\mathrm{Asc}(\sigma)} q^{d-j}}{(1-q)(1-q^2)\cdots(1-q^d)}$$

$$= \frac{\sum_{\sigma\in\mathfrak{S}_d}\prod_{d-j\in\mathrm{Asc}(\sigma)} q^{j}}{(1-q)(1-q^2)\cdots(1-q^d)}. \tag{6.4.3}$$

Here the second equation follows from the observation

$$\{\mathrm{Des}(\sigma) : \sigma\in\mathfrak{S}_d\} = \{\mathrm{Asc}(\sigma) : \sigma\in\mathfrak{S}_d\}.$$

We can make (6.4.3) look even nicer by defining, for each $\sigma\in\mathfrak{S}_d$, the complementary permutation

$$\sigma^{\mathrm{op}}(j) := \sigma(d+1-j).$$

Complementing a permutation switches the roles of ascents and descents in the sense that

$$d-j\in\mathrm{Asc}(\sigma) \quad \text{if and only if} \quad j\in\mathrm{Des}(\sigma^{\mathrm{op}}). \tag{6.4.4}$$

This allows us to rewrite (6.4.3) as

$$\frac{1}{(1-q)^d} = \frac{\sum_{\sigma\in\mathfrak{S}_d}\prod_{j\in\mathrm{Des}(\sigma^{\mathrm{op}})} q^{j}}{(1-q)(1-q^2)\cdots(1-q^d)}$$

$$= \frac{\sum_{\sigma\in\mathfrak{S}_d}\prod_{j\in\mathrm{Des}(\sigma)} q^{j}}{(1-q)(1-q^2)\cdots(1-q^d)}, \tag{6.4.5}$$

where the last equation follows from the fact that $\{\sigma^{\mathrm{op}} : \sigma\in\mathfrak{S}_d\} = \mathfrak{S}_d$. The right-hand side of (6.4.5) motivates the definition

$$\mathrm{maj}(\sigma) := \sum_{j\in\mathrm{Des}(\sigma)} j,$$

the **major index** of σ. Now the summands in the numerator on the right-hand side of (6.4.5) are simply $q^{\mathrm{maj}(\sigma)}$, and in fact (6.4.5) gives a distribution of this statistic:

$$\sum_{\sigma\in\mathfrak{S}_d} q^{\mathrm{maj}(\sigma)} = (1+q)(1+q+q^2)\cdots(1+q+\cdots+q^{d-1}). \tag{6.4.6}$$

The factors on the right are **q-integers**, commonly abbreviated through

$$[n]_q := 1+q+\cdots+q^{n-1}. \tag{6.4.7}$$

We can thus rewrite (6.4.6) compactly as

$$\sum_{\sigma \in \mathfrak{S}_d} q^{\operatorname{maj}(\sigma)} = [1]_q \, [2]_q \cdots [d]_q \,,$$

and the right-hand side is, in turn, (naturally!) often abbreviated as $[d]_q!$, an example of a **q-factorial**.

With a small modification, we can see the permutation statistic from Section 6.3 appearing. Namely, the order cone of an antichain appended by a maximal element is

$$\mathsf{K}_{\widehat{\mathrm{A}}_d} = \left\{ \mathbf{x} \in \mathbb{R}^{d+1} : 0 \le x_1, x_2, \dots, x_d \le x_{d+1} \right\}.$$

The contemplations of the previous two pages apply almost verbatim to this modified situation. The analogues of (6.2.4) and (6.2.5) are the half-open unimodular triangulations

$$\mathsf{K}_{\widehat{\mathrm{A}}_d} = \biguplus_{\sigma \in \mathfrak{S}_d} \left\{ \mathbf{x} : \begin{array}{l} 0 \le x_{\sigma(1)} \le x_{\sigma(2)} \le \cdots \le x_{\sigma(d)} \le x_{d+1} \\ x_{\sigma(j)} < x_{\sigma(j+1)} \text{ if } j \in \operatorname{Des}(\sigma) \end{array} \right\}, \tag{6.4.8}$$

$$\mathsf{K}^{\circ}_{\widehat{\mathrm{A}}_d} = \biguplus_{\sigma \in \mathfrak{S}_d} \left\{ \mathbf{x} : \begin{array}{l} 0 < x_{\sigma(1)} \le x_{\sigma(2)} \le \cdots \le x_{\sigma(d)} < x_{d+1} \\ x_{\sigma(j)} < x_{\sigma(j+1)} \text{ if } j \in \operatorname{Asc}(\sigma) \end{array} \right\}, \tag{6.4.9}$$

and the analogue of (6.4.2), i.e., the integer-point transforms of one of the cones on the right-hand side of (6.4.8), is

$$\frac{\prod_{j \in \operatorname{Des}(\sigma)} \mathbf{z}^{\mathbf{u}_j^\sigma + \mathbf{e}_{d+1}}}{\prod_{j=0}^{d} \left(1 - \mathbf{z}^{\mathbf{u}_j^\sigma + \mathbf{e}_{d+1}}\right)} = \frac{\prod_{j \in \operatorname{Des}(\sigma)} z_{\sigma(j+1)} \cdots z_{\sigma(d)} z_{d+1}}{\prod_{j=0}^{d} \left(1 - z_{\sigma(j+1)} \cdots z_{\sigma(d)} z_{d+1}\right)} . \tag{6.4.10}$$

On the other hand, the integer-point transform of $\mathsf{K}_{\widehat{\mathrm{A}}_d}$ is, practically by definition,

$$\sigma_{\mathsf{K}_{\widehat{\mathrm{A}}_d}}(\mathbf{z}) = \sum_{n \ge 0} \left(1 + z_1 + \cdots + z_1^n\right) \cdots \left(1 + z_d + \cdots + z_d^n\right) z_{d+1}^n \,,$$

and so (6.4.8) implies

$$\begin{aligned} &\sum_{n \ge 0} \left(1 + z_1 + \cdots + z_1^n\right) \cdots \left(1 + z_d + \cdots + z_d^n\right) z_{d+1}^n \\ &= \sum_{\sigma \in \mathfrak{S}_d} \frac{\prod_{j \in \operatorname{Des}(\sigma)} \mathbf{z}^{\mathbf{u}_j^\sigma + \mathbf{e}_{d+1}}}{\prod_{j=0}^{d} \left(1 - \mathbf{z}^{\mathbf{u}_j^\sigma + \mathbf{e}_{d+1}}\right)} . \end{aligned} \tag{6.4.11}$$

It is evident, already from the definition of $\widehat{\mathrm{A}}_d$, that the role of z_{d+1} is different from that of the other variables, and so we now specialize (6.4.11) to $z_{d+1} = z$ and $z_1 = z_2 = \cdots = z_d = q$:

$$\sum_{n \geq 0} (1 + q + \cdots + q^n)^d \, z^n = \frac{\sum_{\sigma \in \mathfrak{S}_d} \prod_{j \in \operatorname{Des}(\sigma)} q^{d-j} z}{(1-z)(1-qz)(1-q^2 z) \cdots (1-q^d z)}$$

$$= \frac{\sum_{\sigma \in \mathfrak{S}_d} \prod_{j \in \operatorname{Des}(\sigma)} q^{j} z}{(1-z)(1-qz)(1-q^2 z) \cdots (1-q^d z)} . \tag{6.4.12}$$

Here the last equation follows with the same change of variables as in (6.4.3) and (6.4.5); you ought to check this in Exercise 6.22. In the numerator of the right-hand side of (6.4) we again see the major index of the permutation σ appearing, alongside its descent number $\operatorname{des}(\sigma) = |\operatorname{Des}(\sigma)|$, which made its first appearance in Section 6.3. With the abbreviation (6.4.7), we can thus write (6.4) compactly in the following form.

Theorem 6.4.2. $\displaystyle\sum_{n \geq 0} [n+1]_q^d \, z^n = \frac{\sum_{\sigma \in \mathfrak{S}_d} q^{\operatorname{maj}(\sigma)} z^{\operatorname{des}(\sigma)}}{(1-z)(1-qz)(1-q^2 z) \cdots (1-q^d z)} .$

Specializing this identity for $q = 1$ recovers the special case $\Pi = \mathrm{A}_d$ of Theorem 6.3.11:

$$\sum_{n \geq 0} (n+1)^d \, z^n = \frac{\sum_{\sigma \in \mathfrak{S}_d} z^{\operatorname{des}(\sigma)}}{(1-z)^{d+1}} ,$$

and the numerator polynomial $\sum_{\sigma \in \mathfrak{S}_d} z^{\operatorname{des}(\sigma)}$ is an Eulerian polynomial. The joint distribution for $(\operatorname{maj}, \operatorname{des})$ given by Theorem 6.4.2 is known as an **Euler–Mahonian statistic**.

The computations for order cones of an antichain, leading up to Theorem 6.4.1 and (6.4.11), are surprisingly close to those for a general finite poset Π, even though an antichain seems to be a rather special poset. The reason for this similarity is geometric: in the general case, the unimodular triangulation (6.2.2)—which resulted in (6.2.4) and thus ultimately in the integer-point transform identities above—gets replaced by the more general Corollary 6.2.3. The "only" difference is that the unions and sums on the right-hand sides of the various decomposition formulas of the previous section will now be taken not over the full set $\mathfrak{S}_d$ of permutations of $[d]$, but rather only over those $\tau \in \mathfrak{S}_d$ that are in the Jordan–Hölder set of Π.

The next step is to make the triangulation

$$\mathsf{K}_\Pi = \bigcup_{\tau \in \operatorname{JH}(\Pi)} \mathsf{K}_\tau$$

of Corollary 6.2.3 disjoint, analogous to (6.2.4). For the rest of this section, except for Theorem 6.4.9 at the very end, we will assume that $\Pi = [d]$ is naturally labelled. This has the convenient consequence that $\phi = (1, 2, \dots, d) \in \mathsf{K}_\Pi^\circ$, which means, in turn, that ascents and descents play exactly the same role in our decomposition formulas as in the previous section, and we can literally repeat their proofs to arrive at the following results.

Theorem 6.4.3. *Let* $\Pi = [d]$ *be a naturally labelled poset. Then*

$$\mathsf{K}_\Pi = \biguplus_{\tau \in \mathrm{JH}(\Pi)} \left\{ \mathbf{x} \in \mathbb{R}^d : \begin{array}{l} 0 \le x_{\tau(1)} \le x_{\tau(2)} \le \cdots \le x_{\tau(d)} \\ x_{\tau(j)} < x_{\tau(j+1)} \text{ if } j \in \mathrm{Des}(\tau) \end{array} \right\},$$

$$\mathsf{K}_\Pi^\circ = \biguplus_{\tau \in \mathrm{JH}(\Pi)} \left\{ \mathbf{x} \in \mathbb{R}^d : \begin{array}{l} 0 < x_{\tau(1)} \le x_{\tau(2)} \le \cdots \le x_{\tau(d)} \\ x_{\tau(j)} < x_{\tau(j+1)} \text{ if } j \in \mathrm{Asc}(\tau) \end{array} \right\},$$

$$\mathsf{K}_{\widehat{\Pi}} = \biguplus_{\tau \in \mathrm{JH}(\Pi)} \left\{ \mathbf{x} \in \mathbb{R}^d : \begin{array}{l} 0 \le x_{\tau(1)} \le x_{\tau(2)} \le \cdots \le x_{\tau(d)} \le x_{d+1} \\ x_{\tau(j)} < x_{\tau(j+1)} \text{ if } j \in \mathrm{Des}(\tau) \end{array} \right\},$$

$$\mathsf{K}_{\widehat{\Pi}}^\circ = \biguplus_{\tau \in \mathrm{JH}(\Pi)} \left\{ \mathbf{x} \in \mathbb{R}^d : \begin{array}{l} 0 < x_{\tau(1)} \le x_{\tau(2)} \le \cdots \le x_{\tau(d)} < x_{d+1} \\ x_{\tau(j)} < x_{\tau(j+1)} \text{ if } j \in \mathrm{Asc}(\tau) \end{array} \right\}.$$

Corollary 6.4.4. *Let* $\Pi = [d]$ *be a naturally labelled poset. Then*

$$\sigma_{\mathsf{K}_\Pi}(\mathbf{z}) = \sum_{\tau \in \mathrm{JH}(\Pi)} \frac{\prod_{j \in \mathrm{Des}(\tau)} \mathbf{z}^{\mathbf{u}_j^\tau}}{\prod_{j=0}^{d-1} \left(1 - \mathbf{z}^{\mathbf{u}_j^\tau}\right)}$$

and

$$\sigma_{\mathsf{K}_{\widehat{\Pi}}}(\mathbf{z}) = \sum_{\tau \in \mathrm{JH}(\Pi)} \frac{\prod_{j \in \mathrm{Des}(\tau)} \mathbf{z}^{\mathbf{u}_j^\tau + \mathbf{e}_{d+1}}}{\prod_{j=0}^{d} \left(1 - \mathbf{z}^{\mathbf{u}_j^\tau + \mathbf{e}_{d+1}}\right)}.$$

The analogous integer-point transforms for K_Π° and $\mathsf{K}_{\widehat{\Pi}}^\circ$ are the subject of Exercise 6.23. Stanley reciprocity (Theorem 5.4.2) gives an identity between the rational functions $\sigma_{\mathsf{K}_{\widehat{\Pi}}}(\mathbf{z})$ and $\sigma_{\mathsf{K}_{\widehat{\Pi}}^\circ}(\mathbf{z})$. In the case of order cones, Stanley reciprocity is equivalent to the following interplay of ascents and descents.

Corollary 6.4.5. *Let* $\Pi = [d]$ *be naturally labelled. Then*

$$\sum_{\tau \in \mathrm{JH}(\Pi)} \frac{\prod_{j \in \mathrm{Des}(\tau)} \mathbf{z}^{-\mathbf{u}_j^\tau}}{\prod_{j=0}^{d-1} \left(1 - \mathbf{z}^{-\mathbf{u}_j^\tau}\right)} = (-1)^d \, z_1 z_2 \cdots z_d \sum_{\tau \in \mathrm{JH}(\Pi)} \frac{\prod_{j \in \mathrm{Asc}(\tau)} \mathbf{z}^{\mathbf{u}_j^\tau}}{\prod_{j=0}^{d-1} \left(1 - \mathbf{z}^{\mathbf{u}_j^\tau}\right)}.$$

A natural next step, parallel to the beginning of this section, is to specialize the variables in the integer-point transforms of order cones. This yields two applications of the theorems in the previous section; the first concerns order polynomials and parallels part of our treatment of order polytopes in

Section 6.3. Given a poset $\Pi = [d]$, we consider

$$\mathsf{K}_{\widehat{\Pi}} = \left\{ \mathbf{x} \in \mathbb{R}^{d+1} : \begin{array}{l} 0 \le x_j \le x_{d+1} \text{ for all } 1 \le j \le d \\ j \preceq k \Rightarrow x_j \le x_k \end{array} \right\}.$$

Integer points in this order cone with a fixed x_{d+1}-coordinate are precisely order-preserving maps $\Pi \to [0, x_{d+1}] \cap \mathbb{Z}$. Thus

$$\begin{aligned} \sigma_{\mathsf{K}_{\widehat{\Pi}}}(1,1,\dots,1,z) &= \sum_{n\ge 0} \#(\text{order-preserving maps } \Pi \to [0,n]\cap\mathbb{Z})\, z^n \\ &= \sum_{n\ge 1} \#(\text{order-preserving maps } \Pi \to [0,n-1]\cap\mathbb{Z})\, z^{n-1} \\ &= \frac{1}{z}\sum_{n\ge 1} \#(\text{order-preserving maps } \Pi \to [1,n]\cap\mathbb{Z})\, z^n \\ &= \frac{1}{z}\sum_{n\ge 1} \Omega_\Pi(n)\, z^n, \end{aligned} \tag{6.4.13}$$

and so the second identity in Corollary 6.4.4 gives another derivation of Corollary 6.3.12. This line of reasoning allows us to give yet another proof of the reciprocity theorem for order polynomials.

Third proof of Theorem 1.3.2. One object of Exercise 6.23 is the integer-point transform of $\mathsf{K}^\circ_{\widehat{\Pi}}$, whose specialization involves the number $\operatorname{asc}(\tau)$ of ascents of a permutation τ:

$$\sigma_{\mathsf{K}^\circ_{\widehat{\Pi}}}(1,1,\dots,1,z) = \frac{\sum_{\tau\in \mathrm{JH}(\Pi)} z^{2+\operatorname{asc}(\tau)}}{(1-z)^{d+1}}.$$

On the other hand, a computation exactly parallel to (6.4.13) gives (Exercise 6.25)

$$\sigma_{\mathsf{K}^\circ_{\widehat{\Pi}}}(1,1,\dots,1,z) = z\sum_{n\ge 0} \Omega^\circ_\Pi(n)\, z^n, \tag{6.4.14}$$

and so we obtain

$$\sum_{n\ge 0} \Omega^\circ_\Pi(n)\, z^n = \frac{\sum_{\tau\in \mathrm{JH}(\Pi)} z^{1+\operatorname{asc}(\tau)}}{(1-z)^{d+1}}.$$

Now $\operatorname{Asc}(\tau) \uplus \operatorname{Des}(\tau) = [d-1]$ for every $\tau \in \mathfrak{S}_d$, and so by Theorem 4.1.6,

$$\begin{aligned} \sum_{n\ge 1} \Omega^\circ(-n)\, z^n &= -\frac{\sum_{\tau\in \mathrm{JH}(\Pi)} z^{-1-\operatorname{asc}(\tau)}}{\left(1-\frac{1}{z}\right)^{d+1}} \\ &= (-1)^d\, \frac{z^{d+1}\sum_{\tau\in \mathrm{JH}(\Pi)} z^{\operatorname{des}(\tau)-d}}{(1-z)^{d+1}} \end{aligned}$$

$$= (-1)^d \frac{\sum_{\tau \in \mathrm{JH}(\Pi)} z^{\mathrm{des}(\tau)+1}}{(1-z)^{d+1}}$$
$$= (-1)^d \sum_{n \ge 1} \Omega_\Pi(n)\, z^n. \qquad \square$$

An important special evaluation at the beginning of this section was $\mathbf{z} = (q, q, \ldots, q)$ in the integer-point transform of $\mathbb{R}^d_{\ge 0}$, the order cone of the antichain A_d. For a general order cone, this gives, with the first identity of Corollary 6.4.4,

$$\begin{aligned} \sigma_{\mathsf{K}_\Pi}(q, q, \ldots, q) &= \frac{\sum_{\tau \in \mathrm{JH}(\Pi)} \prod_{j \in \mathrm{Des}(\tau)} q^{d-j}}{(1-q)(1-q^2)\cdots(1-q^d)} \\ &= \frac{\sum_{\tau \in \mathrm{JH}(\Pi)} \prod_{j \in \mathrm{Asc}(\tau^{\mathrm{op}})} q^j}{(1-q)(1-q^2)\cdots(1-q^d)}\,; \end{aligned} \tag{6.4.15}$$

here the second equation follows from (6.4.4). With the definition of the **comajor index**

$$\mathrm{comaj}(\tau) := \sum_{j \in \mathrm{Asc}(\tau)} j\,,$$

the denominator on the right-hand side of (6.4.15) can be compactly written as $\sum_{\tau \in \mathrm{JH}(\Pi)} q^{\mathrm{comaj}(\tau^{\mathrm{op}})}$.

This brings back memories of Chapter 4, and in fact the generating function on the left-hand side of (6.4.15) enumerates certain compositions:

$$\sigma_{\mathsf{K}_\Pi}(q, q, \ldots, q) = \sum q^{a_1 + a_2 + \cdots + a_d},$$

where the sum is over all $(a_1, a_2, \ldots, a_d) \in \mathbb{Z}^d_{\ge 0}$ such that

$$j \preceq k \quad \Longrightarrow \quad a_j \le a_k\,. \tag{6.4.16}$$

In this case we say that the composition $(a_1, a_2, \ldots, a_d)$ **respects** the poset Π. In this language, a partition is a composition that respects a chain,[1] a general composition (in the sense of Section 4.2) is a composition that respects an antichain (yes, that's as close to an oxymoron as we will get in this book), and the plane partitions of Section 4.3 are situated between these two extremes: in their simplest instance (4.3.1), they are compositions that respect the diamond poset $\diamond$.

Furthermore, if we interpret the composition $(a_1, a_2, \ldots, a_d)$ as a function $\Pi \to \mathbb{Z}^d_{\ge 0}$, then (6.4.16) says that this function is order preserving. The rational functions in (6.4.15) represent the generating function of the counting function for compositions satisfying (6.4.16). Here is what we have proved.

[1] More generally, as long as Π is connected and naturally labelled, compositions that respect Π are *partitions*.

Theorem 6.4.6. *Let $\Pi = [d]$ be naturally labelled, and let $c_\Pi(n)$ be the number of compositions of n that respect Π. Then*

$$\sum_{n \geq 0} c_\Pi(n)\, q^n \;=\; \frac{\sum_{\tau \in \mathrm{JH}(\Pi)} q^{\mathrm{comaj}(\tau^{\mathrm{op}})}}{(1-q)(1-q^2)\cdots(1-q^d)}\,.$$

For example, when $\Pi = \diamond$ is the diamond poset, the two permutations in $\mathfrak{S}_4$ that respect $\diamond$ are the identity $[1234]$ and the permutation $[1324]$ switching 2 and 3. Here

$$\mathrm{comaj}\left([1234]^{\mathrm{op}}\right) = 0 \qquad \text{and} \qquad \mathrm{comaj}\left([1324]^{\mathrm{op}}\right) = 2$$

and so Theorem 6.4.6 gives

$$\sum_{n \geq 0} c_\diamond(n)\, q^n \;=\; \frac{1+q^2}{(1-q)(1-q^2)(1-q^3)(1-q^4)}$$
$$=\; \frac{1}{(1-q)(1-q^2)^2(1-q^3)}\,,$$

confirming once more (4.3.2).

By now you might expect a reciprocity theorem for $c_\Pi(n)$, and the natural candidate for a counting function that is reciprocal to $c_\Pi(n)$ is the number of compositions of n whose parts satisfy

$$j \prec k \qquad \Longrightarrow \qquad a_j < a_k\,.$$

(As above, we allow 0 as a part, i.e., such a composition might have less than d positive parts.) We say that such a composition **strictly respects** Π, and we count them with the function $c_\Pi^\circ(n)$. We define the two accompanying generating functions as

$$C_\Pi(q) \;:=\; \sum_{n \geq 0} c_\Pi(n)\, q^n \qquad \text{and} \qquad C_\Pi^\circ(q) \;:=\; \sum_{n \geq 0} c_\Pi^\circ(n)\, q^n.$$

Theorem 6.4.7. *The rational generating functions $C_\Pi(q)$ and $C_\Pi^\circ(q)$ are related via*

$$C_\Pi\left(\frac{1}{q}\right) \;=\; (-q)^{|\Pi|}\, C_\Pi^\circ(q)\,.$$

Proof. A composition that strictly respects Π, viewed as an integer point $(a_1, a_2, \dots, a_d) \in \mathbb{Z}^d$, satisfies

$$(a_1+1, a_2+1, \dots, a_d+1) \;\in\; \mathsf{K}_\Pi^\circ\,,$$

which directly translates to

$$q^d \sum_{n \geq 0} c_\Pi^\circ(n)\, q^n \;=\; \sigma_{\mathsf{K}_\Pi^\circ}(q, q, \dots, q)\,. \tag{6.4.17}$$

Stanley reciprocity (Theorem 5.4.2) applied to the order cone K_Π specializes to

$$\sigma_{\mathsf{K}_\Pi}\left(\frac{1}{q}, \frac{1}{q}, \dots, \frac{1}{q}\right) = (-1)^d \, \sigma_{\mathsf{K}_\Pi^\circ}(q, q, \dots, q) .$$

Now use (6.4.17) and the fact that $\sigma_{\mathsf{K}_\Pi}(q, q, \dots, q) = C_\Pi(q)$. □

While our composition counting function $c_\Pi(n)$ appears most naturally from the viewpoint of posets and order preserving maps, the right-hand side of the formula in Theorem 6.4.6 is not quite as aesthetic as, say, that of Theorem 6.4.2. The remedy is to instead consider order *reversing* maps, which gives rise to the theory of P-partitions.[2] On the one hand, this point of view is philosophically no different from our treatment of compositions respecting a given poset; on the other hand, its history is older and, again, the ensuing formulas are prettier.

Given a poset $\Pi = [d]$, we call $(a_1, a_2, \dots, a_d) \in \mathbb{Z}_{\geq 0}^d$ a **Π-partition**[3] of n if

$$n = a_1 + a_2 + \cdots + a_d$$

and

$$j \preceq k \qquad \Longrightarrow \qquad a_j \geq a_k ,$$

that is, the map $\Pi \to \mathbb{Z}_{\geq 0}$ given by $(a_1, a_2, \dots, a_d)$ is order reversing. The reciprocal concept is that of a **strict Π-partition** of n, for which the last condition is replaced by

$$j \prec k \qquad \Longrightarrow \qquad a_j > a_k .$$

We count all Π-partitions of n with the function $p_\Pi(n)$ and the strict Π-partitions of n with $p_\Pi^\circ(n)$, with accompanying generating functions $P_\Pi(q)$ and $P_\Pi^\circ(q)$, respectively. The easiest way to state and prove the Π-partition analogues of Theorems 6.4.6 and 6.4.7 is to change our convention that Π is naturally labelled to Π being **dual naturally labelled**, namely, when $i \preceq j$ implies $i \geq j$ for all $i, j \in [d]$. In other words, the elements of Π are now labelled with order-*reversing* indices. We can now safely leave the proof of the following two theorems to you (Exercise 6.31).

Theorem 6.4.8. *Let $\Pi = [d]$ be a dual naturally labelled poset. Then*

$$P_\Pi(q) = \frac{\sum_\tau q^{\operatorname{maj}(\tau)}}{(1-q)(1-q^2)\cdots(1-q^d)} ,$$

where the sum is over all $\tau \in \mathfrak{S}_d$ that satisfy $\tau(j) \preceq_\Pi \tau(k) \implies j \geq k$.

[2] Here P stands for a specific poset—for which we tend to use Greek letters such as Π to avoid confusions with polyhedra.

[3] Despite the name, a Π-partition might not be a partition when Π is not connected.

Theorem 6.4.9. *The rational generating functions $P_\Pi(q)$ and $P_\Pi^\circ(q)$ are related by*

$$P_\Pi\left(\frac{1}{q}\right) = (-q)^{|\Pi|}\, P_\Pi^\circ(q)\,.$$

Notes

The order polytope was first studied by Ladnor Geissinger [**69**] but its full potential was uncovered by Richard Stanley in his seminal *Two poset polytopes* paper [**166**] in 1986. The order cone systematically studied here not only captures the order polytopes but also many variations of it; see [**9, 95, 171**]. We will see order polytopes in a more general context in Section 7.4.

The canonical triangulation of Theorem 6.3.3 has important computational consequences: given a rational polytope in terms of inequalities, is there a polynomial-time algorithm to compute its volume? If so, then we could compute the number of linear extensions of a poset Π in polynomial-time. However, Graham Brightwell and Peter Winkler [**39**] showed that this is not possible.

The Eulerian numbers go back to (surprise!) Leonard Euler. For a bit of history on how Euler got interested in these numbers, see [**86**]. Descent statistics of permutations go back to at least Percy MacMahon, who proved Theorem 6.4.2 and numerous variations of it [**117**]. Back then $\operatorname{maj}(\sigma)$ was called the *greater index* of σ; in the 1970's Dominique Foata suggested to rename it *major index* in honor of MacMahon, who was a major in the British army (and usually published his papers including this title) [**65**]. The geometric treatment of Eulerian polynomials stemming from integer-point transforms of suitable cones was spearheaded in [**20**], which also contains analogous results when $\mathfrak{S}_d$ is replaced by other reflection groups.

Permutation statistics, q-integers, and q-factorials are indispensable in combinatorial number theory, in particular, the theory of partitions; see, e.g., [**4, 5**].

MacMahon's study [**116**] of plane partitions (which we introduced in Section 4.3) can be viewed as a first step towards P-partitions, including his insight that the major index of a permutation is useful here. Much later, Knuth [**101**] used MacMahon's approach to enumerate *solid partitions* (a 3-dimensional variant of plane partitions) by recognizing there is a poset structure underlying MacMahon's computations. The full glory and power of P-partitions—in fact, their generalization treated in Exercise 6.34—were introduced in Richard Stanley's Ph.D. thesis [**160**]; a modern treatment, from a more algebraic perspective than ours, can be found in [**170**]. The main decomposition and reciprocity theorems (Theorems 6.4.8 and 6.4.9) are due to Stanley. An application to the enumeration of Young tableaux is hinted at in Exercise 6.35. For a sampler of other applications and generalizations of P-partitions, see [**3, 10, 47, 64, 67, 85, 118, 128, 132, 135, 177**]. A nice survey about (the history of) P-partitions is [**72**].

Exercises

Throughout the exercises, Π is a finite poset.

6.1 ⬠ Prove Lemma 6.1.3: $\mathsf{K}_{\Pi_1 \uplus \Pi_2} = \mathsf{K}_{\Pi_1} \times \mathsf{K}_{\Pi_2}$.

6.2 ⬠ Verify the claims in the proof of Theorem 6.1.4: The set $\check{\Pi}' := \check{\Pi}/\sim$ equipped with $\preceq'$ is naturally a partially ordered set and the map $\check{\Pi} \to \check{\Pi}'$ is order preserving.

6.3 ⬠ Let $(\Pi, \preceq)$ be a poset and $\mathcal{N} = \mathcal{N}(\Pi, \preceq)$ the poset of refinements of $(\Pi, \preceq)$.
   (a) Show that any two elements in $\mathcal{N}$ have a join.
   (b) Denote by $\check{\mathcal{N}}$ the poset $\mathcal{N}$ extended by a minimal element $\hat{0}$. Is $\check{\mathcal{N}}$ a distributive lattice?

6.4 ⬠ Show that the binary relation $\preceq_\phi$ constructed in the proof of Theorem 6.2.2 is a total order on Π.

6.5 ⬠ Show that the dissection given in Corollary 6.2.3 describes the maximal cells in a triangulation of K_Π. Conversely, show that not every dissection from Theorem 6.2.2 is a subdivision.

6.6 ⬠ Prove (6.4.1): For $\phi = (1, 2, \ldots, d)$,

$$\mathbb{H}_\phi \mathsf{K}_\sigma = \sum_{j \in \mathrm{Des}(\sigma)} \mathbb{R}_{>0}\, \mathbf{u}_j^\sigma + \sum_{j \in \{0,1,\ldots,d-1\} \setminus \mathrm{Des}(\sigma)} \mathbb{R}_{\geq 0}\, \mathbf{u}_j^\sigma ,$$

and conclude from this (6.4.2):

$$\sigma_{\mathbb{H}_\phi \mathsf{K}_\sigma}(\mathbf{z}) = \frac{\prod_{j \in \mathrm{Des}(\sigma)} \mathbf{z}^{\mathbf{u}_j^\sigma}}{\prod_{j=0}^{d-1} \left(1 - \mathbf{z}^{\mathbf{u}_j^\sigma}\right)} = \frac{\prod_{j \in \mathrm{Des}(\sigma)} z_{\sigma(j+1)} \cdots z_{\sigma(d)}}{\prod_{j=0}^{d-1} \left(1 - z_{\sigma(j+1)} \cdots z_{\sigma(d)}\right)} .$$

6.7 ⬠ Verify that $\hom(\mathsf{O}_\Pi) = \mathsf{K}_{\hat{\Pi}}$.

6.8 ⬠ Let $F, F' \subseteq \Pi$ be two distinct filters. Show that $[\mathbf{e}_F, \mathbf{e}_{F'}]$ is an edge of O_Π if and only if $F \subset F'$ and $F' \setminus F$ is a connected poset.

6.9 What is the maximal number of vertices that a 2-face of O_Π can have?

6.10 Let $F, F' \subseteq \Pi$ be two distinct filters. Let $\mathsf{F} \subseteq \mathsf{O}_\Pi$ be the join of $\mathbf{e}_F$ and $\mathbf{e}_{F'}$, that is, the inclusion-minimal face that contains both vertices. Show that F is combinatorially isomorphic to $[0,1]^r$, where $r = |\max(F \setminus F')| + |\max(F' \setminus F)|$.

6.11 Let Π be a poset and $\widehat{\Pi} := \Pi \cup \{\hat{1}\}$. Show that $\mathsf{O}_{\widehat{\Pi}}$ is a pyramid over O_{Π}.

6.12 Let $\mathcal{J}$ be a lattice. A subset $\mathcal{J}' \subseteq \mathcal{J}$ is an **induced sublattice** if $\mathcal{J}'$ is a lattice and meets/joins in $\mathcal{J}'$ agree with the meets/joins in $\mathcal{J}$. An **embedded sublattice** is an induced sublattice $\mathcal{J}' \subseteq \mathcal{J}$ if

$$a \wedge b,\ a \vee b \in \mathcal{J}' \quad \Longrightarrow \quad a, b \in \mathcal{J}' \,.$$

(a) Let $\Phi : \Pi \to \Pi'$ be a surjective order preserving map. Show that the induced map $\Phi^* : \mathcal{J}(\Pi') \to \mathcal{J}(\Pi)$ that sends $F \subseteq \Pi'$ to $\Phi^{-1}(F)$ is injective.
(b) Show that the image of Φ^* is an embedded sublattice of $\mathcal{J}(\Pi)$ and that every embedded sublattice arises this way.
(c) Show that the face lattice of O_{Π} is isomorphic to the poset of embedded sublattices of $\mathcal{J}(\Pi)$ ordered by inclusion.

6.13 ⬠ Show that the pulling triangulation of O_{Π} in Theorem 6.3.3 is independent of the chosen linear extension of $\mathcal{J}(\Pi)$.

6.14 ⬠ Show that if $F' \subset F \subseteq \Pi$ are two filters such that $|F \setminus F'| \geq 2$, then there is a filter F'' with $F' \subset F'' \subset F$.

6.15 ⬠ Show that any pulling triangulation of O_{Π} is unimodular. Are they always isomorphic to the canonical triangulation $\mathcal{T}_{\Pi}$?

6.16 Prove that the number of strictly order-preserving maps of Π into a k-chain are in bijection with chains of filters in Π of length $k-1$.

6.17 Revisiting Exercise 1.19, compute $\Omega^{\circ}_{D_{10}}(n)$ via linear extensions.

6.18 ⬠ Prove (6.3.2): $\Omega^{\circ}_{\Pi}(n) \;=\; \mathrm{ehr}_{\mathsf{O}^{\circ}_{\Pi}}(n+1)\,.$

6.19 ⬠ Theorem 6.4.1 is the generating-function analogue of (6.2.4). Derive the generating-function analogue of (6.2.5).

6.20 Let $\mathrm{inv}(\sigma) := |\{(j,k) : j < k \text{ and } \sigma(j) > \sigma(k)\}|$, the number of **inversions** of $\sigma \in \mathfrak{S}_d$. Show that

$$\sum_{\sigma \in \mathfrak{S}_d} q^{\mathrm{inv}(\sigma)} \;=\; (1+q)(1+q+q^2)\cdots(1+q+\cdots+q^{d-1})\,.$$

(Looking back at (6.4.6), this implies $\sum_{\sigma \in \mathfrak{S}_d} q^{\mathrm{maj}(\sigma)} = \sum_{\sigma \in \mathfrak{S}_d} q^{\mathrm{inv}(\sigma)}$, i.e., the statistics maj and inv are *equidistributed.*)

6.21 Explain how (6.2.4) and (6.2.5), and subsequently Theorem 6.4.3, would change if we had chosen ϕ in Sections 6.2 and 6.4 differently—still with entries $1, 2, \ldots, d$, but now in some other order.

6.22 ⬠ Verify (6.4.10), (6.4.11), and (6.4).

6.23 ◇ Prove the following open analogues of Corollary 6.4.4: Let $\Pi = [d]$ be naturally labelled. Then

$$\sigma_{\mathsf{K}^\circ_\Pi}(\mathbf{z}) = \sum_{\tau \in \mathrm{JH}(\Pi)} \frac{z_1 z_2 \cdots z_d \prod_{j \in \mathrm{Asc}(\tau)} \mathbf{z}^{\mathbf{u}_j^\tau}}{\prod_{j=0}^{d-1} \left(1 - \mathbf{z}^{\mathbf{u}_j^\tau}\right)},$$

$$\sigma_{\mathsf{K}^\circ_{\widehat{\Pi}}}(\mathbf{z}) = \sum_{\tau \in \mathrm{JH}(\Pi)} \frac{z_0^2 z_1 z_2 \cdots z_d \prod_{j \in \mathrm{Asc}(\tau)} \mathbf{z}^{\mathbf{u}_j^\tau + \mathbf{e}_{d+1}}}{\prod_{j=0}^{d} \left(1 - \mathbf{z}^{\mathbf{u}_j^\tau + \mathbf{e}_{d+1}}\right)}.$$

Deduce Corollary 6.4.5.

6.24 Give a direct proof of Corollary 6.4.5 (i.e., one that does not use Theorem 5.4.2).

6.25 ◇ Show (6.4.14): $\sigma_{\mathsf{K}^\circ_{\widehat{\Pi}}}(1, 1, \dots, 1, z) = z \sum_{n \ge 0} \Omega^\circ_\Pi(n)\, z^n$.

6.26 Recompute the generating function for plane partition diamonds from Exercise 4.20 by way of the theorems in this chapter.

6.27 Show that the generating function

$$\sum_{n \ge 1} \Omega_\Pi(n) z^{n-1} = \frac{h_\Pi(z)}{(1-z)^{d+1}}$$

has a palindromic numerator polynomial h_Π (i.e., its coefficients form a symmetric sequence) if and only if Π is a graded poset.

6.28 Show that every Eulerian polynomial is palindromic.

6.29 Compute the order polynomial of $\hat{\mathrm{A}}_d$, an antichain appended by a maximal element. (*Hint:* The Bernoulli polynomials from (4.9.6) will make an appearance.)

6.30 Revisiting Exercise 2.8, let Π be the poset on $2d$ elements $a_1, a_2, \dots, a_d$, $b_1, b_2, \dots, b_d$, defined by the relations

$$a_1 \prec a_2 \prec \cdots \prec a_d \qquad \text{and} \qquad a_j \succ b_j \text{ for } 1 \le j \le d\,.$$

Compute $\sum_{n \ge 0} \Omega_\Pi(n) z^n$. (*Hint:* Consider permutations of the multiset $\{1, 1, 2, 2, \dots, d, d\}$.)

6.31 ◇ Prove Theorems 6.4.8 and 6.4.9: Let $\Pi = [d]$ be dual naturally labelled. Then

$$P_\Pi(q) = \frac{\sum_\tau q^{\mathrm{maj}(\tau)}}{(1-q)(1-q^2)\cdots(1-q^d)},$$

where the sum is over all $\tau \in \mathfrak{S}_d$ that satisfy $\tau(j) \preceq_\Pi \tau(k) \Rightarrow j \ge_\mathbb{R} k$. Furthermore, the rational generating functions $P_\Pi(q)$ and $P^\circ_\Pi(q)$ are related by

$$P_\Pi\left(\frac{1}{q}\right) = (-q)^{|\Pi|}\, P^\circ_\Pi(q)\,.$$

6.32 Derive from Theorem 6.4.9 a reciprocity theorem that relates the quasi-polynomials $p_\Pi(t)$ and $p^\circ_\Pi(t)$.

6.33 Let $\Pi = [d]$ be dual naturally labelled. Prove that

$$P^\circ_\Pi(z) \;=\; \frac{\sum_\tau z^{\operatorname{comaj}\tau}}{(1-z)(1-z^2)\cdots(1-z^d)}\,,$$

where the sum is over all $\tau \in \mathfrak{S}_d$ that satisfy $\tau(j) \preceq_\Pi \tau(k) \Rightarrow j \geq_{\mathbb{R}} k$.

6.34 Given a poset Π on $[d]$, fix a bijection $\omega : [d] \to [d]$ which we call a **labeling** of Π. We say $(a_1, a_2, \dots, a_d) \in \mathbb{Z}^d_{\geq 0}$ is a (Π,ω)**-partition** of n if

$$n \;=\; a_1 + a_2 + \cdots + a_d\,,$$

$$j \preceq k \qquad \Longrightarrow \qquad a_j \geq a_k\,,$$

and

$$j \prec k \text{ and } \omega(j) > \omega(k) \qquad \Longrightarrow \qquad a_j > a_k\,.$$

Thus if ω is natural, then a (Π,ω)-partition is simply a Π-partition. Generalize Theorems 6.4.8 and 6.4.9 to (Π,ω)-partitions. (*Hint:* Exercise 6.21.)

6.35 A **Young diagram** associated to a partition $\lambda = (\lambda_1 \geq \lambda_2 \geq \cdots \geq \lambda_k \geq 1)$ is a collection of rows of left-justified boxes with λ_i boxes in the i-th row for $i = 1, \dots, k$. Figure 6.7 shows a Young diagram for the partition $\lambda = (5,3,3,1)$.

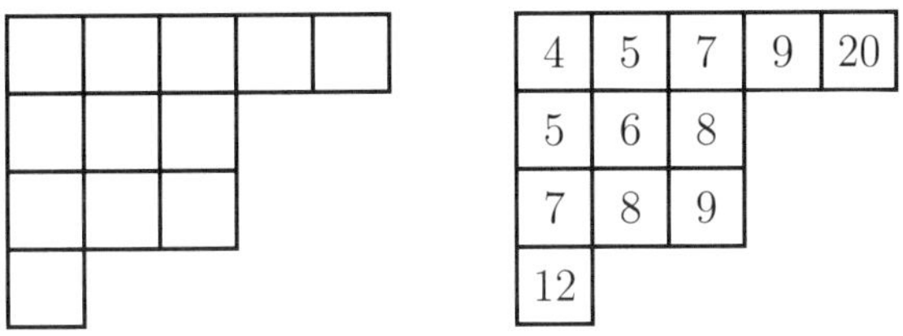

Figure 6.7. Young diagram and a Young tableau for $\lambda = (5,3,3,1)$.

(a) Putting natural numbers in the boxes of a Young diagram is called a **filling** if

- the numbers weakly increase along each row, and
- the numbers strictly increase down each column.

Find a suitable poset and use Exercise 6.34 to give them a geometric incarnation.

(b) Let $n = \lambda_1 + \cdots + \lambda_k$. A filling with numbers in $[n]$ yields a **(standard) Young tableau** if the numbers also strictly increase along rows. Show that standard Young tableaux for λ are in bijection to linear extensions of certain posets.

6.36 Recompute the order polynomials of an antichain, a chain, and the diamond poset $\diamond$ via (6.3.5) and Exercise 6.16.

Chapter 7

Hyperplane Arrangements

The purely formal language of geometry describes adequately the reality of space. We might say, in this sense, that geometry is successful magic. I should like to state a converse: is not all magic, to the extent that it is successful, geometry?
René Thom

We begin the final chapter of this book with a look back to its first. Our very first family of polynomials in Chapter 1 came from proper colorings of graphs, but chromatic polynomials made no real appearance beyond that first chapter—though they have been waiting backstage all along: Corollary 1.3.4 gave an intimate connection between chromatic and order polynomials. The latter have played a prominent (geometric) role in several chapters since, and one of our goals in this chapter is to illustrate how chromatic polynomials exhibit a parallel geometric existence. The second family of polynomials in Chapter 1, related to flows on graphs, have been all but orphaned; they will make a well-deserved comeback here.

In this chapter we will develop a general framework that gives a coherent picture for polynomial counting functions such as chromatic or flow polynomials. The evolving theory will gently force us to revitalize hyperplane arrangements, which we briefly introduced in Section 3.6. These, in turn, will naturally lead us to study two special classes of polytopes: alcoved polytopes and zonotopes.

7.1. Chromatic, Order Polynomials, and Subdivisions Revisited

Our geometric intuition behind order polynomials guided us in Chapter 6 to realize them as Ehrhart polynomials (of order polytopes). While a chromatic polynomial *cannot* be the Ehrhart polynomial of some lattice polytope (e.g., the constant term of a chromatic polynomial is 0, not 1), the geometric picture implicitly painted by Corollary 1.3.4 is attractive: rewritten in the language of Chapter 6, it says that a chromatic polynomial is the sum of Ehrhart polynomials of open (order) polytopes, all of the same dimension and without overlap. We will now continue this painting.

We start by recalling the construction underlying Corollary 1.3.4: an acyclic orientation ρ of a graph $G = (V, E)$ can be viewed as a poset $\Pi({}_\rho G)$ by starting with the binary relations given by ${}_\rho G$ and adding the necessary transitive and reflexive relations. Corollary 1.3.4 states that the chromatic polynomial can be expressed as

$$\chi_G(n) \;=\; \sum_\rho \Omega^\circ_{\Pi({}_\rho G)}(n)\,, \tag{7.1.1}$$

where the sum is over all acyclic orientations ρ of G, which we can rewrite with (6.3.2) as

$$\chi_G(n) \;=\; \sum_\rho \operatorname{ehr}_{\mathcal{O}^\circ_{\Pi({}_\rho G)}}(n+1)\,. \tag{7.1.2}$$

The case $G = K_2$ is depicted in Figure 7.1. The geometric expression (7.1.2)

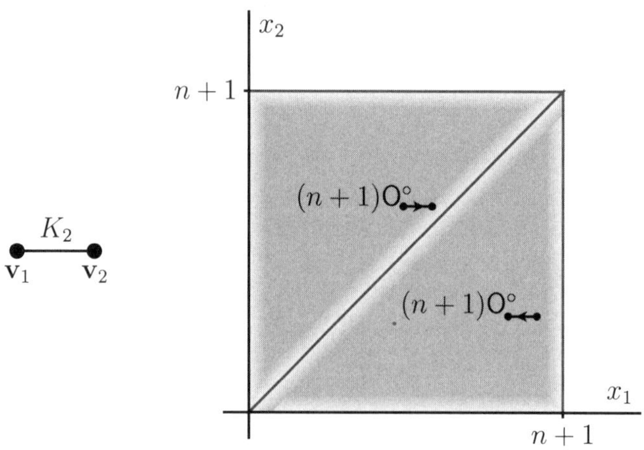

Figure 7.1. The geometry behind (7.1.2) for $G = K_2$.

for the chromatic polynomial, in a sense, actually goes back to basics: what we are counting on the right-hand side are integer lattice points in the open cube

$$(0, n+1)^V = \{\mathbf{c} \in \mathbb{R}^V : 0 < c(v) < n+1 \text{ for } v \in V\}$$

minus those points for which $c(j) = c(k)$ for some edge $jk \in E$; the resulting set naturally represents precisely the proper n-colorings $\mathbf{c}$ of G. The parameter n acts like a dilation factor, and the underlying polytope is the unit cube in $\mathbb{R}^V$, which gets dissected by the hyperplanes $\{c(j) = c(k)\}$, one for each edge $jk \in E$. This gives a subdivision that we already know—at least a refinement of it: Proposition 5.1.9 gives a regular triangulation of $[0,1]^V$ into unimodular simplices, indexed by permutations of V. If G is a complete graph (i.e., E consists of all possible pairs from V), the above dissection is precisely the triangulation from Proposition 5.1.9. In the general case (7.1.2), we essentially use the triangulation from Proposition 5.1.9 but glue some of the (open) simplices along faces; we note that the resulting subdivision consists of *lattice* polytopes.

This motivates the following general geometric construct: for a polyhedral complex $\mathcal{S}$ of dimension d, we write $\mathcal{S}^{(-1)}$ for the **codimension-1 skeleton**, that is, the subcomplex of faces $\mathsf{F} \in \mathcal{S}$ of dimension $\le \dim \mathcal{S} - 1$. We define $\operatorname{ehr}_{\mathcal{S}}^{[1]}(n)$ to count lattice points in the n-th dilate of $|\mathcal{S}| \setminus |\mathcal{S}^{(-1)}|$. That is,

$$\operatorname{ehr}_{\mathcal{S}}^{[1]}(n) = \operatorname{ehr}_{\mathcal{S}}(n) - \operatorname{ehr}_{\mathcal{S}^{(-1)}}(n) = \sum_{\substack{\mathsf{F} \in \mathcal{S} \\ \dim \mathsf{F} = \dim \mathcal{S}}} \operatorname{ehr}_{\mathsf{F}^\circ}(n) . \tag{7.1.3}$$

Figure 7.2 shows an example.

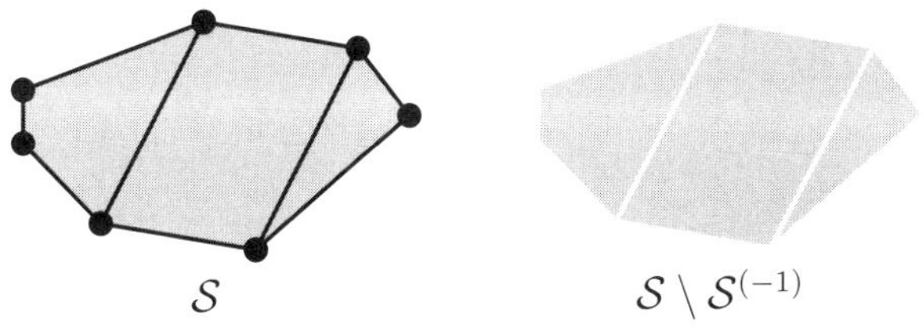

Figure 7.2. Removing the codimension-1 skeleton of a polyhedral complex.

If $\mathcal{S}$ is a complex of lattice polytopes, then $\operatorname{ehr}_{\mathcal{S}}^{[1]}(n)$ agrees with a polynomial of degree $\dim \mathcal{S}$ for all integers $n > 0$.

Proposition 7.1.1. *Let $\mathcal{S}$ be a pure complex of lattice polytopes. Then*

$$(-1)^{\dim \mathcal{S}} \operatorname{ehr}_{\mathcal{S}}^{[1]}(-n)$$

is the number of lattice points in $n|\mathcal{S}|$, each counted with multiplicity equal to the number of maximal cells containing the point.

Proof. By (7.1.3) and Ehrhart–Macdonald reciprocity (Theorem 5.2.3),

$$(-1)^{\dim \mathcal{S}} \operatorname{ehr}_{\mathcal{S}}^{[1]}(-n) \;=\; \sum_{\substack{\mathsf{F} \in \mathcal{S} \\ \dim \mathsf{F} = \dim \mathcal{S}}} \operatorname{ehr}_{\mathsf{F}}(n) \,,$$

and the right-hand side counts lattice points in $\bigcup_{\mathsf{F}} \mathsf{F} = |\mathcal{S}|$, each weighted by the number of facets F containing it. □

Our first concrete goal in this chapter is to adapt the above setup to give a *geometric* proof of the reciprocity theorem for chromatic polynomials (Theorem 1.1.5). We recall its statement: $(-1)^{|V|} \chi_G(-n)$ equals the number of compatible pairs (ρ, c), where c is an n-coloring and ρ is an acyclic orientation.

We start by drawing a picture of (7.1.2) (the case $G = K_2$ is depicted in Figure 7.1): the order polynomials $\Omega^{\circ}_{\Pi(\rho G)}(n)$ on the right-hand side of (7.1.2) are, by (6.3.2), the Ehrhart polynomials of the interiors of the respective order polytopes, evaluated at $n+1$. Since every proper coloring gives rise to a compatible acyclic orientation, these dilated order polytopes combined must contain all proper n-colorings of G as lattice points. In other words,

$$\bigcup_{\rho \text{ acyclic}} (n+1) \mathsf{O}^{\circ}_{\Pi(\rho G)} \tag{7.1.4}$$

equals $(0, n+1)^V$ minus any point $\mathbf{x}$ with $x_j = x_k$ for $jk \in E$.

Proof of Theorem 1.1.5. We consider the subdivision

$$\mathcal{S} \;:=\; \left\{ \mathsf{O}_{\Pi(\rho G)} \,:\, \rho \text{ acyclic} \right\} \tag{7.1.5}$$

of the cube $[0,1]^V$; Exercise 7.1 asks you to verify that this is indeed a subdivision. With (7.1.2), we can now rewrite

$$\chi_G(n) \;=\; \operatorname{ehr}_{\mathcal{S}}^{[1]}(n+1) \,,$$

and so Proposition 7.1.1 says that

$$(-1)^{|V|} \chi_G(-n) \;=\; (-1)^{|V|} \operatorname{ehr}_{\mathcal{S}}^{[1]}(-n+1)$$

counts the number of lattice points in $(n-1)[0,1]^V = [0, n-1]^V$, each weighted by the number of order polytopes $\mathsf{O}_{\Pi(\rho G)}$ containing the point. This count equals the number of n-colorings (where we shifted the color set to $\{0, 1, \ldots, n-1\}$), each weighted by the number of compatible acyclic orientations. □

There is a picture reciprocal to Figure 7.1 that underlies our proof of Theorem 1.1.5. Namely, with Proposition 6.3.2, we may think of

$$\sum_{\rho \text{ acyclic}} \Omega_{\Pi({}_\rho G)}(n) \tag{7.1.6}$$

as a sum of the Ehrhart polynomials of the order polytopes $\mathsf{O}_{\Pi({}_\rho G)}$ evaluated at $n-1$. As in our geometric interpretation of Corollary 1.3.4, we can think of a lattice point in $(n-1)\mathsf{O}_{\Pi({}_\rho G)}$ as an n-coloring of G that is compatible with ρ. Contrary to the picture behind Corollary 1.3.4, here the dilated order polytopes overlap, and so each n-coloring gets counted with multiplicity equal to the number of its compatible acyclic orientations (see Figure 7.3).

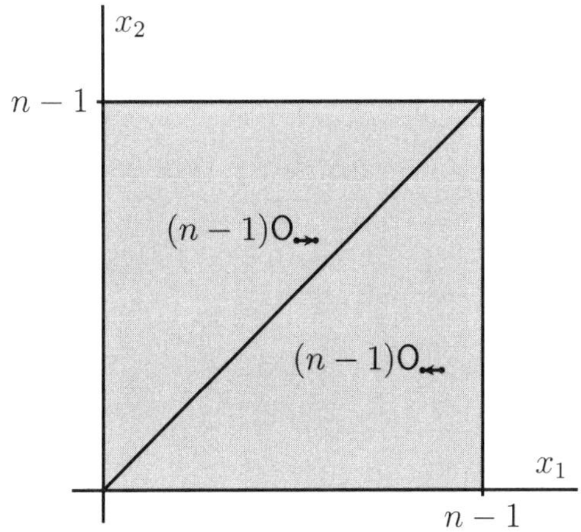

Figure 7.3. The geometry behind our proof of Theorem 1.1.5 for $G = K_2$.

There is an alternative, and more general, way of looking at the above setup and the combinatorial reciprocity exhibited by (our geometric proof of) Theorem 1.1.5. Before we can introduce it, we need to take another look at the combinatorics of hyperplane arrangements.

7.2. Flats and Regions of Hyperplane Arrangements

We recall from Section 3.4 that a hyperplane arrangement is a collection $\mathcal{H} = \{\mathsf{H}_1, \dots, \mathsf{H}_k\}$ of affine hyperplanes in $\mathbb{R}^d$. For $I \subseteq [k]$, we write H_I for the affine subspace obtained by intersecting $\{\mathsf{H}_i : i \in I\}$. In Section 3.4 we also defined the intersection poset $\mathcal{L}(\mathcal{H})$ of $\mathcal{H}$ as

$$\mathcal{L}(\mathcal{H}) = \{\mathsf{H}_I : \mathsf{H}_I \neq \varnothing,\ I \subseteq [k]\},$$

ordered by reverse inclusion, with minimum $\hat{0} = \mathbb{R}^d$. This poset has a maximum precisely when all hyperplanes have a common point or, equivalently, $\mathcal{H}$ is the translate of a central arrangement.

Our motivating and governing example in this section is an arrangement stemming from a given simple graph $G = (V, E)$. To an edge $ij \in E$ we associate the hyperplane $\mathsf{H}_{ij} := \{\mathbf{x} \in \mathbb{R}^V : x_i = x_j\}$. The **graphical arrangement** of G is then

$$(7.2.1) \qquad \mathcal{H}_G := \{\mathsf{H}_{ij} : ij \in E\}.$$

This is a central but not essential arrangement in $\mathbb{R}^V$: in Exercise 7.2 you will investigate the lineality spaces of graphical arrangements, and your findings will imply the following.

Proposition 7.2.1. *Let $G = (V, E)$ be a simple and connected graph. The lineality space of $\mathcal{H}_G$ is*

$$\text{lineal}(\mathcal{H}_G) = \mathbb{R}\,\mathbf{1}.$$

Here $\mathbf{1} \in \mathbb{R}^V$ denotes a vector all of whose entries are 1.

Our main motivation for considering this special class of hyperplane arrangements is that they geometrically carry quite a trove of information about the underlying graph. Let's take a closer look at the flats of a graphical arrangement. The flat corresponding to $S \subseteq E$ is

$$\mathsf{H}_S = \bigcap_{ij \in S} \mathsf{H}_{ij} = \big\{\mathbf{x} \in \mathbb{R}^V : x_i = x_j \text{ for all } ij \in S\big\}.$$

There are potentially many different sets $S' \subseteq E$ with $\mathsf{H}_S = \mathsf{H}_{S'}$. Indeed, let $G[S] := (V, S)$, the graph with the same node set V as G but with the (smaller) edge set S. If $ij \in E \setminus S$ is an edge such that the nodes i and j are in the same connected component of $G[S]$, then $x_i = x_j$ for all $\mathbf{x} \in \mathsf{H}_S$ and hence $\mathsf{H}_{S \cup \{ij\}} = \mathsf{H}_S$.

In general, there is always a unique inclusion-maximal $\bar{S} \subseteq E$ with $\mathsf{H}_{\bar{S}} = \mathsf{H}_S$. What does $\bar{S}$ look like? For any edge $ij \in E \setminus \bar{S}$, the endpoints i and j should be in different connected components of $G[\bar{S}]$; that is, $G[\bar{S} \cup \{ij\}]$ has strictly fewer connected components than $G[\bar{S}]$. We already came across this back in Section 2.4 where we showed that $S = \bar{S}$ if and only if $S \subseteq E$ is a flat of G. This way, we can extend (2.4.5) from colorings to all $\mathbf{p} \in \mathbb{R}^V$. We define

$$S_G(\mathbf{p}) := \{ij \in E : p_i = p_j\}.$$

Then $\mathsf{H}_{S_G(\mathbf{p})}$ is the maximal element in $\mathcal{L}(\mathcal{H}_G)$ (i.e., the inclusion-minimal affine subspace of $\mathcal{H}_G$) that contains $\mathbf{p}$. Our way of describing the elements of $\mathcal{L}(\mathcal{H}_G)$ will lead you (in Exercise 7.5) to realize that flats of $\mathcal{H}_G$ can be naturally identified with flats of G defined in Section 2.4.

Proposition 7.2.2. *Let $G = (V, E)$ be a graph. Then $\mathcal{L}(G)$ and $\mathcal{L}(\mathcal{H}_G)$ are canonically isomorphic as posets.*

The description of flats is reminiscent of a construction on graphs that we introduced, somewhat informally, in Section 1.1, namely, the **contraction** of an edge. To reintroduce this construct thoroughly, for $e = ij \in E$ we define G/e to be the graph $(\widetilde{V}, \widetilde{E})$, where $\widetilde{V} := V \setminus \{j\}$ and $\widetilde{E}$ consists of $E \setminus \{e\}$ plus all edges $\{ik : jk \in E,\ k \neq i\}$. We chose to define G/e so that no new loops are created; all edges between i and j disappear in G/e.

It is a short step to extend contraction from one to several edges: for $S \subseteq E$, we define G/S to be the graph resulting from consecutively contracting all edges in S; Exercise 7.3 makes sure that this definition is independent on the order with which we contract. In particular, $G/S \cong G/\bar{S}$, where $\bar{S}$ is the smallest flat of G that contains S. Figure 7.4 shows a contraction at an edge. Of course, G/S also comes with a graphical hyperplane arrangement and we can actually find it within $\mathcal{H}_G$.

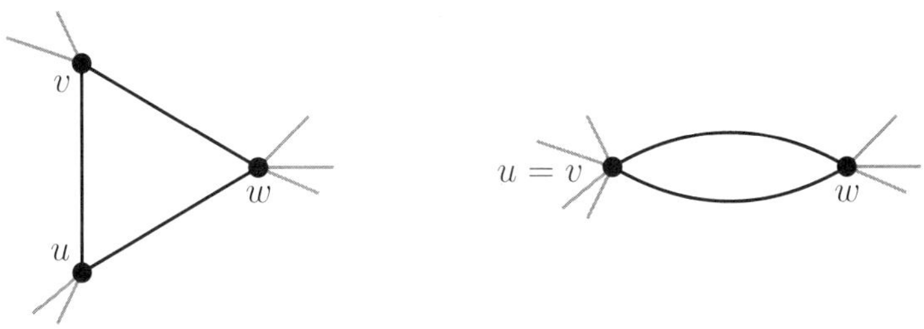

Figure 7.4. Contracting the edge uv.

Let $F \in \mathcal{L}(\mathcal{H})$ be a flat of an arrangement $\mathcal{H}$. We define the **restriction** of $\mathcal{H}$ to F as

$$\mathcal{H}|_F := \{\mathsf{H} \cap F : \mathsf{H} \in \mathcal{H},\ \varnothing \subsetneq \mathsf{H} \cap F \subsetneq F\}.$$

Since $F \cap \mathsf{H} \neq F$ and $F \cap \mathsf{H} \neq \varnothing$, we know that $F \cap \mathsf{H}$ is an affine subspace of F of dimension $\dim F - 1$. Hence $\mathcal{H}|_F$ is an arrangement of hyperplanes in $F \cong \mathbb{R}^{\dim F}$, and we invite you to establish the following in Exercise 7.6.

Proposition 7.2.3. *Let $G = (V, E)$ be a graph, $S \subseteq E$ a flat of G, and $F \in \mathcal{L}(\mathcal{H}_G)$ the corresponding flat of $\mathcal{H}_G$. Then*

$$\mathcal{H}_G|_F \cong \mathcal{H}_{G/S} \quad \textit{and} \quad \mathcal{L}(\mathcal{H}_G|_F) \cong \mathcal{L}(\mathcal{H}_{G/S}),$$

in the sense that there is a bijection between $\mathcal{H}_G|_F$ and $\mathcal{H}_{G/S}$ that induces an order-preserving bijection between $\mathcal{L}(\mathcal{H}_G|_F)$ and $\mathcal{L}(\mathcal{H}_{G/S})$.

Any hyperplane arrangement $\mathcal{H}$ dissects $\mathbb{R}^d$ into regions, and we studied the number of regions $r(\mathcal{H})$ and the number of relative bounded regions $b(\mathcal{H})$ in Section 3.6. The arrangement $\mathcal{H}_G$ is central and hence has no relatively

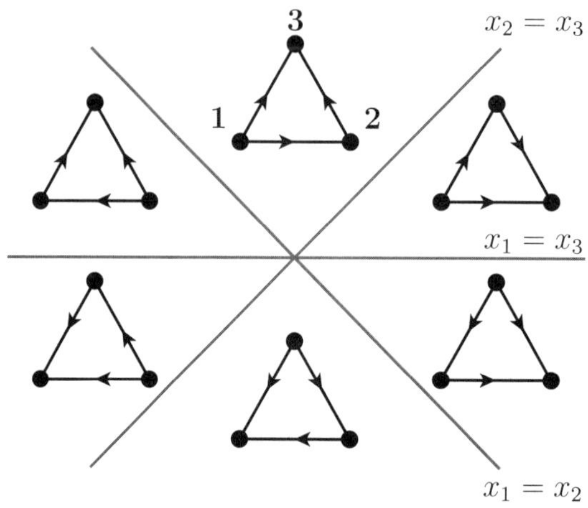

Figure 7.5. The regions of $\mathcal{H}_{K_3}$ (projected to the plane $x_1+x_2+x_3=0$) and their corresponding acyclic orientations.

bounded regions. The number of all regions of $\mathcal{H}_G$ turns out to be a familiar face.

We study the regions of $\mathcal{H}_G$ based on the methodology developed in Section 3.6. If $\mathbf{p} \in \mathbb{R}^V$ is not contained in any of the hyperplanes of $\mathcal{H}_G$, then for each edge $ij \in E$, we have $p_i > p_j$ or $p_i < p_j$. Each of these inequalities, in turn, can be thought of as orienting the edge: if $p_j > p_i$, then orient ij from i to j, and if $p_j < p_i$ then orient ij from j to i. That is, we orient an edge so that the edge $ij \in E$ is oriented towards the node having the *larger* of the two distinct values p_i and p_j. In this way, each region of $\mathcal{H}_G$ gives rise to an orientation of G; see Figure 7.5 for an illustration in the case $G = K_3$, the complete graph on three nodes. The orientations that we can associate with the regions of $\mathcal{H}_G$ are precisely the *acyclic* ones—a directed circle would correspond to a sequence of the nonsensical inequalities $x_{i_1} > x_{i_2} > \cdots > x_{i_k} > x_{i_1}$. We summarize:

Lemma 7.2.4. *Let $G = (V, E)$ be a graph. The regions of $\mathcal{H}_G$ are in one-to-one correspondence with the acyclic orientations of G. Moreover, for any flat $S \subseteq E$ of G with corresponding flat $F \in \mathcal{L}(\mathcal{H}_G)$, the region count $r(\mathcal{H}_G|_F)$ equals the number of acyclic orientations of G/S.*

Proof. We still need to argue that every acyclic orientation actually determines a region of $\mathcal{H}_G$. Let ρ be an acyclic orientation of G. A **source** of ${}_\rho G$ is a node v with no oriented edges entering v. Pick $M > 1$ and define $\mathbf{p} \in \mathbb{R}^V$ iteratively as follows. Set $p_v = M$ if v is a source of ${}_\rho G$. As the next step, we remove all sources from G, which leaves us with a subgraph $G' = (V', E')$ with $V' \subset V$, and G' still carries an induced acyclic orientation. We now repeat the procedure with M^2 instead of M, etc. At some point, when we removed the last node, we are left with a well-defined point $\mathbf{p} \in \mathbb{R}^V$. If ρ orients an edge $ij \in E$ from i to j, then i was removed before j and hence $p_j \geq M \cdot p_i > p_i$, which proves the first claim.

For the second claim, we simply appeal to Proposition 7.2.3. □

The intersection poset $\mathcal{L}(\mathcal{H})$ of any arrangement in $\mathbb{R}^d$ is graded with rank function

$$\mathrm{rk}_{\mathcal{L}(\mathcal{H})}(F) \;=\; d - \dim \mathrm{lineal}(\mathcal{H}) - \dim F\,,$$

and the characteristic polynomial of $\mathcal{H}$, as defined in Section 3.6, is

$$\chi_{\mathcal{H}}(n) \;=\; \sum_{F \in \mathcal{L}(\mathcal{H})} \mu_{\mathcal{L}(\mathcal{H})}(\hat{0}, F)\, n^{\dim F}.$$

Our central result in Section 3.6 was Zaslavsky's Theorem 3.6.4: if $\mathcal{H}$ is an arrangement in $\mathbb{R}^d$ with e-dimensional lineality space $\mathrm{lineal}(\mathcal{H})$, then

$$r(\mathcal{H}) \;=\; (-1)^d\, \chi_{\mathcal{H}}(-1) \qquad \text{and} \qquad b(\mathcal{H}) \;=\; (-1)^{d-e}\, \chi_{\mathcal{H}}(1)\,.$$

This connects geometric and combinatorial quantities. Next we show how we can use region counts to determine the Möbius function of $\mathcal{L}(\mathcal{H})$.

Complementary to the restriction, we define the **localization** of $\mathcal{H}$ at a flat $F \in \mathcal{L}(\mathcal{H})$ by

$$\mathcal{H}|^F \;:=\; \{\mathsf{H} \in \mathcal{H} \,:\, F \subseteq \mathsf{H}\}\,.$$

So $\mathcal{H}|^F$ is the largest subarrangement of $\mathcal{H}$ that contains F. The upshot is that, if $S \preceq_{\mathcal{L}(\mathcal{H})} T$ are two flats, then $T \subseteq S$ and the interval $[S, T]_{\mathcal{L}(\mathcal{H})}$ is canonically isomorphic to $\mathcal{L}(\mathcal{H}|_S^T)$. This brings us closer to the Möbius function of intersection posets of central hyperplane arrangements.

Let $\mathcal{H} = \{\mathsf{H}_1, \dots, \mathsf{H}_m\}$ be a central hyperplane arrangement in $\mathbb{R}^d$. An affine hyperplane $\mathsf{H} \subset \mathbb{R}^d$ is **in general position relative to $\mathcal{H}$** if $\mathrm{lineal}(\mathcal{H}) \cap \mathsf{H} = \varnothing$ and

$$\dim F \cap \mathsf{H} \;=\; \dim F - 1 \quad \text{for every} \quad F \in \mathcal{L}(\mathcal{H}) \setminus \{\mathrm{lineal}(\mathcal{H})\}\,.$$

Proposition 7.2.5. *Let $\mathcal{H}$ be a central hyperplane arrangement and H_0 a hyperplane in general position relative to $\mathcal{H}$. Then*

$$\mathcal{L}(\mathcal{H}|_{\mathsf{H}_0}) \;\cong\; \mathcal{L}(\mathcal{H}) \setminus \{\mathrm{lineal}(\mathcal{H})\}$$

via $F \mapsto F \cap \mathsf{H}_0$.

With this result in hand (whose proof we leave to Exercise 7.7), we can give an interpretation of the values of the Möbius function of $\mathcal{L}(\mathcal{H})$.

Theorem 7.2.6. *Let $\mathcal{H}$ be a central hyperplane arrangement and H_0 a hyperplane in general position relative to $\mathcal{H}$. Then H_0 meets all but*

$$(-1)^{d-\dim \operatorname{lineal}(\mathcal{H})}\,\mu_{\mathcal{L}(\mathcal{H})}(\hat{0},\hat{1})$$

many regions of $\mathcal{H}$.

Proof. We observe that the regions of $\mathcal{H}$ that meet H_0 are in bijection to the regions of $\mathcal{H}|_{\mathsf{H}_0}$ in $\mathsf{H}_0 \cong \mathbb{R}^{d-1}$ under the map that takes a region R to $R \cap \mathsf{H}_0$. Hence the number of regions of $\mathcal{H}$ missed by H_0 is

$$r(\mathcal{H}) - r(\mathcal{H}|_{\mathsf{H}_0}) \;=\; (-1)^d \chi_{\mathcal{H}}(-1) - (-1)^{d-1}\chi_{\mathcal{H}|_{\mathsf{H}_0}}(-1)\,, \tag{7.2.2}$$

by Zaslavsky's Theorem 3.6.4. Using the definition of the characteristic polynomial together with Proposition 7.2.5, the right-hand side in (7.2.2) can be rewritten as

$$\sum_{\hat{0}\preceq F\preceq\hat{1}} \mu_{\mathcal{L}(\mathcal{H})}(\hat{0},F)(-1)^{d-\dim F} - \sum_{\hat{0}\preceq F\prec\hat{1}} \mu_{\mathcal{L}(\mathcal{H})}(\hat{0},F)(-1)^{(d-1)-(\dim F-1)}\,.$$

All terms cancel except for $(-1)^{d-\dim \operatorname{lineal}(\mathcal{H})}\mu_{\mathcal{L}(\mathcal{H})}(\hat{0},\hat{1})$. □

We outsource the interpretation of the full Möbius function of $\mathcal{L}(\mathcal{H})$ to Exercise 7.12.

Theorem 7.2.6 gives a pleasing interpretation of $|\mu_{\mathcal{L}(\mathcal{H})}(\hat{0},\hat{1})|$ but the quite fascinating implication is that the number of regions missed by a hyperplane in relative general position is always the same! For graphs this has the following interpretation, which is not so easy to prove without appealing to Theorem 7.2.6; do try in Exercise 7.13.

Theorem 7.2.7. *Let $G = (V, E)$ be a connected graph and let $\mathcal{L} = \mathcal{L}(G)$ be its lattice of flats. Let $u \in V$ be an arbitrary but fixed node. Then $(-1)^{|V|}\mu_{\mathcal{L}}(\hat{0},\hat{1})$ is the number of acyclic orientations for which u is the unique source.*

Proof. Using Proposition 7.2.1, we may restrict $\mathcal{H}_G$ to the hyperplane

$$U \;:=\; \left\{\mathbf{x}\in\mathbb{R}^V \;:\; x_u = 0\right\}.$$

This makes $\mathcal{H}_G$ central and essential restricted to $U \cong \mathbb{R}^{|V|-1}$. Exercise 7.14 verifies that the hyperplane

$$\mathsf{H}_0 \;:=\; \left\{\mathbf{x}\in U \;:\; \sum_{v\in V} x_v = -1\right\}$$

is in general position relative to $\mathcal{H}_G$.

Proposition 7.2.2 and Theorem 7.2.6 now say that $(-1)^{|V|}\mu_{\mathcal{L}}(\hat{0},\hat{1})$ is the number of regions missed by H_0. By Lemma 7.2.4, the regions R of $\mathcal{H}_G$ correspond exactly to the acyclic orientations of G. Pick a region R such that the corresponding acyclic orientation ρ has u as its unique source. Thus, for any $v \in V$, there is a directed path $u = v_0 v_1 \cdots v_k = v$, and so

$$0 \; = \; x_u \; = \; x_{v_0} \; < \; x_{v_1} \; < \; \cdots \; < \; x_{v_k} \; = \; x_u$$

for every $\mathbf{x} \in R$. It follows that $x_u \geq 0$ for all $u \in V$ and hence

$$R \; \subseteq \; \left\{ \mathbf{x} \in U \, : \, \sum_{v \in V} x_v \geq 0 \right\} .$$

So R cannot meet H_0.

Conversely, let R be a region such that the corresponding orientation has a source $w \neq u$. Then $p_w < p_v$ for all $uw \in E$ and all $\mathbf{p} \in R$. In particular, for $\mathbf{p} \in R$, the point $\mathbf{q} := \mathbf{p} - \lambda \mathbf{e}_w \in R$ for any $\lambda > 0$. For sufficiently large $\lambda > 0$,

$$\sum_{v \in V} q_v \; < \; 0 \, ,$$

and scaling $\mathbf{q}$ by a positive scalar shows that $R \cap \mathsf{H}_0 \neq \varnothing$. □

7.3. Inside-out Polytopes

Now we pick up the thread from the end of Section 7.1. We already remarked that (7.1.4) can be interpreted as the open cube $(0, n+1)^V$ minus any point $\mathbf{x}$ with $x_j = x_k$ for $jk \in E$; these latter points form precisely the hyperplanes in the graphic arrangement of G.

Our next goal is to study an analogue of this setup for a general rational d-polytope $\mathsf{P} \subset \mathbb{R}^d$ and a general rational[1] hyperplane arrangement $\mathcal{H}$ in $\mathbb{R}^d$. (The case that P is not full dimensional is a little subtle; for a flavor, see Exercises 7.40 and 7.43.) Let

$$I_{\mathsf{P},\mathcal{H}}(n) \; := \; \left| n \left(\mathsf{P} \setminus \bigcup \mathcal{H} \right) \cap \mathbb{Z}^d \right| \; = \; \left| \left(\mathsf{P} \setminus \bigcup \mathcal{H} \right) \cap \frac{1}{n} \mathbb{Z}^d \right| , \tag{7.3.1}$$

in words: $I_{\mathsf{P},\mathcal{H}}(n)$ counts those points in $\frac{1}{n}\mathbb{Z}^d$ that are in the polytope P but off the hyperplanes in $\mathcal{H}$. For example, the chromatic polynomial $\chi_G(n)$ of a given graph G can thus be written as

$$\chi_G(n) \; = \; I_{(0,1)^V, \mathcal{H}_G}(n+1) \, , \tag{7.3.2}$$

that is, the polytope in question is the open unit cube in $\mathbb{R}^V$ and the hyperplane arrangement is the one associated with G.

In the absence of $\mathcal{H}$, our definition (7.3.1) matches (4.6.1) giving the Ehrhart quasipolynomial of P counting lattice points as we dilate P or, equivalently, shrink the integer lattice $\mathbb{Z}^d$. We refer to $(\mathsf{P}, \mathcal{H})$ as an **inside-out**

[1] A hyperplane arrangement is **rational** if all of its hyperplanes can be described by linear equalities with rational (or, equivalently, integer) coefficients.

polytope because we think of the hyperplanes in $\mathcal{H}$ as additional boundary inside P.

The counting function $I_{\mathsf{P},\mathcal{H}}(n)$ can be computed through the Möbius function $\mu(\mathbb{R}^d, F)$ of $\mathcal{L}(\mathcal{H})$ if we know the Ehrhart quasipolynomial of $\mathsf{P} \cap F$ for each flat F.

Theorem 7.3.1. *Suppose $\mathsf{P} \subset \mathbb{R}^d$ is a rational d-polytope and $\mathcal{H}$ is a rational hyperplane arrangement in $\mathbb{R}^d$ with Möbius function $\mu = \mu_{\mathcal{L}(\mathcal{H})}$. Then*

$$I_{\mathsf{P},\mathcal{H}}(n) \;=\; \sum_{F \in \mathcal{L}(\mathcal{H})} \mu(\mathbb{R}^d, F)\, \mathrm{ehr}_{\mathsf{P}\cap F}(n)\,.$$

Proof. Given a flat $F \in \mathcal{L}(\mathcal{H})$, we can compute $\mathrm{ehr}_{\mathsf{P}\cap F}(n)$ by counting the lattice points in the faces of $\mathcal{H}|_F$ in the following way: each lattice point is in the interior of a unique minimal face Q of $\mathcal{H}|_F$, and Q is a region of $\mathcal{H}|_G$ for some flat $G \subseteq F$ (more precisely, G is the affine span of Q). Thus our lattice point is one of the points counted by $I_{\mathsf{P}\cap G,\mathcal{H}|_G}(n)$, and grouping the lattice points in $\mathsf{P} \cap F$ according to this scheme yields

$$\mathrm{ehr}_{\mathsf{P}\cap F}(n) \;=\; \sum_{G \subseteq F} I_{\mathsf{P}\cap G,\mathcal{H}|_G}(n)\,.$$

By Möbius inversion (Theorem 2.4.2),

$$I_{\mathsf{P}\cap F,\mathcal{H}|_F}(n) \;=\; \sum_{G \subseteq F} \mu(F,G)\; \mathrm{ehr}_{\mathsf{P}\cap G}(n)\,,$$

and so in particular for $F = \mathbb{R}^d$,

$$I_{\mathsf{P},\mathcal{H}}(n) \;=\; \sum_{G \in \mathcal{L}(\mathcal{H})} \mu(\mathbb{R}^d, G)\; \mathrm{ehr}_{\mathsf{P}\cap G}(n)\,. \qquad \square$$

With Theorem 5.2.4 we conclude effortlessly:

Corollary 7.3.2. *If $\mathsf{P} \subset \mathbb{R}^d$ is a rational polytope and $\mathcal{H}$ is a rational hyperplane arrangement in $\mathbb{R}^d$, then $I_{\mathsf{P},\mathcal{H}}(n)$ is a quasipolynomial in n.*

You have undoubtedly noticed the similarities shared by the definition of the characteristic polynomial $\chi_{\mathcal{H}}(n)$ and the formula for $I_{\mathsf{P},\mathcal{H}}(n)$ given in Theorem 7.3.1. In fact, the two functions are intimately connected, as the following first application shows.

Corollary 7.3.3. *Let $(\mathsf{P},\mathcal{H})$ be an inside-out polytope for which there exists a function $\phi(n)$ such that $\mathrm{ehr}_{\mathsf{P}\cap F}(n) = \phi(n)^{\dim F}$ for any flat $F \in \mathcal{L}(\mathcal{H})$. Then*

$$I_{\mathsf{P},\mathcal{H}}(n) \;=\; \chi_{\mathcal{H}}(\phi(n))\,.$$

Proof. If $(\mathsf{P}, \mathcal{H})$ is an inside-out polytope satisfying the conditions of Corollary 7.3.3 then, by Theorem 7.3.1,

$$\begin{aligned} I_{\mathsf{P},\mathcal{H}}(n) &= \sum_{F \in \mathcal{L}(\mathcal{H})} \mu(\mathbb{R}^d, F)\, \mathrm{ehr}_{\mathsf{P} \cap F}(n) = \sum_{F \in \mathcal{L}(\mathcal{H})} \mu(\mathbb{R}^d, F)\, \phi(n)^{\dim F} \\ &= \chi_{\mathcal{H}}(\phi(n)) . \end{aligned}$$ □

Corollary 7.3.3 allows us to compute the characteristic polynomial of certain arrangements by counting lattice points. Here is a sample.

Corollary 7.3.4. *Let $\mathcal{H} = \{\{x_j = 0\} : 1 \le j \le d\}$, the Boolean arrangement in $\mathbb{R}^d$. Then $\chi_{\mathcal{H}}(n) = (n-1)^d$.*

Proof. Let P be the d-dimensional unit cube $[0,1]^d$. Then $I_{\mathsf{P},\mathcal{H}}(n) = n^d$, since $(n\mathsf{P} \setminus \bigcup \mathcal{H}) \cap \mathbb{Z}^d$ contains precisely those lattice points in $n\mathsf{P}$ that have nonzero coordinates. On the other hand, a k-dimensional flat $F \in \mathcal{L}(\mathcal{H})$ is defined by $d-k$ equations of the form $x_j = 0$, and so

$$\mathrm{ehr}_{\mathsf{P} \cap F}(n) = (n+1)^k .$$

Thus $\phi(n) = n+1$ in Corollary 7.3.3 gives $\chi_{\mathcal{H}}(n+1) = I_{\mathsf{P},\mathcal{H}}(n) = n^d$. □

Note that, with Zaslavsky's Theorem 3.6.4, Corollary 7.3.4 implies Exercise 3.65, namely, that the Boolean arrangement in $\mathbb{R}^d$ has 2^d regions.

Corollary 7.3.5. *Let $\mathcal{H} = \{\{x_j = x_k\} : 1 \le j < k \le d\}$, the real braid arrangement in $\mathbb{R}^d$. Then*

$$\chi_{\mathcal{H}}(n) = n(n-1)(n-2)\cdots(n-d+1) .$$

Proof. Again let P be the d-dimensional unit cube $[0,1]^d$. We can pick a point $(x_1, x_2, \dots, x_d) \in (n\mathsf{P} \setminus \bigcup \mathcal{H}) \cap \mathbb{Z}^d$ by first choosing x_1 (for which there are $n+1$ choices), then choosing $x_2 \ne x_1$ (for which there are n choices), etc., down to choosing $x_d \ne x_1, x_2, \dots, x_{d-1}$ (for which there are $n-d+2$ choices), and so

$$I_{\mathsf{P},\mathcal{H}}(n) = (n+1)\, n\, (n-1) \cdots (n-d+2) .$$

On the other hand, a k-dimensional flat $F \in \mathcal{L}(\mathcal{H})$ is defined by $d-k$ equations of the form $x_i = x_j$, and so again

$$\mathrm{ehr}_{\mathsf{P} \cap F}(n) = (n+1)^k .$$

Thus we take $\phi(n) = n+1$ in Corollary 7.3.3, whence

$$\chi_{\mathcal{H}}(n+1) = I_{\mathsf{P},\mathcal{H}}(n) = (n+1)\, n\, (n-1) \cdots (n-d+2) .$$ □

Again we can use Zaslavsky's Theorem 3.6.4, with which Corollary 7.3.5 implies Exercise 3.66.

Corollary 7.3.6. *Let* $\mathcal{H} = \{\{x_j = \pm x_k\}, \{x_j = 0\} : 1 \le j < k \le d\}$. *Then*

$$\chi_{\mathcal{H}}(n) \ = \ (n-1)(n-3)\cdots(n-2d+1)\,.$$

Proof. Let P be the cube $[-1,1]^d$. We can pick a point $(x_1, x_2, \dots, x_d) \in (n\mathsf{P} \setminus \bigcup \mathcal{H}) \cap \mathbb{Z}^d$ by first choosing $x_1 \neq 0$ (for which there are $2n$ choices), then choosing $x_2 \neq \pm x_1, 0$ (for which there are $2n-2$ choices), etc., down to choosing $x_d \neq \pm x_1, \pm x_2, \dots, \pm x_{d-1}, 0$ (for which there are $2n-2d+2$ choices), and so

$$I_{\mathsf{P},\mathcal{H}}(n) \ = \ 2n(2n-2)\cdots(2n-2d+2)\,.$$

On the other hand, a k-dimensional flat $F \in \mathcal{L}(\mathcal{H})$ is defined by $d-k$ equations of the form $x_i = \pm x_j$ or $x_j = 0$, and so $\operatorname{ehr}_{\mathsf{P}\cap F}(n) = (2n+1)^k$. Thus we take $\phi(n) = 2n+1$ in Corollary 7.3.3, whence

$$\chi_{\mathcal{H}}(2n+1) \ = \ I_{\mathsf{P},\mathcal{H}}(n) \ = \ 2n(2n-2)\cdots(2n-2d+2)\,. \qquad \square$$

It's time to return to proper colorings of a graph $G = (V, E)$. Let $\mathsf{P} = [0,1]^V$, the unit cube in $\mathbb{R}^V$. Just like in the case of real braid arrangements—i.e., our proof of Corollary 7.3.5—, any k-dimensional flat F of the graphical arrangement $\mathcal{H}_G$ will give rise to the Ehrhart polynomial $\operatorname{ehr}_{\mathsf{P}\cap F}(n) = (n+1)^k$, and with Corollary 7.3.3 we obtain

$$I_{\mathsf{P},\mathcal{H}_G}(n) \ = \ \chi_{\mathcal{H}_G}(n+1)\,.$$

We can play the same game with the *open* unit cube $\mathsf{P}^\circ = (0,1)^V$: if F is a k-dimensional flat of $\mathcal{H}_G$, then $\operatorname{ehr}_{\mathsf{P}^\circ\cap F}(n) = (n-1)^k$, and so

$$I_{\mathsf{P}^\circ,\mathcal{H}_G}(n) \ = \ \chi_{\mathcal{H}_G}(n-1)\,. \tag{7.3.3}$$

The left-hand side already appeared in (7.3.2) in connection with the chromatic polynomial of G, and so combining (7.3.2) and (7.3.3) yields something you might have suspected by now.

Corollary 7.3.7. *For any graph G we have* $\chi_G(n) = \chi_{\mathcal{H}_G}(n)$.

In particular, this reconfirms that $\chi_G(n)$ is a polynomial, as we have known since Section 1.1. There are a few more immediate consequences, reproving parts of Theorem 1.1.5.

Corollary 7.3.8. *The chromatic polynomial $\chi_G(n)$ of G is monic, has degree $|V|$, and constant term 0. Its evaluation $(-1)^{|V|}\chi_G(-1)$ equals the number of acyclic orientations of G.*

Proof. Since the unit cube $\mathsf{P} = [0,1]^V$ has volume one, the polynomial $\chi_G(n) = I_{\mathsf{P}^\circ,\mathcal{H}_G}(n+1)$ is monic, of degree $|V|$, and has constant term $\chi_G(0) = I_{\mathsf{P}^\circ,\mathcal{H}_G}(1) = 0$. The final statement in Corollary 7.3.8 follows from Lemma 7.2.4 whose last part can now be restated as: $(-1)^{|V|}\chi_G(-1)$ equals the number of acyclic orientations of G. $\square$

We recall that Lemma 7.2.4 seemingly effortlessly gave us an instance of the reciprocity theorem for chromatic polynomials (Theorem 1.1.5). In fact, we can give a second proof of this reciprocity theorem with the machinery developed in this chapter. We'll do so for the more general class of inside-out Ehrhart quasipolynomials and specialize to chromatic polynomials soon thereafter.

Let $\mathsf{P} \subset \mathbb{R}^d$ be a rational polytope, and let $\mathcal{H}$ be a rational hyperplane arrangement in $\mathbb{R}^d$. Then $\mathsf{P}^\circ \setminus \bigcup \mathcal{H}$ is the union of the (relative) interiors of some rational polytopes, each of dimension $\dim(\mathsf{P})$:

$$\mathsf{P}^\circ \setminus \bigcup \mathcal{H} \ = \ \mathsf{Q}_1^\circ \cup \mathsf{Q}_2^\circ \cup \cdots \cup \mathsf{Q}_k^\circ \, .$$

(In the graphic case, this leads to (7.1.2).) In the language of Chapter 5, the Q_js and their faces form a subdivision of P—one that is induced by $\mathcal{H}$. Thus $I_{\mathsf{P}^\circ, \mathcal{H}}(n) = \sum_{j=1}^k \operatorname{ehr}_{\mathsf{Q}_j^\circ}(n)$, and Ehrhart–Macdonald reciprocity (Theorem 5.2.4) gives

$$I_{\mathsf{P}^\circ, \mathcal{H}}(-n) \ = \ (-1)^{\dim(\mathsf{P})} \sum_{j=1}^{k} \operatorname{ehr}_{\mathsf{Q}_j}(n) \, . \tag{7.3.4}$$

The sum on the right can be interpreted purely in terms of the inside-out polytope $(\mathsf{P}, \mathcal{H})$. Namely, each point in $\frac{1}{n}\mathbb{Z}^d$ that is counted by $\sum_{j=1}^k \operatorname{ehr}_{\mathsf{Q}_j}(n)$ lies in P, and it gets counted with multiplicity equal to the number of closed regions of $(\mathsf{P}, \mathcal{H})$ that contain it. Here, by analogy with our hyperplane arrangement terminology from Section 3.6, a **(closed) region** of $(\mathsf{P}, \mathcal{H})$ is (the closure of) a connected component of $\mathsf{P} \setminus \bigcup \mathcal{H}$.

These observations motivate the following definitions. The **multiplicity of $\mathbf{p} \in \mathbb{R}^d$ with respect to $(\mathsf{P}, \mathcal{H})$** is[2]

$$\operatorname{mult}_{\mathsf{P}, \mathcal{H}}(\mathbf{p}) \ := \ \# \text{ closed regions of } (\mathsf{P}, \mathcal{H}) \text{ that contain } \mathbf{p} \, .$$

Note that this definition implies $\operatorname{mult}_{\mathsf{P}, \mathcal{H}}(\mathbf{p}) = 0$ if $\mathbf{p} \notin \mathsf{P}$. Thus

$$O_{\mathsf{P}, \mathcal{H}}(n) \ := \sum_{\mathbf{p} \in \frac{1}{n}\mathbb{Z}^d} \operatorname{mult}_{\mathsf{P}, \mathcal{H}}(\mathbf{p})$$

equals the sum $\sum_{j=1}^k \operatorname{ehr}_{\mathsf{Q}_j}(n)$ in (7.3.4), which gives the following reciprocity theorem.

Theorem 7.3.9. *Suppose $\mathsf{P} \subset \mathbb{R}^d$ is a rational d-polytope, and $\mathcal{H}$ is a rational hyperplane arrangement in $\mathbb{R}^d$. Then $O_{\mathsf{P}, \mathcal{H}}(n)$ and $I_{\mathsf{P}^\circ, \mathcal{H}}(n)$ are quasipolynomials that satisfy*

$$I_{\mathsf{P}^\circ, \mathcal{H}}(-n) \ = \ (-1)^d \, O_{\mathsf{P}, \mathcal{H}}(n) \, .$$

[2] In the definition of $\operatorname{mult}_{\mathsf{P}, \mathcal{H}}(\mathbf{p})$, we can replace the phrase *closed regions of* $(\mathsf{P}, \mathcal{H})$ by *closed regions of* $\mathcal{H}$ if P and $\mathcal{H}$ are **transversal**, i.e., if every flat of $\mathcal{L}(\mathcal{H})$ that intersects P also intersects P°.

There is a version of this theorem when P is not full dimensional, but one has to be a bit careful; see Exercise 7.40.

We can say more: first, the periods of the quasipolynomials $O_{\mathsf{P},\mathcal{H}}(n)$ and $I_{\mathsf{P}^\circ,\mathcal{H}}(n)$ divide the least common multiple of the denominators of the vertex coordinates of all of the regions of $(\mathsf{P},\mathcal{H})$, by Theorem 5.2.4. Second, the leading coefficient of both $O_{\mathsf{P},\mathcal{H}}(n)$ and $I_{\mathsf{P}^\circ,\mathcal{H}}(n)$ equals the volume of P, by Exercise 5.10. Finally, since $\mathrm{ehr}_{\mathsf{Q}_j}(0) = 1$ (Exercise 5.15), we conclude the following.

Corollary 7.3.10. *The constant term $O_{\mathsf{P},\mathcal{H}}(0)$ equals the number of regions of $(\mathsf{P},\mathcal{H})$.*

Now we apply the inside-out machinery to graph coloring.

Second proof of Theorem 1.1.5. We recall that (7.3.2) said

$$\chi_G(n) \quad = \quad I_{\mathsf{P}^\circ,\mathcal{H}_G}(n+1)\,,$$

where $\mathsf{P} = [0,1]^V$. With Theorem 7.3.9 we thus obtain

$$(-1)^{|V|}\,\chi_G(-n) \quad = \quad O_{\mathsf{P},\mathcal{H}_G}(n-1)\,.$$

In the absence of the graphical arrangement $\mathcal{H}_G$, the function on the right counts the integer lattice points in the cube $[0,n-1]^V$; each such lattice point can naturally be interpreted as an n-coloring. Now taking $\mathcal{H}_G$ into account, $O_{\mathsf{P},\mathcal{H}_G}(n-1)$ counts each such lattice point with multiplicity equal to the number of closed regions the point lies in. But Lemma 7.2.4 asserts that these regions correspond exactly to the acyclic orientations of G. So if we think of a lattice point in $[0,n-1]^V$ as an n-coloring, the region multiplicity gives the number of compatible acyclic orientations. Because $O_{\mathsf{P},\mathcal{H}_G}(n-1)$ takes these multiplicities into account, the theorem follows. □

7.4. Alcoved Polytopes

Many of the combinatorially-rich polytopes that we encountered throughout can be put on common ground. One way to construct inside-out polytopes is from hyperplane arrangements: given a hyperplane arrangement $\mathcal{H}$, its regions are convex polyhedra and, similar to what we did in Section 3.4, we can form polyhedra from unions of regions. An **$\mathcal{H}$-polytope** is a convex polytope that is a union of closed (and bounded) regions of $\mathcal{H}$. Such polytopes automatically come with a dissection by an arrangement of hyperplanes, that is, $\mathcal{H}$-polytopes are inside-out polytopes.

The focus here will be on one particularly interesting class of hyperplane arrangements. For fixed $d \geq 1$, we label the coordinates of $\mathbb{R}^{d+1}$ by

$x_0, x_1, \ldots, x_d$. For $0 \le i < j \le d$ and $a \in \mathbb{Z}$, we define

$$\mathsf{H}_{ij}^a := \left\{ \mathbf{x} \in \mathbb{R}^{d+1} : x_j - x_i = a \right\},$$

giving rise to an infinite hyperplane arrangement in $\mathbb{R}^{d+1}$. (We should not worry that this is an infinite arrangement; at any moment we will only work with a finite subarrangement.) Note that each hyperplane is parallel to the line $\mathbb{R}\mathbf{1}$. So, this arrangement is not essential but we can make it essential by restricting it to the hyperplane $\{\mathbf{x} \in \mathbb{R}^{d+1} : x_0 = 0\}$. We define $\mathcal{A}_d$ to be the essential (but still infinite) arrangement of hyperplanes H_{ij}^a for $1 \le i \le j \le d$ and $a \in \mathbb{Z}$ as well as the hyperplanes

$$\mathsf{H}_i^a := \left\{ \mathbf{x} \in \mathbb{R}^d : x_i = a \right\}$$

for $1 \le i \le d$ and $a \in \mathbb{Z}$; see Figure 7.6 for the two-dimensional picture. Note that we will keep in mind that here $\mathbb{R}^d$ corresponds to the hyperplane $\{x_0 = 0\}$ in $\mathbb{R}^{d+1}$, and hence for every point $\mathbf{p} \in \mathbb{R}^d$, we will conveniently add $p_0 = 0$.

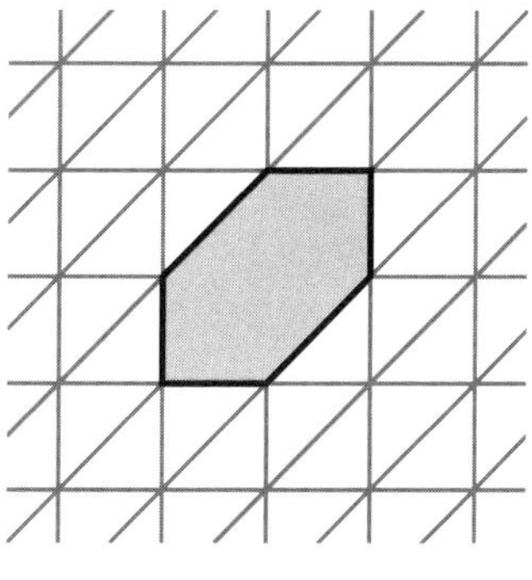

Figure 7.6. The alcoved arrangement in dimension 2 and an alcoved hexagon.

What are the regions of $\mathcal{A}_d$? We pick an arbitrary point $\mathbf{p} \in \mathbb{R}^d$ that is not contained in any of the hyperplanes of $\mathcal{A}_d$. Hence, $p_i \notin \mathbb{Z}$ for $i = 1, \ldots, d$ and we can write $\mathbf{p} = \mathbf{q} + \mathbf{r}$ with $\mathbf{q} \in \mathbb{Z}^d$ and $\mathbf{r} \in (0,1)^d$. Since $\mathbf{p}$ also misses all of the hyperplanes H_{ij}^a, we note that $r_i \ne r_j$ for all $i \ne j$. We have seen this before: as in Section 5.1, there is a unique permutation $\tau \in \mathfrak{S}_d$ such that

$$0 < r_{\tau^{-1}(1)} < r_{\tau^{-1}(2)} < \cdots < r_{\tau^{-1}(d)} < 1 .$$

The hyperplanes bounding the simplex

$$\triangle^\tau := \left\{ \mathbf{x} \in \mathbb{R}^d : 0 \le x_{\tau^{-1}(1)} \le x_{\tau^{-1}(2)} \le \cdots < x_{\tau^{-1}(d)} \le 1 \right\}$$

are contained in $\mathcal{A}_d$ and the unique closed (and bounded) region of $\mathcal{A}_d$ that contains $\mathbf{p}$ is given by $\mathbf{q} + \triangle^{\tau^{-1}}$. Of course, $\triangle^\tau$ is exactly the simplex

$\triangle_{\tau^{-1}}$ defined in Section 5.1, but you will soon see that it is nicer to work with τ^{-1} instead of τ, and our new notation reflects that. Since all points $\mathbf{p} \in \mathbb{R}^d \setminus \bigcup \mathcal{A}_d$ are of the form $\mathbf{p} = \mathbf{q} + \mathbf{r}$ as above, our reasoning proves the following fact.

Proposition 7.4.1. *The closed regions of $\mathcal{A}_d$, called* **alcoves**, *are given by $\mathbf{q} + \triangle^\tau$ for $\mathbf{q} \in \mathbb{Z}^d$ and $\tau \in \mathfrak{S}_d$. In particular all regions of $\mathcal{A}_d$ are bounded.*

An **alcoved polytope** is a polytope $\mathsf{P} \subset \mathbb{R}^d$ that is the union of alcoves. In particular, P is of the form

$$\mathsf{P} = \left\{ \mathbf{x} \in \mathbb{R}^d : \begin{array}{rcll} \alpha_i \le & x_i & \le \beta_i & \text{for } 1 \le i \le d \\ \alpha_{ij} \le & x_j - x_i & \le \beta_{ij} & \text{for } 1 \le i < j \le d \end{array} \right\}, \tag{7.4.1}$$

for some $\alpha_i, \beta_i, \alpha_{ij}, \beta_{ij} \in \mathbb{Z}$. Since alcoves are lattice polytopes and the vertices of P are among the vertices of the alcoves it contains, we put the following on record.

Corollary 7.4.2. *Alcoved polytopes are lattice polytopes.*

From their very definition (6.3.1) we remark that order polytopes O_Π and thus cubes $[0,1]^d$ are alcoved polytopes. What is less obvious is that the hypersimplices that we considered in Section 5.7 are up to a change of coordinates alcoved polytopes as well. For $0 \le k \le d-1$, we recall that the $(d+1, k+1)$-hypersimplex $\triangle(d+1, k+1)$ is the d-dimensional polytope given by all points $\mathbf{s} \in [0,1]^{d+1}$ with $s_1 + \cdots + s_{d+1} = k+1$. The linear equation says that if we know any d coordinates of $\mathbf{s}$, then we can infer the last one. Hence, we might as well project $\triangle(d+1, k)$ onto the first d coordinates $(s_1, \ldots, s_d)$ to get a full-dimensional embedding. In the following we will identify

$$\triangle(d+1, k+1) = \left\{ \mathbf{y} \in \mathbb{R}^d : \begin{array}{c} 0 \le y_i \le 1 \text{ for } i = 1, \ldots, d \\ k \le y_1 + \cdots + y_d \le k+1 \end{array} \right\}.$$

To unmask $\triangle(d+1, k+1)$ as an alcoved polytope, we define the linear transformation $T : \mathbb{R}^d \to \mathbb{R}^d$ by

$$T(\mathbf{y})_i := y_1 + y_2 + \cdots + y_i \quad \text{for } i = 1, \ldots, d. \tag{7.4.2}$$

Following Exercise 7.41, the map T is invertible, lattice-preserving (i.e., $T(\mathbb{Z}^d) = \mathbb{Z}^d$), and

$$\begin{aligned} \widetilde{\triangle}(d+1, k+1) &:= T(\triangle(d+1, k+1)) \\ &= \left\{ \mathbf{p} \in \mathbb{R}^d : \begin{array}{rcll} 0 \le & p_1 & \le 1, & \\ k \le & p_d & \le k+1, & \text{and} \\ 0 \le & p_i - p_{i-1} & \le 1 & \text{for } 1 < i \le d \end{array} \right\}. \end{aligned} \tag{7.4.3}$$

This is an alcoved polytope *par excellence.*

An attractive feature of alcoved polytopes is that they automatically come equipped with a (regular) unimodular triangulation—the **alcoved triangulation**—with maximal cells corresponding exactly to the alcoves they contain. The combinatorics of these triangulations is intimately related to that of permutations. To illustrate, we give a more combinatorially-flavored proof of Theorem 5.7.7.

Theorem 7.4.3. *For $1 \le k < d$, the number of unimodular simplices (i.e., alcoves) in the alcoved triangulation of $\widetilde{\triangle}(d+1, k+1)$ equals the number of permutations $\tau \in \mathfrak{S}_d$ with k descents.*

To set the stage for the proof, let $\mathsf{P} \subset \mathbb{R}^d$ be a general full-dimensional alcoved polytope. Since P is by definition a union of alcoves, we note that an alcove $\mathbf{q} + \triangle^\tau$ is part of the alcoved triangulation of P if and only if $\mathbf{q} + \triangle^\tau \subseteq \mathsf{P}$. By definition, no two alcoves intersect in their interiors and hence $\mathbf{q} + \triangle^\tau \subseteq \mathsf{P}$ if and only if $\mathbf{q} + \mathbf{b} \in \mathsf{P}$ for some $\mathbf{b} \in (\triangle^\tau)^\circ$ in the interior. A canonical choice for such a point $\mathbf{b}$ is the **barycenter** of $\triangle^\tau$,

$$\mathbf{b}^\tau := \tfrac{1}{d+1}(\mathbf{v}_0 + \mathbf{v}_1 + \cdots + \mathbf{v}_d) = \tfrac{1}{d+1}\left(\tau(1), \tau(2), \ldots, \tau(d)\right),$$

where $\operatorname{vert}(\triangle^\tau) = \{\mathbf{v}_0, \ldots, \mathbf{v}_d\}$. This, at least in principle, leads to a counting formula for the number of alcoves in a given alcoved polytope P.

Corollary 7.4.4. *Let $\mathsf{P} \subset \mathbb{R}^d$ be an alcoved polytope. For $\tau \in \mathfrak{S}_d$, let*

$$I_\tau(\mathsf{P}) := \mathbb{Z}^d \cap (-\mathbf{b}^\tau + P), \tag{7.4.4}$$

the lattice points in the rational polytope $-\mathbf{b}^\tau + \mathsf{P}$. Then the number of full-dimensional unimodular simplices in the alcoved triangulation of P is given by $\sum_{\tau \in \mathfrak{S}_d} |I_\tau(\mathsf{P})|$.

Proof. We have

$$\mathbf{q} + \triangle^\tau \subseteq \mathsf{P} \iff \mathbf{q} + \mathbf{b}^\tau \in \mathsf{P} \iff \mathbf{q} \in -\mathbf{b}^\tau + \mathsf{P}. \qquad \square$$

The question, of course, is how to determine $I_\tau(\mathsf{P})$. At least for hypersimplices this can be done explicitly.

Proof of Theorem 7.4.3. Let $\tau \in \mathfrak{S}_d$ and $\mathbf{q} \in \mathbb{Z}^d$. As argued above,

$$\mathbf{q} + \triangle^\tau \subseteq \widetilde{\triangle}(d+1, k+1) \iff \mathbf{q} + \mathbf{b}^\tau \in \widetilde{\triangle}(d+1, k+1).$$

So we need to check when the defining inequalities (7.4.3) are satisfied. The first inequality in (7.4.3) yields

$$0 \le (\mathbf{q} + \mathbf{b}^\tau)_1 = q_1 + \tfrac{\tau(1)}{d+1} \le 1.$$

But $\mathbf{q}$ is an integer vector, and so the inequality is satisfied if and only if $q_1 = 0$. For $1 \le i < d$, we compute

$$0 \le (\mathbf{q} + \mathbf{b}^\tau)_{i+1} - (\mathbf{q} + \mathbf{b}^\tau)_i = q_{i+1} - q_i + \tfrac{\tau(i+1) - \tau(i)}{d+1} \le 1.$$

With $\delta_i := \frac{1}{d+1}(\tau(i+1) - \tau(i))$, we rewrite this as

$$q_i - \delta_i \ \leq \ q_{i+1} \ \leq \ q_i + 1 - \delta_i \, ,$$

and again because $\mathbf{q}$ is an integer vector, this set of inequalities is satisfied if and only if

$$q_{i+1} \ = \ \begin{cases} q_i + 1 & \text{if } \delta_i < 0, \\ q_i & \text{otherwise.} \end{cases}$$

Now $(d+1)\,\delta_i = \tau(i+1) - \tau(i) < 0$ if i is a descent of τ. Thus, q_i is the number of descents $j \in \mathrm{Des}(\tau)$ with $j \leq i$.

The final inequality for $\widetilde{\triangle}(d,k)$ in (7.4.3) now states that

$$k \ \leq \ q_d + \frac{\tau(d)}{d+1} \ \leq \ k+1 \, ,$$

which holds if and only if τ has exactly k descents. Thus the number of alcoves in $\widetilde{\triangle}(d+1, k+1)$ equals the number of permutations $\tau \in \mathfrak{S}_d$ with k descents. □

Let's see one more alcoved polytope in action. Let Π be a finite poset. We will assume that Π is naturally labelled, that is, $\Pi = \{1, \dots, d\}$ and $i \prec_\pi j$ implies $i < j$. Much of what we have done in this book is centered around the notion of order-preserving maps $\phi : \Pi \to \mathbb{Z}_{\geq 0}$. Looking back at Chapter 6, we counted those by bounding

$$\max\{\phi(a) \, : \, a \in \Pi\} \, ,$$

which led us to order polynomials and order polytopes, or we fixed

$$\phi(1) + \phi(2) + \cdots + \phi(d) \, ,$$

which gave us the notion of Π-partitions. In both cases, this led to a plethora of combinatorial results, most of which related to permutations. We now explore one more variation. To motivate this, we appeal to analysis: a function $f : \mathbb{R}^d \to \mathbb{R}$ is **k-Lipschitz continuous** in a given metric D on $\mathbb{R}^d$ if

$$|f(\mathbf{y}) - f(\mathbf{x})| \ \leq \ k \cdot \mathrm{D}(\mathbf{x}, \mathbf{y})$$

for all $\mathbf{x}, \mathbf{y} \in \mathbb{R}^d$. If we set $\mathrm{D}_\Pi(a,b)$ to be the length of a shortest saturated chain $a = a_0 \prec\!\cdot \cdots \prec\!\cdot\, a_s = b$, then this defines not a metric on Π but a *quasimetric*, i.e., $\mathrm{D}_\Pi(a,b)$ satisfies all requirements of a metric except for symmetry. Quasimetrics are fine, as long as we measure distance in the right way. This perfectly fits with order-preserving functions.

An order-preserving function $f : \Pi \to \mathbb{R}_{\geq 0}$ is **k-Lipschitz** if

$$f(b) - f(a) \ \leq \ k \cdot \mathrm{D}_\Pi(a,b)$$

for all $a, b \in \Pi$. We simply say that f is **Lipschitz** if $k = 1$. Since we assume Π to be finite, the collection of k-Lipschitz functions is a polyhedron in $\mathbb{R}^{\Pi}$ but it is not bounded. Indeed, if $g : \Pi \to \mathbb{R}_{\geq 0}$ is any constant (and thus order-preserving!) function, then $f + g$ is again k-Lipschitz. We take the following measures. We recall that $\check{\Pi} := \Pi \cup \{\hat{0}\}$ is the poset obtained from Π by adjoining a minimal element, irrespective of whether Π already had a unique minimal element. We define Lip_Π to be the collection of all order-preserving Lipschitz functions $f : \check{\Pi} \to \mathbb{R}_{\geq 0}$ such that $f(\hat{0}) = 0$. A little thinking (to be done in Exercise 7.17) reveals that

$$\mathsf{Lip}_\Pi = \left\{ \mathbf{x} \in \mathbb{R}^d : \begin{array}{rcll} 0 \leq & x_i & \leq 1 & \text{for } i \in \min(\Pi) \\ 0 \leq & x_j - x_i & \leq 1 & \text{for } i \prec\!\cdot\, j \end{array} \right\}. \tag{7.4.5}$$

Thus $k \cdot \mathsf{Lip}_\Pi$ captures exactly the k-Lipschitz functions up to adding constant functions supported on connected components of Π. We call Lip_Π the **Lipschitz polytope** of Π. In particular, looking at (7.4.5), it is crystal clear that Lip_Π is an alcoved polytope.

If Π is the antichain A_d on d elements, then $\mathsf{D}_\Pi = [0,1]^d$, the standard cube. On the other hand, if $\Pi = [d]$ is the d-chain, then you should verify that $\mathsf{D}_{[d]} = T([0,1]^d)$ and hence is linearly isomorphic to a cube. This is illustrated in Figure 7.7. In fact, in Exercise 7.18 you will learn about a class of posets all of whose Lipschitz polytopes are cubes.

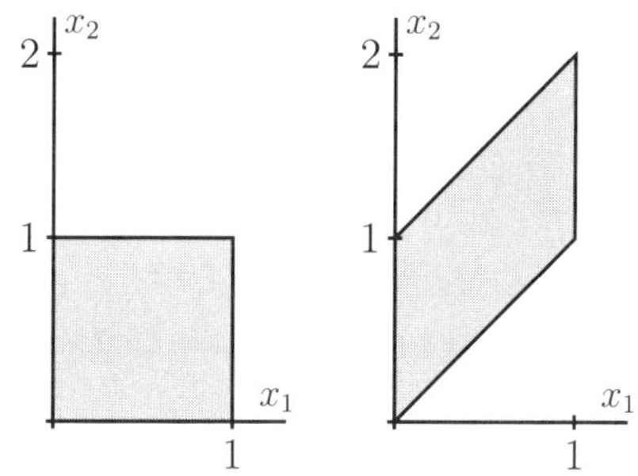

Figure 7.7. Applying $T(\mathbf{x})$ to the unit cube.

Exercise 7.19 says that the inequalities given in (7.4.5) are facet defining. What we do not yet know are the vertices of Lip_Π. Note that if $\Pi = \Pi_1 \uplus \Pi_2$, then $\mathsf{Lip}_\Pi = \mathsf{Lip}_{\Pi_1} \times \mathsf{Lip}_{\Pi_2}$. Thus, we need to only worry about connected posets and, henceforth, we assume that Π is connected. Let $F \subseteq \Pi$ be a filter. We denote by

$$N(F) := \left\{ a \in \check{\Pi} \setminus F : a \prec\!\cdot\, b \text{ for some } b \in F \right\}$$

the **neighborhood** of F in $\check{\Pi}$. A chain of filters $\varnothing \neq F_m \subset \cdots \subset F_1 \subseteq \Pi$ is **neighbor closed** if $F_{i+1} \cup N(F_{i+1}) \subseteq F_i$; that is, there is no cover relation $a \prec\!\cdot\, b$ in $\check{\Pi}$ such that $b \in F_i$ and $a \notin F_{i-2}$. By Corollary 7.4.2, Lip_Π is a lattice polytope and we only have to determine which lattice points are vertices.

Theorem 7.4.5. *Let* Π *be a finite poset and* $\mathbf{v} \in \mathbb{Z}^d$. *Then* $\mathbf{v}$ *is a vertex of* Lip_Π *if and only if there is a neighbor-closed chain of nonempty filters* $F_m \subset \cdots \subset F_1 \subseteq \Pi$ *such that*

$$\mathbf{v} \ = \ \mathbf{e}_{F_m} + \cdots + \mathbf{e}_{F_1} \,.$$

Proof. We first observe that if $\mathbf{p} \in \mathsf{Lip}_\Pi \cap \mathbb{Z}^d$, then

$$p_b - p_a \ = \ 0 \qquad \text{or} \qquad p_b - p_a \ = \ 1$$

for every $a \prec\!\cdot\, b$ in $\check{\Pi}$. Hence, every lattice point can be recovered from the knowledge of which linear inequalities of (7.4.5) are satisfied with equality. It follows that the only lattice points in Lip_Π are its vertices. Since points in Lip_Π are, in particular, order-preserving maps, by Theorem 6.1.6 every vertex is of the form

$$\mathbf{v} \ = \ \mathbf{e}_{F_m} + \cdots + \mathbf{e}_{F_1}$$

for some chain of nonempty filters $F_m \subset \cdots \subset F_1 \subseteq \Pi$. Now, for $a, b \in \check{\Pi}$ with $a \prec\!\cdot\, b$, observe that $v_b - v_a \le 1$ if and only if there is at most one i such that $b \in F_i$ and $a \notin F_i$. But since $F_i \subseteq F_{i-1}$, this is the case if and only if the chain is neighbor closed. $\square$

To determine the alcoves that comprise Lip_Π, we can reuse the arguments in our proof of Theorem 7.4.3. This yields an amalgamation of posets and permutations not unlike those of Section 6.4. To make the description more palpable, we need some definitions. As a reminder, $\check{\Pi}$ is a poset whose elements we identified with $\hat{0} = 0, 1, \dots, d$ using a linear extension. For a permutation $\tau \in \mathfrak{S}_d$, we use the convention $\tau(\hat{0}) = \tau(0) := 0$. Any chain $C = \{c_1 \prec c_2 \prec \cdots \prec c_k\} \subseteq \check{\Pi}$ yields a subword of τ via

$$\tau|_C \ := \ \tau(c_1)\tau(c_2)\cdots\tau(c_k)\,.$$

As before, a **descent** of $\tau|_C$ is an index $1 \le i \le k-1$ such that $\tau(c_i) > \tau(c_{i+1})$, and we write $\operatorname{des}(\tau|_C)$ for the number of descents. A permutation $\tau \in \mathfrak{S}_d$ is **descent compatible** with Π if for each $a \in \Pi$ the number of descents $\operatorname{des}(\tau|_C)$ for $C = \{\hat{0} = c_1 \prec\!\cdot\, c_2 \prec\!\cdot \cdots \prec\!\cdot\, c_k = a\}$ is independent of C. In this case, we write $\operatorname{des}_{\Pi,\tau}(a) := \operatorname{des}(\tau|_C)$ for any such chain C. The collection of descent-compatible permutations of Π is denoted by $\mathrm{DC}(\Pi) \subseteq \mathfrak{S}_d$.

Theorem 7.4.6. *Let* $\Pi = ([d], \preceq)$ *be a partially ordered set. An alcove* $\mathbf{q} + \triangle^\tau$ *with* $\mathbf{q} \in \mathbb{Z}^d$ *and* $\tau \in \mathfrak{S}_d$ *is contained in* Lip_Π *if and only if* $\tau \in \mathrm{DC}(\Pi)$ *and* $q_a = \operatorname{des}_{\Pi,\tau}(a)$ *for all* $a \in \Pi$.

Proof. Looking back at our proof of Theorem 7.4.3, we now realize that the methodology can, in fact, handle Lip_Π. We can follow the arguments to infer that if $\mathbf{q} + \mathbf{b}^\tau \in \mathsf{Lip}_\Pi$, then $\mathbf{q}$ is uniquely determined by

$$q_b = \begin{cases} q_a & \text{if } a \prec\!\cdot\, b \text{ and } \tau(a) < \tau(b), \\ q_a + 1 & \text{if } a \prec\!\cdot\, b \text{ and } \tau(a) > \tau(b). \end{cases}$$

Together with $q_{\hat{0}} = 0$, this shows that $\mathbf{q} + \mathbf{b}^\tau \in \mathsf{Lip}_\Pi$ if and only if $q_b = \operatorname{des}(\tau|_C)$ for any saturated chain ending in b. □

Lipschitz polytopes are particularly nice when Π is a poset for which $\check{\Pi}$ is **ranked**, that is, for $a, b \in \check{\Pi}$ with $a \preceq b$, any two maximal chains in $[a, b]_{\check{\Pi}}$ have the same length. Graded posets are ranked but not vice versa. If $\check{\Pi}$ is ranked, then there is a unique function $\rho : \Pi \to \mathbb{Z}_{\geq 0}$, called the **rank function**, such that $\rho(\hat{0}) = 0$ and $\rho(b) = \rho(a) + 1$ for $a \prec\!\cdot\, b$. A polytope $\mathsf{P} \subset \mathbb{R}^d$ is **centrally symmetric** if there is a point $\mathbf{q} \in \mathbb{R}^d$ such that $\mathsf{P} = \mathbf{q} - \mathsf{P}$.

Proposition 7.4.7. *If Π is a poset for which $\check{\Pi}$ is ranked, then Lip_Π is centrally symmetric.*

Proof. If $\check{\Pi}$ is ranked, we define $\mathbf{r} \in \mathbb{R}^d$ by $r_i = \rho(i)$, where ρ is the rank function of $\check{\Pi}$. We claim that

$$\mathsf{Lip}_\Pi = \mathsf{K}_\Pi \cap (\mathbf{r} - \mathsf{K}_\Pi),$$

where K_Π is the order cone of Π. A point $\mathbf{q} \in \mathbb{R}^d$ is contained in $\mathbf{r} - \mathsf{K}_\Pi$ if and only if $\mathbf{r} - \mathbf{q} \in \mathsf{K}_\Pi$. This is the case if and only if for $a, b \in \check{\Pi}$ with $a \prec\!\cdot\, b$

$$r_a - q_a \leq r_b - q_b \iff q_b - q_a \leq r_b - r_a = 1.$$

Since $q_{\hat{0}} = r_{\hat{0}} = 0$, this also implies that $q_b \leq 1$ for all $b \in \min(\Pi)$. Thus a point $\mathbf{q} \in \mathsf{K}_\Pi$ is contained in Lip_Π if and only if $\mathbf{q} \in \mathbf{r} - \mathsf{K}_\Pi$. With this representation, central symmetry with respect to the point $\mathbf{p} = \frac{1}{2}\mathbf{r}$ is apparent. □

It follows from our proof of Proposition 7.4.7 that $2\,\mathsf{Lip}_\Pi$ is centrally-symmetric with respect to the point $\mathbf{r} = \rho$ and $\mathbf{r}$ is the only lattice point in $2\,\mathsf{Lip}_\Pi^\circ$.

Proposition 7.4.8. *Let Π be a poset on d elements for which $\check{\Pi}$ is ranked. Then for any $k \geq 0$,*

$$(k+2)\,\mathsf{Lip}_\Pi^\circ \cap \mathbb{Z}^d = (\mathbf{r} + k\,\mathsf{Lip}_\Pi) \cap \mathbb{Z}^d$$

and, consequently,

$$(-1)^d \operatorname{ehr}_{\mathsf{Lip}_\Pi}(-(k+2)) = \operatorname{ehr}_{\mathsf{Lip}_\Pi}(k).$$

Proof. Let $\mathbf{r} = \rho$ be the point representing the rank function of $\check{\Pi}$. A point $\mathbf{q} \in \mathbb{Z}^d$ is contained in $(k+2)\,\mathsf{Lip}^\circ_\Pi$ if and only if

$$\begin{aligned} & 0 \;<\; q_b - q_a \;<\; k+2 \\ \iff\; & 1 \;\le\; q_b - q_a \;\le\; k+1 \\ \iff\; & 0 \;\le\; q_b - q_a - 1 \;\le\; k \\ \iff\; & 0 \;\le\; q_b - q_a - (r_b - r_a) \;\le\; k \end{aligned}$$

for all $a, b \in \check{\Pi}$ with $a \prec\!\cdot\, b$, that is, if and only if $\mathbf{q} - \mathbf{r} \in k\,\mathsf{Lip}_\Pi$. The second claim follows from Theorem 5.2.3. □

In the language of Exercise 5.13, this means that if $\check{\Pi}$ is ranked, then Lip_Π is a Gorenstein polytope and

$$h_i^*(\mathsf{Lip}_\Pi) \;=\; h_{d-i-1}^*(\mathsf{Lip}_\Pi) \tag{7.4.6}$$

for all $0 \le i \le \lfloor \frac{d-1}{2} \rfloor$. If $\Pi = \mathrm{A}_d$, an antichain, and so $\mathsf{Lip}_\Pi = [0,1]^d$, then $h^*(\mathsf{Lip}_\Pi)$ is an Eulerian polynomial, and Proposition 7.4.8 together with (7.4.6) yields an alternative solution to Exercise 6.28: the number of permutations of $[d]$ with i descents is equal to the number of permutations of $[d]$ with $d-1-i$ descents.

This brings us to the question of how to compute h^*-polynomials of alcoved polytopes. The most elegant way to do this is to determine a half-open decomposition of the infinite subdivision of $\mathbb{R}^d$ by alcoves. We will compute the half-open decomposition with respect to the point $\mathbf{w} = \frac{1}{d+1}(1, 2, \ldots, d)$ and the next lemma states that this point is good enough.

Lemma 7.4.9. *Let $\mathsf{P} \subset \mathbb{R}^d$ be a full-dimensional alcoved polytope. Then there are a lattice translation and a relabeling of coordinates such that*

$$\triangle^{\mathrm{id}} \;\subseteq\; \mathsf{P} \;\subset\; \mathbb{R}^d_{\ge 0}\,.$$

Proof. For $\mathbf{p}, \mathbf{q} \in \mathsf{P}$ define the point $\mathbf{r} \in \mathbb{R}^d$ by $r_i = \min(p_i, q_i)$ for $1 \le i \le d$. We claim that $\mathbf{r} \in \mathsf{P}$ as well. Indeed, assume that $x, x', y, y' \in \mathbb{R}$ are real numbers such that $\alpha \le x - x' \le \beta$ and $\alpha \le y - y' \le \beta$ for some $\alpha, \beta \in \mathbb{R}$; then

$$\alpha \;\le\; \min(x,y) - \min(x',y') \;\le\; \beta\,.$$

Hence, if $\mathbf{p}$ and $\mathbf{q}$ satisfy the inequalities given in (7.4.1), then so does $\mathbf{r}$. This implies that there is a point $\mathbf{b} \in \mathsf{P}$ such that $b_i \le q_i$ for all $\mathbf{q} \in \mathsf{P}$ and $1 \le i \le d$. Thus,

$$\mathsf{P}' \;:=\; \mathsf{P} - \mathbf{b} \;\subseteq\; \mathbb{R}^d_{\ge 0}$$

differs from P by a lattice translation. The origin is contained in P' and any alcove in P' that contains $\mathbf{0}$ is of the form $\triangle^\tau$ for some $\tau \in \mathfrak{S}_d$. Relabeling coordinates turns $\triangle^\tau$ into $\triangle^{\mathrm{id}}$ and finishes the proof. □

The benefit of using $\mathbf{w} = \frac{1}{d+1}(1, 2, \ldots, d)$ for the half-open decomposition is captured by the following slightly technical lemma. We recall that the h^*-polynomial of a (half-open) polytope $\mathsf{Q} \subset \mathbb{R}^d$ is given by

$$h^*_{\mathsf{Q}}(z) \ = \ h^*_0(\mathsf{Q}) + h^*_1(\mathsf{Q})z + \cdots + h^*_d(\mathsf{Q})z^d .$$

Also note that by our convention, $0 \in \operatorname{Asc}(\tau)$ for all $\tau \in \mathfrak{S}_d$.

Lemma 7.4.10. *For $\tau \in \mathfrak{S}_d$ and $\mathbf{q} \in \mathbb{Z}^d_{\geq 0}$, let*

$$a(\tau, \mathbf{q}) \ := \ \left|\left\{ i \in \operatorname{Asc}(\tau^{-1}) \ : \ q_{\tau^{-1}(i)} < q_{\tau^{-1}(i+1)} \right\}\right| ,$$
$$d(\tau, \mathbf{q}) \ := \ \left|\left\{ i \in \operatorname{Des}(\tau^{-1}) \ : \ q_{\tau^{-1}(i)} \leq q_{\tau^{-1}(i+1)} \right\}\right| .$$

Then $h^*_{\mathbb{H}_{\mathbf{w}}(\mathbf{q}+\triangle^\tau)}(z) \ = \ z^{a(\tau,\mathbf{q})+d(\tau,\mathbf{q})}$.

One proof consists of checking which inequalities of $\mathbf{q} + \triangle^\tau$ are violated for the point $\mathbf{w} = \frac{1}{d+1}(1, 2, \ldots, d)$; we leave it for Exercise 7.22.

If Π is a finite poset and $\tau \in \operatorname{DC}(\Pi)$ is a descent-compatible permutation, we define $\operatorname{ldes}_\Pi(\tau)$ to be number of pairs $(a, b) \in \check{\Pi} \times \check{\Pi}$ such that $\tau(a) = \tau(b) - 1$ and

$$\operatorname{des}_{\Pi,\tau}(a) \ < \ \operatorname{des}_{\Pi,\tau}(b) \quad \text{or} \quad (\operatorname{des}_{\Pi,\tau}(a) \ = \ \operatorname{des}_{\Pi,\tau}(b) \quad \text{and} \quad a \ > \ b) .$$

Theorem 7.4.11. *Let $\Pi = ([d], \preceq)$ be a naturally labelled poset. Then*

$$h^*_{\mathsf{Lip}_\Pi}(z) \ = \ \sum_{\tau \in \operatorname{DC}(\Pi)} z^{\operatorname{ldes}_\Pi(\tau)} .$$

Proof. Using the additivity of h^*-polynomials stated in (5.5.2),

$$h^*_{\mathsf{Lip}_\Pi}(z) \ = \ \sum_{\tau \in \operatorname{DC}(\Pi)} h^*_{\mathbb{H}_{\mathbf{w}}(\mathbf{q}^\tau+\triangle^\tau)}(z) \ = \ \sum_{\tau \in \operatorname{DC}(\Pi)} z^{a(\tau,\mathbf{q}^\tau)+d(\tau,\mathbf{q}^\tau)} .$$

To complete the proof, let $0 \leq i < d$ and $(a, b) \in \check{\Pi} \times \check{\Pi}$ such that $\tau(a) = i$ and $\tau(b) = i + 1$. Then i is counted by $a(\tau, \mathbf{q}^\tau) + d(\tau, \mathbf{q}^\tau)$ if and only if

$$q^\tau_a \ = \ q^\tau_{\tau^{-1}(i)} \ < \ q^\tau_{\tau^{-1}(i)} \ = \ q^\tau_b ,$$

or $q^\tau_a = q^\tau_b$ and $a = \tau^{-1}(i) > \tau^{-1}(i + 1) = b$. □

For the case that $\Pi = [d]$ is a chain, this gives quite a nice result. For a permutation $\tau \in \mathfrak{S}_d$, we call an index $0 \leq i < d$ a **big ascent** or **2-ascent** if $\tau(i + 1) - \tau(i) \geq 2$. In particular, $i = 0$ is a big ascent if $\tau(1) > 1$. We write $\operatorname{asc}^{(2)}(\tau)$ for the number of big ascents of τ.

Theorem 7.4.12. *Let $\Pi = [d]$ be the d-chain. Then $\operatorname{DC}(\Pi) = \mathfrak{S}_d$ and*

$$h^*_{\mathsf{Lip}_{[d]}}(z) \ = \ \sum_{\tau \in \mathfrak{S}_d} z^{\operatorname{asc}^{(2)}(\tau)} .$$

Proof. What we will actually prove is that $\operatorname{ldes}_{[d]}(\tau) = \operatorname{asc}^{(2)}(\tau^{-1})$. But since $\operatorname{DC}([d]) = \mathfrak{S}_d$ and we are thus summing over all permutations of d to compute the h^*-polynomial of $\mathsf{Lip}_{[d]}$, that's fine.

For the case of the chain, we note that $q_i^\tau = \operatorname{des}_{[d],\tau}(i)$ is the number of descents in the sequence $\tau(0)\tau(1)\tau(2)\cdots\tau(i)$. In particular, $q_a^\tau \le q_b^\tau$ for $a < b$ and $q_a^\tau = q_b^\tau$ if there is no descent in $\tau(a)\tau(a+1)\cdots\tau(b)$. Now pick $(a,b) \in \{0,1,\dots,d\}^2$ such that $\tau(a) = i$ and $\tau(b) = i+1$. If $a > b$, then the sequence $i+1 = \tau(b)\tau(b+1)\cdots\tau(a) = i$ inevitably contains a descent and thus will not be counted by $\operatorname{ldes}_{[d]}(\tau)$. If $a < b$, then the pair is counted if and only if $i = \tau(a)\tau(a+1)\cdots\tau(b) = i+1$ contains a descent. For that to be even possible, we need at least

$$2 \ \le\ b - a \ =\ \tau^{-1}(i+1) - \tau^{-1}(i)\,,$$

that is, i is a 2-ascent of τ^{-1}. This is also sufficient: in every sequence of length ≥ 3 that starts with i and ends with $i+1$, there has to be a descent. □

Big ascents might seem to be less natural to consider than ordinary ascents but they are not that outlandish.

Corollary 7.4.13. *Let $d \ge 1$. Then*

$$\sum_{\tau\in\mathfrak{S}_d} z^{\operatorname{des}(\tau)} \ =\ \sum_{\tau\in\mathfrak{S}_d} z^{\operatorname{asc}^{(2)}(\tau)}\,.$$

Proof. As part of Exercise 7.18, you verified that $T([0,1]^d) = \mathsf{Lip}_{[d]}$. Since the linear transformation T satisfies that $T(\mathbf{q}) \in \mathbb{Z}^d$ if and only if $\mathbf{q} \in \mathbb{Z}^d$, we conclude that $\operatorname{ehr}_{[0,1]}(n) = \operatorname{ehr}_{\mathsf{Lip}_{[d]}}(n)$ for all $n \ge 0$. In particular

$$\sum_{\tau\in\mathfrak{S}_d} z^{\operatorname{des}(\tau)} \ =\ h^*_{[0,1]^d}(z) \ =\ h^*_{\mathsf{Lip}_{[d]}}(z) \ =\ \sum_{\tau\in\mathfrak{S}_d} z^{\operatorname{asc}^{(2)}(\tau)}\,,$$

where the first equality stems from Corollary 6.3.13 and the last equality is Theorem 7.4.12. □

This again looks very much like what we did at the beginning of the section when dealing with hypersimplices. In fact, we can define a version of hypersimplices for all sorts of posets. Let Π be a poset with $\hat{1}$. For a descent-compatible permutation $\tau \in \operatorname{DC}(\Pi)$, the number of descents along any maximal chain, which necessarily has to end in $\hat{1}$, is the same and we define the number of **Π-descents** of τ as $\operatorname{des}_\Pi(\tau) := \operatorname{des}_{\Pi,\tau}(\hat{1})$. The **height** $\operatorname{ht}(\Pi)$ of Π is the number of elements in a maximal chain of Π. For $1 \le k \le \operatorname{ht}(\Pi)$, we define the **$(\Pi, k)$-hypersimplex** as

$$\Delta(\Pi,k) \ :=\ \left\{ f \in \mathsf{Lip}_\Pi \,:\, k-1 \le f(\hat{1}) \le k \right\}.$$

You might want to check that $\Delta([d],k) = \widetilde{\triangle}(d+1,k)$ and that

$$\Delta(\Pi,1) = \mathsf{O}_\Pi$$

is the order polytope of Π. Note that we get a dissection

$$\mathsf{Lip}_\Pi = \Delta(\Pi,1) \cup \Delta(\Pi,2) \cup \cdots \cup \Delta(\Pi,\mathrm{ht}(\Pi))$$

and, assuming that Π still adheres to our labeling convention,

$$\mathsf{Lip}_\Pi = \mathbb{H}_\mathbf{w}\Delta(\Pi,1) \uplus \mathbb{H}_\mathbf{w}\Delta(\Pi,2) \uplus \cdots \uplus \mathbb{H}_\mathbf{w}\Delta(\Pi,\mathrm{ht}(\Pi)).$$

For $k=1$, we have $\mathbf{w} \in \Delta(\Pi,1)$ and $\mathbb{H}_\mathbf{w}\Delta(\Pi,1) = \Delta(\Pi,1)$. For $1 < k \le \mathrm{ht}(\Pi)$,

$$\mathbb{H}_\mathbf{w}\Delta(\Pi,k) = \left\{ f \in \mathsf{Lip}_\Pi : k-1 < f(\hat{1}) \le k \right\}.$$

Using this half-open decomposition into hypersimplices, we can get a more refined picture of descent-compatible permutations. For the d-chain we record the following.

Corollary 7.4.14. *For $0 \le k < d$,*

$$h^*_{\mathbb{H}_\mathbf{q}\widetilde{\triangle}(d+1,k+1)}(z) = \sum_{\substack{\tau \in \mathfrak{S}_d \\ \mathrm{des}(\tau)=k}} z^{\mathrm{asc}^{(2)}(\tau^{-1})}.$$

7.5. Zonotopes and Tilings

The line segment between two points $\mathbf{a},\mathbf{b} \in \mathbb{R}^d$ is the set

$$[\mathbf{a},\mathbf{b}] := \{(1-\lambda)\,\mathbf{a} + \lambda\,\mathbf{b} : 0 \le \lambda \le 1\}.$$

We recall from Section 3.1 that the Minkowski sum of two convex sets $\mathsf{K}_1, \mathsf{K}_2 \subseteq \mathbb{R}^d$ is

$$\mathsf{K}_1 + \mathsf{K}_2 := \{\mathbf{p} + \mathbf{q} : \mathbf{p} \in \mathsf{K}_1,\ \mathbf{q} \in \mathsf{K}_2\}.$$

This section is devoted to the class of polytopes we obtain from Minkowski sums of line segments. As we will see, these polytopes are quite sympathetic to combinatorics. A **zonotope** $\mathsf{Z} \subset \mathbb{R}^d$ is a polytope of the form

$$\mathsf{Z} = \mathsf{Z}(\mathbf{a}_1,\dots,\mathbf{a}_m;\mathbf{b}_1,\dots,\mathbf{b}_m) := [\mathbf{a}_1,\mathbf{b}_1] + \cdots + [\mathbf{a}_m,\mathbf{b}_m] \tag{7.5.1}$$

for some $\mathbf{a}_1,\mathbf{b}_1,\dots,\mathbf{a}_m,\mathbf{b}_m \in \mathbb{R}^d$. Figure 7.8 shows an example.

If $\mathbf{b}_1-\mathbf{a}_1,\dots,\mathbf{b}_m-\mathbf{a}_m$ are linearly independent (and hence $m \le d$), then Z is simply a parallelepiped. In Exercise 7.23 you will show that in this case Z is affinely isomorphic to the cube $[0,1]^m$, and hence parallelepipeds are the simplest examples of zonotopes, akin to simplices in relation to general polytopes. The analogy is not at all far-fetched: we will soon see that parallelepipeds indeed play the role of building blocks for zonotopes.

To ease notation, we abbreviate

$$\mathsf{Z}(\mathbf{z}_1,\dots,\mathbf{z}_m) := \mathsf{Z}(\mathbf{0},\dots,\mathbf{0};\mathbf{z}_1,\dots,\mathbf{z}_m)$$

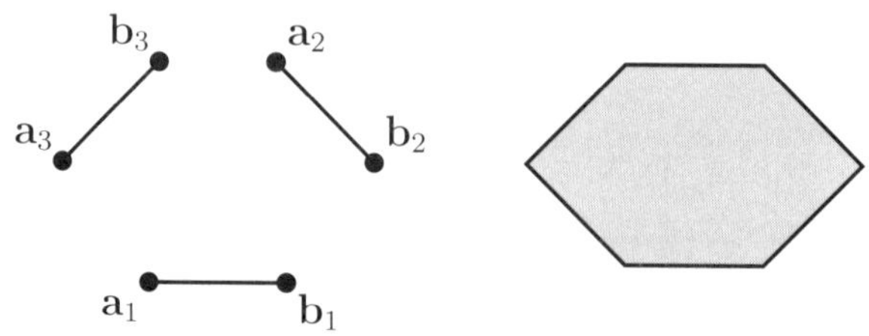

Figure 7.8. A hexagon is a zonotope generated by three line segments.

for a collection $\mathbf{z}_1, \ldots, \mathbf{z}_m \in \mathbb{R}^d$. Since $[\mathbf{a}, \mathbf{b}] = \mathbf{a} + [\mathbf{0}, \mathbf{b} - \mathbf{a}]$, every zonotope is of the form $\mathbf{t} + \mathsf{Z}(\mathbf{z}_1, \ldots, \mathbf{z}_m)$ for some $\mathbf{t}, \mathbf{z}_1, \ldots, \mathbf{z}_m \in \mathbb{R}^d$. As a running example, let $G = ([d], E)$ be a graph, possibly with parallel edges and loops. We define the **graphical zonotope** of G as

$$\mathsf{Z}_G \ := \ \sum_{ij \in E} [\mathbf{e}_i, \mathbf{e}_j]$$

where, as usual, $\mathbf{e}_j$ denotes the j-th unit vector in $\mathbb{R}^d$. We note that loops only contribute translations to the above sum and hence do not change the combinatorics or geometry of Z_G.

As a first step to getting a feel for the class of zonotopes, we consider the faces of a zonotope. To this end, we note the following general fact about Minkowski sums of polytopes. For $\mathbf{w} \in \mathbb{R}^d$ and a polyhedron $\mathsf{Q} \subset \mathbb{R}^d$, let

$$\mathsf{F}_{\mathbf{w}}\mathsf{Q} \ := \ \{\mathbf{y} \in \mathsf{Q} : \langle \mathbf{w}, \mathbf{y} \rangle \geq \langle \mathbf{w}, \mathbf{x} \rangle \text{ for all } \mathbf{x} \in \mathsf{Q}\},$$

the face of Q that maximizes the linear function $\mathbf{x} \mapsto \langle \mathbf{w}, \mathbf{x} \rangle$ over Q.

Lemma 7.5.1. *Let $\mathsf{Q}_1, \mathsf{Q}_2 \subset \mathbb{R}^d$ be polytopes and $\mathbf{w} \in \mathbb{R}^d$. Then*

$$\mathsf{F}_{\mathbf{w}}(\mathsf{Q}_1 + \mathsf{Q}_2) \ = \ \mathsf{F}_{\mathbf{w}}\mathsf{Q}_1 + \mathsf{F}_{\mathbf{w}}\mathsf{Q}_2 .$$

Proof. Let $\mathbf{p} \in \mathsf{F}_{\mathbf{w}}(\mathsf{Q}_1 + \mathsf{Q}_2)$. Then $\mathbf{p} = \mathbf{p}_1 + \mathbf{p}_2$ for some $\mathbf{p}_1 \in \mathsf{Q}_1$ and $\mathbf{p}_2 \in \mathsf{Q}_2$. If, say, $\mathbf{p}_1 \not\in \mathsf{F}_{\mathbf{w}}\mathsf{Q}_1$, then there is some $\mathbf{p}_1' \in \mathsf{Q}_1$ with $\langle \mathbf{w}, \mathbf{p}_1' \rangle > \langle \mathbf{w}, \mathbf{p}_1 \rangle$. But then

$$\mathbf{p}_1' + \mathbf{p}_2 \ \in \ \mathsf{Q}_1 + \mathsf{Q}_2 \qquad \text{and} \qquad \langle \mathbf{w}, \mathbf{p}_1' + \mathbf{p}_2 \rangle \ > \ \langle \mathbf{w}, \mathbf{p}_1 + \mathbf{p}_2 \rangle,$$

a contraction. □

For this section, the upshot of Lemma 7.5.1 is that zonotopes are closed under taking faces.

Corollary 7.5.2. *Every face of a zonotope is a zonotope.*

Proof. Let $\mathsf{Z} = \mathsf{Z}(\mathbf{a}_1, \ldots, \mathbf{a}_m; \mathbf{b}_1, \ldots, \mathbf{b}_m)$ be a zonotope and $\mathbf{w} \in \mathbb{R}^d$. By Lemma 7.5.1,

$$\mathsf{F}_{\mathbf{w}}\mathsf{Z} \;=\; \sum_{i=1}^{m} \mathsf{F}_{\mathbf{w}}[\mathbf{a}_i, \mathbf{b}_i]$$

and since $\mathsf{F}_{\mathbf{w}}[\mathbf{a}_i, \mathbf{b}_i]$ is either $\{\mathbf{a}_i\}$, $\{\mathbf{b}_i\}$, or $[\mathbf{a}_i, \mathbf{b}_i]$, it follows that $\mathsf{F}_{\mathbf{w}}\mathsf{Z}$ is a zonotope. □

Our proof of Corollary 7.5.2 suggests a natural encoding for the faces of a zonotope $\mathsf{Z} = \mathsf{Z}(\mathbf{a}_1, \ldots, \mathbf{a}_m; \mathbf{b}_1, \ldots, \mathbf{b}_m)$. For $\mathbf{w} \in \mathbb{R}^d$, we define a vector $\sigma \in \{<, =, >\}^m$ by setting, for example, σ_i to $<$ if $\langle \mathbf{w}, \mathbf{b}_i \rangle < \langle \mathbf{w}, \mathbf{a}_i \rangle$. To get the most of our notation, we write

$$\sigma^{>} \;:=\; \{i \,:\, \langle \mathbf{w}, \mathbf{b}_i \rangle > \langle \mathbf{w}, \mathbf{a}_i \rangle\}$$

and define $\sigma^{=}$ and $\sigma^{<}$ analogously. Thus, if σ is defined with respect to $\mathbf{w}$, then our proof of Corollary 7.5.2 yields

$$\mathsf{F}_{\mathbf{w}}\mathsf{Z} \;=\; \mathsf{Z}_{\sigma} \;:=\; \sum_{i \in \sigma^{<}} \mathbf{a}_i + \sum_{i \in \sigma^{>}} \mathbf{b}_i + \sum_{i \in \sigma^{=}} [\mathbf{a}_i, \mathbf{b}_i]\,. \tag{7.5.2}$$

For example, for a graph $G = ([d], E)$ and $\mathbf{w} \in \mathbb{R}^d$, let $G' = ([d], E')$ be the subgraph with $E' := \{ij \in E \,:\, w_i = w_j\}$. Then (7.5.2) says that $\mathsf{F}_{\mathsf{Z}_G}\mathbf{w}$ is a translate of $\mathsf{Z}_{G'}$. Does that look familiar when you think back to the graphical hyperplane arrangements of Section 7.2? Let's determine the vertices of a graphical zonotope.

Proposition 7.5.3. *Let $G = ([d], E)$ be a graph, possibly with parallel edges but without loops. The vertices of Z_G are in bijection with the acyclic orientations of G.*

Proof. Let $\mathbf{w} \in \mathbb{R}^d$ such that $\mathsf{F}_{\mathbf{w}}\mathsf{Z}_G$ is a vertex. From (7.5.2), we infer that

$$\langle \mathbf{w}, \mathbf{e}_i \rangle \;\neq\; \langle \mathbf{w}, \mathbf{e}_j \rangle$$

for any edge $ij \in E$. We orient ij from i to j if $w_j > w_i$ and from j to i otherwise. As in our proof of Lemma 7.2.4, the resulting orientation is acyclic, and every acyclic orientation arises this way. □

The similarity of our treatment of zonotopes and hyperplane arrangements is uncanny. To get the full picture, we start afresh with a collection $\mathbf{a}_1, \ldots, \mathbf{a}_m, \mathbf{b}_1, \ldots, \mathbf{b}_m \in \mathbb{R}^d$ such that $\mathbf{a}_i \neq \mathbf{b}_i$ for all i. In addition to the zonotope $\mathsf{Z} = \mathsf{Z}(\mathbf{a}_1, \ldots, \mathbf{a}_m; \mathbf{b}_1, \ldots, \mathbf{b}_m)$, we associate to Z a central hyperplane arrangement $\mathcal{H}(\mathsf{Z}) := \{\mathsf{H}_1, \ldots, \mathsf{H}_m\}$ with

$$\mathsf{H}_i \;:=\; \left\{\mathbf{x} \in \mathbb{R}^d \,:\, \langle \mathbf{b}_i - \mathbf{a}_i, \mathbf{x} \rangle = 0\right\}$$

for $i = 1, \ldots, m$. As introduced in (3.4.3), every central arrangement of hyperplanes $\mathcal{H} = \{\mathsf{H}_1, \ldots, \mathsf{H}_m\}$ decomposes $\mathbb{R}^d$ into relatively open cones. For any such relatively open cone C°, there is a unique $\sigma \in \{<, =, >\}^m$ with

$$\mathsf{C}^\circ = \mathsf{H}_\sigma := \bigcap_{i \in \sigma^<} \mathsf{H}_i^< \cap \bigcap_{i \in \sigma^=} \mathsf{H}_i^= \cap \bigcap_{i \in \sigma^>} \mathsf{H}_i^> .$$

Proposition 7.5.4. *Let* $\mathsf{Z} = \mathsf{Z}(\mathbf{a}_1, \ldots, \mathbf{a}_m; \mathbf{b}_1, \ldots, \mathbf{b}_m) \subset \mathbb{R}^d$ *with associated hyperplane arrangement* $\mathcal{H}(\mathsf{Z})$. *For* $\sigma \in \{<, =, >\}^m$, Z_σ *is a face of* Z *of dimension* k *if and only if* H_σ *is a face of* $\mathcal{H}$ *of dimension* $d - k$.

The correspondence mentioned in this proposition is illustrated in Figure 7.9.

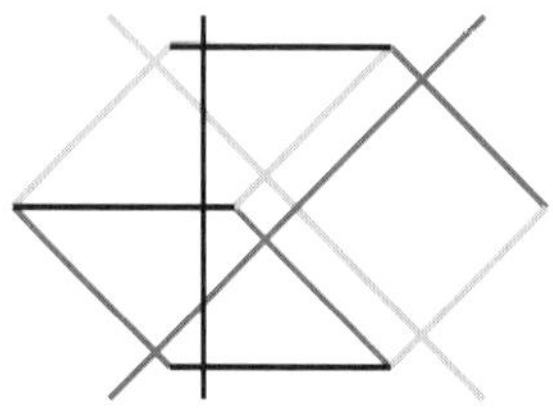

Figure 7.9. The correspondence of Proposition 7.5.4.

Proof. If Z_σ is a face, then there is some $\mathbf{w} \in \mathbb{R}^d$ such that $\mathsf{Z}_\sigma = \mathsf{F}_\mathbf{w}\mathsf{Z}$. That is,

$$\langle \mathbf{w}, \mathbf{b}_i - \mathbf{a}_i \rangle \begin{cases} > 0 & \text{if } i \in \sigma^>, \\ = 0 & \text{if } i \in \sigma^=, \\ < 0 & \text{if } i \in \sigma^<, \end{cases}$$

for all $1 \le i \le m$. This is the case if and only if $\mathbf{w} \in \mathsf{H}_i^{\sigma_i}$ for all $i \in [m]$ or, in compact notation, $\mathbf{w} \in \mathsf{H}_\sigma$. This shows that Z_σ is a face if and only if H_σ is a face.

For the statements about dimensions, we note that the linear span of H_σ is given by $\{\mathbf{x} : \langle \mathbf{b}_i - \mathbf{a}_i, \mathbf{x} \rangle = 0 \text{ for } i \in \sigma^=\}$ whereas the affine hull of Z_σ is a translate of the span of $\{\mathbf{b}_i - \mathbf{a}_i : i \in \sigma^=\}$. These define complementary and even orthogonal subspaces, and a dimension count completes the argument. □

We can considerably strengthen Proposition 7.5.4. For a hyperplane arrangement $\mathcal{H}$, let $\Phi(\mathcal{H})$ be the set of closed faces $\overline{\mathsf{H}_\sigma}$ for $\sigma \in \{<, =, >\}^m$ ordered by inclusion, which we call the **face poset** of $\mathcal{H}$. We call two posets Π, Π' **anti-isomorphic** if there is a bijection $\phi : \Pi \to \Pi'$ such that $a \preceq_\Pi b$ if and only if $\phi(a) \succeq_{\Pi'} \phi(b)$. We leave the following result as Exercise 7.28.

Theorem 7.5.5. *Let* Z *by a zonotope with associated central hyperplane arrangement* $\mathcal{H} = \mathcal{H}(\mathsf{Z})$*. Then* $\Phi(\mathsf{Z})$ *and* $\Phi(\mathcal{H})$ *are anti-isomorphic.*

In particular, Proposition 7.2.2 gives the complete face lattice of the graphical zonotope Z_G.

We can associate a zonotope to any central arrangement $\mathcal{H}$ by picking a nonzero normal vector $\mathbf{z}_{\mathsf{H}}$ for every hyperplane $\mathsf{H} \in \mathcal{H}$, and then $\mathsf{Z} = \sum_{\mathsf{H}\in\mathcal{H}}[-\mathbf{z}_{\mathsf{H}}, \mathbf{z}_{\mathcal{H}}]$ recovers $\mathcal{H}$ by way of Proposition 7.5.4. But this already shows that Z is not unique, even up to translation.

We can also *see* the lattice of flats of $\mathcal{H}$ by looking at Z.

Proposition 7.5.6. *Let* Z *be a zonotope. For two faces* F, F' *of* Z *the following are equivalent:*

(a) *The affine hulls* $\operatorname{aff}(F)$ *and* $\operatorname{aff}(F')$ *are translates of each other.*

(b) *The faces* F *and* F' *are translates of each other.*

Proof. Let $\sigma, \sigma' \in \{<,=,>\}^m$ such that $F = \mathsf{Z}_\sigma$ and $F' = \mathsf{Z}_{\sigma'}$. In the proof of Proposition 7.5.4 we saw that the affine hull of F is uniquely determined by $\sigma^=$. Hence, $\operatorname{aff}(F) = \mathbf{t} + \operatorname{aff}(F')$ for some $\mathbf{t} \in \mathbb{R}^d$ if and only if $\sigma^= = (\sigma')^=$. On the other hand, it follows from (7.5.2) that this happens if and only if F and F' are translates of each other. □

Proposition 7.5.6 prompts an equivalence relation on $\Phi(\mathsf{Z})$: two faces F and F' of Z are equivalent if and only if they are translates of each other. In Exercise 7.29 you will show that the collection of equivalence classes of faces yields a poset that is anti-isomorphic to $\mathcal{L}(\mathcal{H})$.

For now we have exhausted the combinatorics of zonotopes and turn to ways to subdivide them. We could use the methods of Chapter 5 to triangulate zonotopes but, as we shall see in a second, it is more appropriate to subdivide zonotopes into zonotopes. A **zonotopal tiling** of a zonotope $\mathsf{Z} = \mathsf{Z}(\mathbf{z}_1, \ldots, \mathbf{z}_m)$ is a subdivision[3] $\mathcal{P}$ of Z such that the maximal cells (and hence all faces) of $\mathcal{P}$ are translates of zonotopes defined from subsets of $\mathbf{z}_1, \ldots, \mathbf{z}_m$. Figure 7.10 shows a sample tiling. The finest among such tilings decompose Z into parallelepipeds, and we call them **fine** or **cubical** zonotopal tilings for that reason.

The trick in obtaining zonotopal tilings lies in the simple observation that projections of line segments are line segments. Together with the fact that linear maps distribute over Minkowski sums, this proves the following.

Corollary 7.5.7. *Projections of zonotopes are zonotopes.*

[3]As the letter $\mathcal{T}$ is taken up by triangulations, we use $\mathcal{P}$ as in *Pflasterung*, the German word for *tiling*.

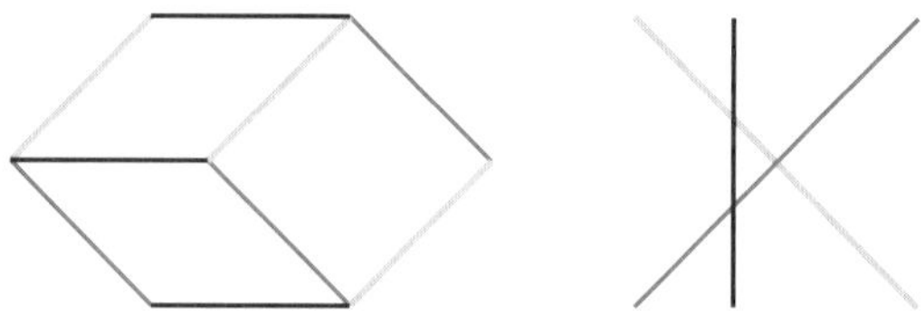

Figure 7.10. Tiling of a zonotope.

Given $\mathbf{z}_1, \ldots, \mathbf{z}_m \in \mathbb{R}^d$ and $\boldsymbol{\delta} = (\delta_1, \ldots, \delta_m) \in \mathbb{R}^m$, let $\widehat{\mathbf{z}}_i := (\mathbf{z}_i, \delta_i)$ for $i = 1, \ldots, m$ and define the zonotope

$$\widehat{\mathsf{Z}} := \mathsf{Z}(\widehat{\mathbf{z}}_1, \ldots, \widehat{\mathbf{z}}_m) \subset \mathbb{R}^{d+1}.$$

Borrowing from Section 5.1, we write $\pi : \mathbb{R}^{d+1} \to \mathbb{R}^d$ for the linear projection that forgets the $(d+1)$-st coordinate, and we reuse $\uparrow_{\mathbb{R}} := \{\mathbf{0}\} \times \mathbb{R}_{\geq 0} \subset \mathbb{R}^{d+1}$. Thus

$$\pi\left(\widehat{\mathsf{Z}} + \uparrow_{\mathbb{R}}\right) = \pi\left(\widehat{\mathsf{Z}}\right) = \mathsf{Z}(z_1, \ldots, z_m),$$

and using Lemma 7.5.1, we infer that every bounded face of $\widehat{\mathsf{Z}} + \uparrow_{\mathbb{R}}$ is a face of $\widehat{\mathsf{Z}}$ and hence a zonotope. Revisiting the arguments used in our proof of Theorem 5.1.5, which we invite you to do in Exercise 7.31, gives the following result.

Theorem 7.5.8. *Let $\mathsf{Z} = \mathsf{Z}(\mathbf{z}_1, \ldots, \mathbf{z}_m) \subset \mathbb{R}^d$ and $\boldsymbol{\delta} \in \mathbb{R}^m$ with associated zonotope $\widehat{\mathsf{Z}} \subset \mathbb{R}^{d+1}$. Then*

$$\mathcal{P}(\mathsf{Z}, \boldsymbol{\delta}) := \left\{\pi(\mathsf{F}) : \mathsf{F} \textit{ bounded face of } \widehat{\mathsf{Z}} + \uparrow_{\mathbb{R}}\right\}$$

is a zonotopal tiling of Z.

For reasons pertaining to Section 5.1, we call $\mathcal{P}$ a **regular tiling** of Z if $\mathcal{P} = \mathcal{P}(\mathsf{Z}, \boldsymbol{\delta})$ for some $\boldsymbol{\delta}$. For a counterpart to Corollary 5.1.6 we first need the following. We call a collection of vectors $\mathbf{z}_1, \ldots, \mathbf{z}_m \in \mathbb{R}^d$ **in general position** if no subset of $k+1$ of them is contained in a linear subspace of dimension k for any $1 \leq k \leq d$.

Proposition 7.5.9. *Let $\mathbf{z}_1, \ldots, \mathbf{z}_m \in \mathbb{R}^d$ be vectors in general position. Then all proper faces of $\mathsf{Z}(\mathbf{z}_1, \ldots, \mathbf{z}_m)$ are parallelepipeds.*

Proof. Without loss of generality, we may assume that $\mathsf{Z} = \mathsf{Z}(\mathbf{z}_1, \ldots, \mathbf{z}_m)$ is full dimensional. As faces of parallelepipeds are parallelepipeds, we only need to show that all *facets* of Z are parallelepipeds.

Let $\mathbf{w} \in \mathbb{R}^d$ such that $\mathsf{F} = \mathsf{F}_{\mathbf{w}}\mathsf{Z}$ is a facet. Following the proof of Corollary 7.5.2, we observe that F is a translate of the zonotope generated by those $\mathbf{z}_i$ with $\langle \mathbf{w}, \mathbf{z}_i \rangle = 0$. There are at least $d-1$ such vectors, as otherwise $\dim \mathsf{F} < d-1$. On the other hand, there are exactly $d-1$ vectors, as our general position assumption forbids the hyperplane $\{\mathbf{x} \in \mathbb{R}^d : \langle \mathbf{w}, \mathbf{x} \rangle = 0\}$ to contain more than $d-1$ vectors among $\mathbf{z}_1, \ldots, \mathbf{z}_m$. □

Corollary 7.5.10. *Every zonotope has a fine zonotopal tiling.*

Proof. For a generic choice of $\boldsymbol{\delta}$, the vectors $\widehat{\mathbf{z}}_1, \ldots, \widehat{\mathbf{z}}_m$ are in general position, and Proposition 7.5.9 finishes the proof. □

As with our construction of regular subdivisions of polytopes in Chapter 5, the above construction of regular zonotopal tilings is elegant but hard to analyze. However, there is a quite charming way to *view* regular zonotopal tilings. Let $\mathcal{H}$ be a central hyperplane arrangement with hyperplanes

$$\mathsf{H}_i \;=\; \{\mathbf{x} : \langle \mathbf{z}_i, \mathbf{x} \rangle = 0\}$$

for $i = 1, \ldots, m$. For $\boldsymbol{\delta} \in \mathbb{R}^m$, we define the **affinization** of $\mathcal{H}$ with respect to $\boldsymbol{\delta}$ as the affine hyperplane arrangement $\mathcal{H}^{\boldsymbol{\delta}}$ with hyperplanes

$$\mathsf{H}_i^{\delta_i} \;=\; \{\mathbf{x} : \langle \mathbf{z}_i, \mathbf{x} \rangle = \delta_i\}$$

for $i = 1, \ldots, m$. Note that $\mathcal{H} = \mathcal{H}^{\mathbf{0}}$ is an affinization and if $\delta_i = \langle \mathbf{z}_i, \mathbf{t} \rangle$ for some $\mathbf{t} \in \mathbb{R}^d$, then $\mathcal{H}^{\boldsymbol{\delta}}$ is simply a translation of $\mathcal{H}$. Nothing exciting so far. In particular, if you look at $\mathcal{H}^{\boldsymbol{\delta}}$ from far, far away (by formally replacing $\mathbf{x}$ with $\nu\mathbf{x}$ for some large number ν), then there is hardly any difference between H_i and $\mathsf{H}_i^{\delta_i}$ and we basically see $\mathcal{H}$, as illustrated in Figure 7.11.

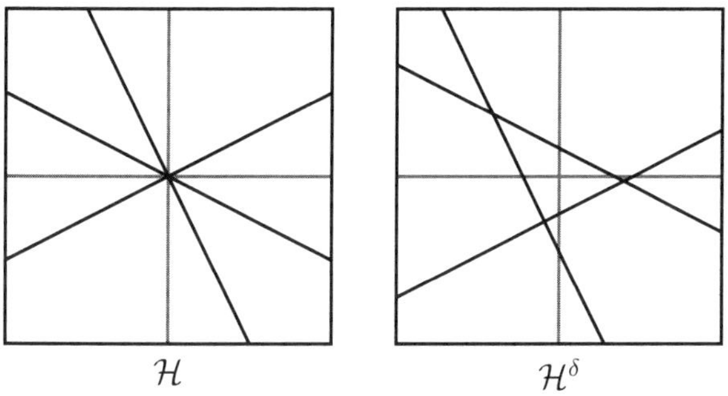

Figure 7.11. A sample affinization.

On the other hand, $\mathcal{H}^{\boldsymbol{\delta}}$ has an interesting local structure. A minimal face $\mathsf{F} \in \Phi(\mathcal{H}^{\boldsymbol{\delta}})$ with corresponding $\sigma \in \{<, =, >\}^m$ is of the form $\mathbf{t} + \mathrm{lineal}(\mathcal{H})$. The subarrangement of $\mathcal{H}$

$$\mathcal{H}' \;:=\; \left\{\mathsf{H}_i^{\delta_i} - \mathbf{t} \,:\, \mathbf{t} + \mathrm{lineal}(\mathcal{H}) \subseteq \mathsf{H}_i^{\delta_i},\ i = 1, \ldots, m\right\} \;=\; \{\mathsf{H}_i : i \in \sigma^=\}$$

is the central arrangement of the zonotope $\mathsf{Z}' := \mathsf{Z}(\mathbf{z}_i : i \in \sigma^=)$. Moreover, setting $\mathbf{s} := \sum_{i \in \sigma^>} \mathbf{z}_i$, the collection of zonotopes $\mathbf{s} + \mathsf{Z}'$ for all σ magically fit together to a zonotopal tiling of Z; see Figure 7.12.

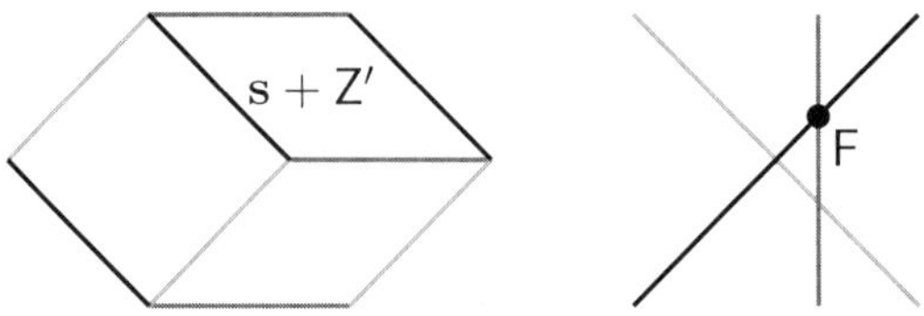

Figure 7.12. Tiling of a zonotope and the corresponding affinization.

The rigorous statement behind these musings is as follows.

Theorem 7.5.11. *Let* $\mathsf{Z} = \mathsf{Z}(\mathbf{z}_1, \dots, \mathbf{z}_m)$ *with associated central hyperplane arrangement* $\mathcal{H} = \{\mathsf{H}_1, \dots, \mathsf{H}_m\}$*. For* $\boldsymbol{\delta} \in \mathbb{R}^m$*, let* $\mathcal{P} = \mathcal{P}(\mathsf{Z}, \boldsymbol{\delta})$ *be the zonotopal tiling corresponding to* $\mathcal{H}^{\boldsymbol{\delta}}$*. Then* $\mathcal{P}$ *is anti-isomorphic to* $\Phi(\mathcal{H}^{\boldsymbol{\delta}})$*.*

Proof. Let $\widehat{\mathbf{w}} = (\mathbf{w}, w_{d+1}) \in \mathbb{R}^{d+1}$ be such that $\mathsf{F}_{\widehat{\mathbf{w}}}(\widehat{\mathsf{Z}} + \uparrow_{\mathbb{R}})$ is a bounded face of Z. As in Proposition 5.1.4, we notice that $w_{d+1} < 0$ and hence we can assume that $w_{d+1} = -1$. In particular,

$$\mathsf{F}_{\widehat{\mathbf{w}}}\left(\widehat{\mathsf{Z}} + \uparrow_{\mathbb{R}}\right) \; = \; \mathsf{F}_{\widehat{\mathbf{w}}}\widehat{\mathsf{Z}} \; = \; \widehat{\mathsf{Z}}_{\sigma}$$

for some $\sigma \in \{<, =, >\}^m$. For $1 \le i \le m$,

$$\langle \widehat{\mathbf{z}}_i, \widehat{\mathbf{w}} \rangle \; = \; \langle \mathbf{z}_i, \mathbf{w} \rangle - \delta_i \,.$$

This translates into the fact that $\mathbf{w}$ is contained in the face of $\mathcal{H}^{\boldsymbol{\delta}}$ uniquely determined by σ. Playing this argument backwards by starting with a face of $\mathcal{H}^{\boldsymbol{\delta}}$ shows that there is a canonical bijection between $\mathcal{P}(\mathsf{Z}, \boldsymbol{\delta})$ and $\mathcal{H}^{\boldsymbol{\delta}}$.

To see that this bijection is inclusion reversing, let $\tau \in \{<, =, >\}^m$ be such that $\widehat{\mathsf{Z}}_{\tau}$ is a face of $\widehat{\mathsf{Z}}_{\sigma}$. Then necessarily $\tau^{=} \subseteq \sigma^{=}$, which implies that the inclusion for the corresponding faces of $\mathcal{H}^{\boldsymbol{\delta}}$ is exactly the other way. □

To reap some of the benefits of the correspondence given in Theorem 7.5.11 with affinizations of $\mathcal{H}$, let $\mathsf{Z} = \mathsf{Z}(\mathbf{z}_1, \dots, \mathbf{z}_m) \subset \mathbb{R}^d$ be a full-dimensional zonotope, i.e., $\mathbf{z}_1, \dots, \mathbf{z}_m$ are spanning $\mathbb{R}^d$ and hence $\mathcal{H}$ is essential.

Proposition 7.5.12. *Let* $\mathcal{P}$ *be a fine regular tiling of the full-dimensional zonotope* $\mathsf{Z}(\mathbf{z}_1, \dots, \mathbf{z}_m) \subset \mathbb{R}^d$*. Then the number of parallelepipeds equals the number of bases of* $\mathbb{R}^d$ *among* $\mathbf{z}_1, \dots, \mathbf{z}_m$*.*

Proof. Let $\mathcal{H}$ be the central hyperplane arrangement corresponding to $\mathsf{Z}(\mathbf{z}_1, \dots, \mathbf{z}_m)$, and let $\mathcal{H}^{\boldsymbol{\delta}}$ be the affinization of $\mathcal{H}$ corresponding to $\mathcal{P}$, courtesy of Theorem 7.5.11. Since $\mathsf{Z}(\mathbf{z}_1, \dots, \mathbf{z}_m)$ is full dimensional, $\mathcal{H}$ is essential and the cells of $\mathcal{P}$ correspond to the vertices of $\mathcal{H}^{\boldsymbol{\delta}}$. Since the maximal cells of $\mathcal{P}$ are parallelepipeds, each vertex of $\mathcal{H}^{\boldsymbol{\delta}}$ is the intersection of exactly d

hyperplanes of $\mathcal{H}^{\boldsymbol{\delta}}$. In linear algebra terms, these correspond exactly to the sets $I = \{i_1 < i_2 < \cdots < i_d\} \subseteq [m]$ such that there is a unique $\mathbf{x} \in \mathbb{R}^d$ with

$$\langle \mathbf{z}_{i_1}, \mathbf{x} \rangle = \delta_{i_1}, \quad \langle \mathbf{z}_{i_2}, \mathbf{x} \rangle = \delta_{i_2}, \quad \dots, \quad \langle \mathbf{z}_{i_d}, \mathbf{x} \rangle = \delta_{i_d} .$$

This, however, happens if and only if $\mathbf{z}_{i_1}, \dots, \mathbf{z}_{i_d}$ are linearly independent. In particular, the linear independence of the vectors indexed by I is independent of the choice of a generic $\boldsymbol{\delta}$. □

The situation for graphical zonotopes is particularly appealing: in Exercise 7.27, you will show that for a graph $G = ([d], E)$ and a subset $E' \subseteq E$, the set $\{\mathbf{e}_i - \mathbf{e}_j : ij \in E'\}$ is linearly independent if and only if the induced graph $G[E'] := ([d], E')$ has no cycle. A **forest** is a graph that has no cycles. A **tree** is a connected forest. Given a connected graph G, a **spanning tree** is an inclusion-maximal cycle-free subgraph of G. We have thus proved the following.

Corollary 7.5.13. *Let $G = ([d], E)$ be a connected graph and $\mathcal{P}$ a fine regular tiling of Z_G into parallelepipeds. Then the number of parallelepipeds in $\mathcal{P}$ equals the number of spanning trees of G.*

If $S = \{\mathbf{z}_1, \dots, \mathbf{z}_m\} \subset \mathbb{Z}^d$, then $\mathsf{Z} = \mathsf{Z}(\mathbf{z}_1, \dots, \mathbf{z}_m)$ is a lattice polytope, which we call a **lattice zonotope**. We can use Theorem 7.5.11 to determine the Ehrhart polynomial of Z. If $S' = \{\mathbf{z}_{i_1}, \dots, \mathbf{z}_{i_k}\} \subseteq S$ is a linearly independent subset, let

$$\widehat{\square}(S') := \{\lambda_1 \mathbf{z}_{i_1} + \cdots + \lambda_k \mathbf{z}_{i_k} : 0 \le \lambda_1, \dots, \lambda_k < 1\},$$

the half-open parallelepiped spanned by $\mathbf{z}_{i_1}, \dots, \mathbf{z}_{i_k}$. For $S' = \varnothing$ we set $\widehat{\square}(S') = \{\mathbf{0}\}$. The following result is illustrated in Figure 7.13.

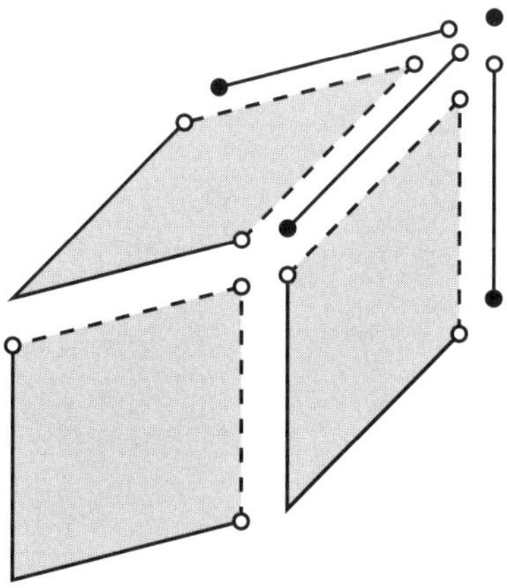

Figure 7.13. A fine zonotopal tiling of a hexagon.

Lemma 7.5.14. *Let $S \subset \mathbb{R}^d$ be a collection of (lattice) vectors. Then for every linearly independent subset $S' \subseteq S$, there is a (lattice) vector $\mathbf{t}_{S'}$ such that*

$$\mathsf{Z}(S) \;=\; \biguplus \left(\mathbf{t}_{S'} + \widehat{\square}(S')\right),$$

where the union is over all linearly independent $S' \subseteq S$.

Proof. We adopt an optimization perspective. Let $S = \{\mathbf{z}_1, \dots, \mathbf{z}_m\}$. Then every point $\mathbf{p} \in \mathsf{Z}$ is of the form

$$\text{(7.5.3)} \quad \mathbf{p} \;=\; \mu_1\mathbf{z}_1 + \mu_2\mathbf{z}_2 + \cdots + \mu_m\mathbf{z}_m \qquad \text{for some} \qquad 0 \le \mu_1, \dots, \mu_m \le 1\,.$$

The parameters $\mu = (\mu_1, \dots, \mu_m)$ are in general not unique. In fact, (7.5.3) shows that for fixed $\mathbf{p}$, the set of all possible μ is a polytope $\mathsf{Q}_\mathbf{p}$. We want to construct a canonical choice of μ. For $\varepsilon > 0$ we define the linear function

$$\ell_\varepsilon(\mathbf{x}) \;:=\; \varepsilon x_1 + \varepsilon^2 x_2 + \cdots + \varepsilon^m x_m\,.$$

The polytope $\mathsf{Q}_\mathbf{p}$ has finitely many vertices. Exercise 7.24 asserts that there is an $\varepsilon_\mathbf{p} > 0$ such that for all $0 < \varepsilon < \varepsilon_\mathbf{p}$, the minimum of $\ell_\varepsilon(\mathbf{x})$ is attained at the same vertex $\mu^* = (\mu_1^*, \dots, \mu_m^*)$ of $\mathsf{Q}_\mathbf{p}$. Let $I := \{i : 0 < \mu_i^* < 1\}$. Then

$$\sum_{i \in I} \mu_i^* \mathbf{z}_i \;=\; \mathbf{p} - \mathbf{t}\,,$$

where $\mathbf{t} := \sum_{i : \mu_i^* = 1} \mathbf{z}_i$. We claim that the set $\{\mathbf{z}_i : i \in I\}$ is linearly independent. If this was not true, then there is some $\eta \in \mathbb{R}^m$ such that $\eta_i = 0$ for all $i \notin I$ and $\sum_i \eta_i \mathbf{z}_i = 0$. But then for $\lambda > 0$ sufficiently small, $\mu^* \pm \lambda\eta \in \mathsf{Q}_\mathbf{p}$, and this would contradict the fact that μ^* is a vertex. Thus for $S' = \{\mathbf{z}_i : i \in I\}$ and $\mathbf{t}_{S'} := \mathbf{t}$, we obtain $\mathbf{p} \in \mathbf{t}_{S'} + \widehat{\square}(S')$.

Exercise 7.25 shows that the choice of μ^* is consistent for all points $\mathbf{p} \in \mathsf{Z}$ provided we choose $\varepsilon > 0$ sufficiently small. This means that for every point $\mathbf{p}$ there is a unique linearly independent $S' \subseteq S$ and a point $\mathbf{t}_{S'}$ such that $\mathbf{p} \in \mathbf{t}_{S'} + \widehat{\square}(S')$. □

If $S = \{\mathbf{z}_1, \dots, \mathbf{z}_k\} \subset \mathbb{Z}^d$ is linearly independent, then

$$\operatorname{ehr}_{\widehat{\square}(S)}(n) \;=\; \left|n\widehat{\square}(\mathbf{z}_1, \dots, \mathbf{z}_k) \cap \mathbb{Z}^d\right| \;=\; \left|\widehat{\square}(n\mathbf{z}_1, \dots, n\mathbf{z}_k) \cap \mathbb{Z}^d\right|,$$

a polynomial in n of degree k. We can write down this polynomial explicitly. Consider $S \subset \mathbb{Z}^d$ as a $d \times k$-matrix, and let $\det(\mathrm{S_J})$ be the determinant of the $k \times k$-submatrix of S obtained by selecting the rows indexed by a given k-subset $J \subseteq [d]$. We define $\operatorname{gd}(S)$ as the greatest common divisor of $\det(\mathrm{S_J})$, where J ranges over all k-subsets of $[d]$.

Proposition 7.5.15. *Let $S = \{\mathbf{z}_1, \dots, \mathbf{z}_k\} \subset \mathbb{Z}^d$ be linearly independent. Then*

$$\operatorname{ehr}_{\widehat{\square}(S)}(n) \;=\; n^k\,|\operatorname{gd}(S)|\,.$$

Proof. We prove only the case $k = d$ and leave the gory details for the general case to you as Exercise 7.26. The same arguments as in our proof of Lemma 7.5.14 show that

$$\widehat{\square}(n\mathbf{z}_1, \dots, n\mathbf{z}_d) \;=\; \biguplus_{0 \le \lambda_1, \dots, \lambda_d < n} (\lambda_1 \mathbf{z}_1 + \cdots + \lambda_d \mathbf{z}_d) + \widehat{\square}(\mathbf{z}_1, \dots, \mathbf{z}_d) \tag{7.5.4}$$

and hence

$$\mathrm{ehr}_{\widehat{\square}(S)}(n) \;=\; n^d \left| \widehat{\square}(\mathbf{z}_1, \dots, \mathbf{z}_d) \cap \mathbb{Z}^d \right| .$$

Lemma 7.5.14 yields that the Ehrhart polynomial of $\mathsf{Z}(S) \backslash \widehat{\square}(S)$ has degree $<$ d and hence $\mathrm{ehr}_{\mathsf{Z}(S)}(n)$ and $\mathrm{ehr}_{\widehat{\square}(S)}(n)$ have the same leading coefficient. By Exercise 5.10, the leading coefficient equals the volume of the parallelepiped $\mathsf{Z}(S)$, which is given by $|\det(\mathrm{S})|$. □

Combining Lemma 7.5.14 with Proposition 7.5.15 gives the Ehrhart polynomial of any lattice zonotope.

Corollary 7.5.16. *Let $S \subset \mathbb{Z}^d$ be a finite collection of lattice vectors and $\mathsf{Z} = \mathsf{Z}(S)$ the corresponding zonotope. Then*

$$\mathrm{ehr}_{\mathsf{Z}}(n) \;=\; \sum_{S' \subseteq S} n^{|S'|} \left| \mathrm{gd}(S') \right| .$$

Proof. Note that $\mathrm{gd}(S') = 0$ whenever $S' \subset \mathbb{Z}^d$ is a finite collection of linearly dependent vectors. Hence, the sum in the statement is actually over all linearly independent subsets $S' \subseteq S$. By Proposition 7.5.15 the corresponding summand is the Ehrhart polynomial of $\widehat{\square}(S')$, and Lemma 7.5.14 says that the sum is the Ehrhart polynomial of Z. □

Again, for graphical zonotopes, the identity in Corollary 7.5.16 is most charming and also follows from Exercise 7.27.

Corollary 7.5.17. *Let $G = ([d], E)$ be a graph. Then*

$$\mathrm{ehr}_{\mathsf{Z}_G}(n) \;=\; \sum_{i=0}^{d-1} b_i(G)\, n^i ,$$

where $b_i(G)$ is the number of induced forests in G with i edges.

As a final thought on the relation between zonotopes and hyperplane arrangements, we briefly discuss one more amazing connection that relies on Theorem 7.5.11. An affine hyperplane arrangement in $\mathbb{R}^d$ is **simple** if every vertex is contained in exactly d hyperplanes.

Theorem 7.5.18. *Let $\mathcal{H} = \{\mathsf{H}_1, \dots, \mathsf{H}_m\}$ be a central and essential arrangement in $\mathbb{R}^d$. Then there is a central arrangement $\widehat{\mathcal{H}}$ in $\mathbb{R}^m$ whose regions are in bijection with the simple affinizations of $\mathcal{H}$.*

With Theorem 7.5.11, this implies in particular:

Corollary 7.5.19. *Let $\mathsf{Z} = \mathsf{Z}(\mathbf{z}_1, \dots, \mathbf{z}_m) \subset \mathbb{R}^d$ be a full-dimensional zonotope. Then there is a hyperplane arrangement $\widehat{\mathcal{H}} \subset \mathbb{R}^m$ such that the regions of $\widehat{\mathcal{H}}$ are in bijection with the fine regular tilings of Z.*

Let $\mathcal{H}^{\boldsymbol{\delta}}$ be a given affinization of $\mathcal{H}$ and let $\mathbf{v}_1, \dots, \mathbf{v}_r$ be the vertices of $\mathcal{H}^{\boldsymbol{\delta}}$. There are unique $\sigma_1, \dots, \sigma_r \in \{<, =, >\}^m$ such that $\{\mathbf{v}_i\} = \mathsf{H}^{\boldsymbol{\delta}}_{\sigma_i}$ for $1 \le i \le r$. We claim that $\sigma_1, \dots, \sigma_r$ together with $\mathcal{H}$ completely determines the face poset of $\mathcal{H}^{\boldsymbol{\delta}}$. Let's explain this for the bounded regions.

If $\mathsf{F} \in \Phi(\mathcal{H}^{\boldsymbol{\delta}})$ is a bounded region of $\mathcal{H}^{\boldsymbol{\delta}}$, then it is determined by its vertices, which are a subset of the vertices of $\mathcal{H}^{\boldsymbol{\delta}}$. Hence, all we have to do is to determine the inclusion-maximal subsets $V \subseteq \{\mathbf{v}_1, \dots, \mathbf{v}_r\}$ whose convex hulls yield bounded regions of $\mathcal{H}^{\boldsymbol{\delta}}$. The Minkowski–Weyl theorem (Theorem 3.2.5) assures us that every F is the bounded intersection of halfspaces and a moment's thought reveals that these halfspaces are among the halfspaces induced by the hyperplanes in $\mathcal{H}^{\boldsymbol{\delta}}$. The bounded regions are $\mathcal{H}^{\boldsymbol{\delta}}$-polytopes in the language of Section 7.4. In particular, no two elements in V are separated by a hyperplane in $\mathcal{H}^{\boldsymbol{\delta}}$. This gives the following combinatorial description of bounded regions: V is the set of vertices of a bounded region F of $\mathcal{H}^{\boldsymbol{\delta}}$ if and only if it is inclusion maximal with the property that for any $\mathbf{v}_s, \mathbf{v}_t \in V$, there is no $1 \le i \le m$ such that $i \in {\sigma_s}^<$ and $i \in {\sigma_t}^>$. However, there are also unbounded regions (and faces) of $\mathcal{H}^{\boldsymbol{\delta}}$ and so this is not a rigorous proof. Hence, we have Exercise 7.32 to guide you to a proof of the following result.

Lemma 7.5.20. *Let $\mathcal{H}^{\boldsymbol{\delta}}$ be an essential affine hyperplane arrangement with vertices $\mathbf{v}_1, \dots, \mathbf{v}_r$ given by $\sigma_1, \dots, \sigma_r \in \{<, =, >\}^m$. Then $\Phi(\mathcal{H}^{\boldsymbol{\delta}})$ is determined by $\mathcal{H}$ and $\sigma_1, \dots, \sigma_r$.*

If $\mathcal{H}^{\boldsymbol{\delta}}$ is simple, then its vertices are easy to determine. Each hyperplane is of the form $\mathsf{H}_i = \{\mathbf{x} : \langle \mathbf{z}_i, \mathbf{x} \rangle = \delta_i\}$. Let $Z \in \mathbb{R}^{m \times d}$ be the matrix whose i-th row is $\mathbf{z}_i$. For $J \subseteq [m]$ we again denote by Z_J the submatrix with rows indexed by J. If $J \subseteq [m]$ such that Z_J is a nonsingular $d \times d$ matrix, then

$$\mathbf{v} \;=\; Z_J^{-1} \boldsymbol{\delta}_J$$

is a vertex of $\mathcal{H}^{\boldsymbol{\delta}}$, and every vertex is of that form. For each vertex $\mathbf{v}$ identified with $J \subseteq [m]$ and any $i \in [m] \setminus J$,

$$\text{(7.5.5)} \qquad \mathbf{v} \in \left(\mathsf{H}_i^{\delta_i}\right)^< \qquad \text{if and only if} \qquad \langle \mathbf{z}_i, Z_J^{-1} \boldsymbol{\delta}_J \rangle \;>\; \delta_i\,.$$

Proof of Theorem 7.5.18. Let $\mathcal{I}$ be the collection of pairs (J, i) with $J \subseteq [m]$ such that Z_J is a nonsingular $d \times d$-matrix and $i \notin J$. Define the hyperplane

$$\mathsf{H}_{(J,i)} \;:=\; \left\{\boldsymbol{\delta} \in \mathbb{R}^m \,:\, \langle \mathbf{z}_i, Z_J^{-1} \boldsymbol{\delta}_J \rangle = \delta_i\right\}$$

and $\widehat{\mathcal{H}} = \{\mathsf{H}_{(J,i)} : (J,i) \in \mathcal{I}\}$. Now any simple affinization $\mathcal{H}^{\boldsymbol{\delta}}$ with $\boldsymbol{\delta} \in \mathbb{R}^m$ determines a partition $\mathcal{I} = \mathcal{I}_< \uplus \mathcal{I}_>$ and a nonempty region

$$\bigcap_{(J,i)\in\mathcal{I}_<} (\mathsf{H}_{(J,i)})^< \cap \bigcap_{(J,i)\in\mathcal{I}_>} (\mathsf{H}_{(J,i)})^> .$$

It follows from Lemma 7.5.20 that $\Phi(\mathcal{H}^{\boldsymbol{\delta}}) = \Phi(\mathcal{H}^{\boldsymbol{\delta}'})$ for each $\boldsymbol{\delta}'$ in this region. And since each $\boldsymbol{\delta}$ in general position determines a simple affinization, this proves the claim. □

Much more can be done. Similar to our considerations about refinements of dissections of order cones in Section 6.1, the collection of all regular tilings of a zonotope Z is a poset $\mathfrak{T}(\mathsf{Z})$ with respect to refinement: a tiling $\mathcal{P}$ is finer than a tiling $\mathcal{P}'$ if for every $\mathsf{Q} \in \mathcal{P}$ there is some $\mathsf{Q}' \in \mathcal{P}'$ with $\mathsf{Q} \subseteq \mathsf{Q}'$. With some more work, you can prove the following consequence.

Corollary 7.5.21. *Let* $\mathsf{Z} = \mathsf{Z}(\mathbf{z}_1, \dots, \mathbf{z}_m)$ *be a full-dimensional zonotope in* $\mathbb{R}^d$. *Then there is an* $(m-d)$*-dimensional zonotope* $\widehat{\mathsf{Z}} \subset \mathbb{R}^m$ *such that* $\Phi(\widehat{\mathsf{Z}}) \cong \mathfrak{T}(\mathsf{Z})$.

Exercise 7.33 gives a different construction of $\widehat{\mathsf{Z}}$.

7.6. Graph Flows and Totally Cyclic Orientations

We finally come to the second family of polynomials and combinatorial reciprocities that we considered in Chapter 1: the nowhere-zero flow polynomials $\varphi_G(n)$ of Section 1.2. Since we learned about flows on graphs almost a whole book ago, we recapitulate the setup and the statement of Theorem 1.2.5, whose proof we still owe. Throughout, let $G = (V, E)$ be a graph with a fixed but arbitrary orientation, which we will refer to as the **base orientation**. We write $u \to v$ if the edge $uv \in E$ is oriented from u to v. A $\mathbb{Z}_n$**-flow** is a function $f : E \to \mathbb{Z}_n = \mathbb{Z}/n\mathbb{Z}$ such that for every node $v \in V$

$$\sum_{u \to v} f(uv) = \sum_{v \leftarrow u} f(uv) .$$

This is an equality in the finite Abelian group $\mathbb{Z}_n$. In this wording, it seems rather difficult to relate the functions f to the lattice points in some polytope and, in fact, there is no single polytope whose Ehrhart polynomial counts **nowhere-zero flows**, i.e., $f(e) \neq 0$ for all $e \in E$ for varying n.

To cut right to the chase, we identify $\mathbb{Z}_n$ with $\{0, 1, \dots, n-1\}$. Then nowhere-zero flows are those functions $f : E \to \mathbb{Z}$ such that $0 < f(e) < n$ for all $e \in E$ and

$$\sum_{u \to v} f(uv) - \sum_{v \to u} f(uv) \quad \text{is divisible by } n$$

for every node $v \in V$. To see why this gets us what we want, we rewrite the conditions one more time: we seek functions $f \in n(0,1)^E \cap \mathbb{Z}^E$ such that for every $v \in V$

$$\sum_{u \to v} f(uv) - \sum_{v \to u} f(uv) \; = \; n\, b_v \tag{7.6.1}$$

for some $b_v \in \mathbb{Z}$. For $\mathbf{b} \in \mathbb{Z}^V$, let $\mathcal{F}_G(\mathbf{b}) \subseteq \mathbb{R}^E$ be the affine subspace of all $f \in \mathbb{R}^E$ satisfying (7.6.1) with $n = 1$. The number $\varphi_G(n)$ of nowhere-zero $\mathbb{Z}_n$-flows of G is the number of lattice points in

$$n\,(0,1)^E \cap \bigcup_{\mathbf{b} \in \mathbb{Z}^V} \mathcal{F}_G(n\mathbf{b})\,. \tag{7.6.2}$$

We note that $\mathcal{F}_G(\mathbf{b}) \cap \mathcal{F}_G(\mathbf{b}') = \varnothing$ whenever $\mathbf{b} \neq \mathbf{b}'$. Since the cube $[0,1]^E$ is compact, there are only finitely many $\mathbf{b} \in \mathbb{Z}^V$ such that

$$(0,1)^E \cap \mathcal{F}_G(\mathbf{b}) \; \neq \; \varnothing$$

and we denote the set of these $\mathbf{b}$ by $\mathcal{C}(G) \subset \mathbb{Z}^V$. For $\mathbf{b} \in \mathbb{Z}^V$, let

$$\mathsf{P}_G(\mathbf{b}) \; := \; [0,1]^E \cap \mathcal{F}_G(\mathbf{b})\,; \tag{7.6.3}$$

as an intersection of halfspaces and hyperplanes, $\mathsf{P}_G(\mathbf{b})$ is a polytope. We remark that our construction of both $\mathcal{F}_G(\mathbf{b})$ and $\mathsf{P}_G(\mathbf{b})$ depends on the base orientation of G.

Iterating Exercise 3.37 shows that

$$\dim \mathsf{P}_G(\mathbf{b}) \; \le \; \dim \mathcal{F}_G(\mathbf{b})\,,$$

and equality is attained exactly for $\mathbf{b} \in \mathcal{C}(G)$. We note that $\mathcal{F}_G(\mathbf{b}) = \mathbf{t} + \mathcal{F}_G(\mathbf{0})$ for some $\mathbf{t} \in \mathbb{R}^E$ and hence $\dim \mathcal{F}_G(\mathbf{b})$ is independent of $\mathbf{b}$.

The linear subspace $\mathcal{F}_G(\mathbf{0}) \subseteq \mathbb{R}^E$ plays a prominent role. It is the linear space of real-valued flows on G, i.e., the functions $f \in \mathbb{R}^E$ such that

$$\sum_{u \to v} f(uv) \; = \; \sum_{v \to u} f(uv) \tag{7.6.4}$$

for all $v \in V$. The condition (7.6.4) is called *conservation of flow* at v: it states that what flows into v also has to flow out of it.[4] We call $\mathcal{F}_G(\mathbf{0})$ the **flow space** of G and simply denote it by $\mathcal{F}_G := \mathcal{F}_G(\mathbf{0})$. Again, the flow space depends on the base orientation of G.

We recall from Section 1.2 that the cyclotomic number of G is $\xi(G) = |E| - |V| + c$, where c is the number of connected components of G.

Proposition 7.6.1. *Let $G = (V, E)$ be a graph with a fixed base orientation. Then* $\dim \mathcal{F}_G = \xi(G)$.

[4] We are stretching this physical interpretation a bit since $f(e)$ can be negative.

It is not difficult to produce nonzero elements in $\mathcal{F}_G$ provided G has cycles. Let C be a cycle of G, i.e., C consists of distinct nodes $v_1, \ldots, v_k$ such that $v_{i-1}v_i \in E$ for $i = 1, \ldots, k$ with $v_0 := v_k$. We have implicitly given C an orientation by labeling its nodes. The orientation from v_{i-1} to v_i might or might not agree with the base orientation. We define a function $f_C : E \to \mathbb{Z}$ through

$$f_C(v_{i-1}v_i) := \begin{cases} 1 & \text{if } v_{i-1} \to v_i\,, \\ -1 & \text{if } v_{i-1} \leftarrow v_i\,, \end{cases} \tag{7.6.5}$$

and $f_C(e) := 0$ for each edge $e \in E$ that is not part of the cycle. Actually, f_C depends on C as well as the choice of orientation of C. However, choosing the other orientation around C would replace f_C by $-f_C$ and does not change the statement of the following lemma, whose verification we leave to Exercise 7.34.

Lemma 7.6.2. *The function f_C is an integer-valued flow for any undirected cycle C of G.*

Proof of Proposition 7.6.1. We will construct a basis with $\xi(G)$ elements. We first observe that, if G is the disjoint union of G_1 and G_2, then $\mathcal{F}_G = \mathcal{F}_{G_1} \times \mathcal{F}_{G_2}$ and hence $\dim \mathcal{F}_G = \dim \mathcal{F}_{G_1} + \dim \mathcal{F}_{G_2}$. We will therefore assume that G is connected.

Let $T \subseteq G$ be a spanning tree, i.e., $T = (V, E')$ for some $E' \subseteq E$, that is connected and without cycles. Let $e = uv \in E \setminus E'$ with orientation $u \to v$. Since T is connected, there is a path $v_1, \ldots, v_k$ in T that connects v to u. In particular, $u =: v_0, v_1, \ldots, v_k$ is a cycle C_e in $T \cup \{e\} \subseteq G$, called the **fundamental cycle** with respect to T and e. The function f_{C_e} defined by (7.6.5) is a nonzero element of $\mathcal{F}_G$ and we claim that $\{f_{C_e} : e \in E \setminus E'\}$ is a basis of $\mathcal{F}_G$. Note that the elements in this collection are linearly independent and hence we only need to show that they are spanning.

For a given $f \in \mathcal{F}_G$, we can add suitable scalar multiples of f_{C_e} to it and can assume that $f(e) = 0$ for all $e \in E \setminus E'$. Arguing by contradiction, let's assume that $f \neq 0$. Then $E_f := \{e \in E : f(e) \neq 0\}$ is a subset of E'. However, for (7.6.4) to be satisfied at a node v, there have to be either zero or at least two edges in E_f incident to w. Exercise 7.35 now shows that the graph $(V, E_f) \subseteq T$ contains a cycle which contradicts our assumption on T.

Since T is a spanning tree, $|E'| = |V| - 1$ and so $\mathcal{F}_G$ is of dimension $|E \setminus E'| = \xi(G)$. □

For a fixed spanning tree $T \subseteq G$, the basis of $\mathcal{F}_G$ constructed in the course of the above proof is called a **cycle basis**. Next, we show that all polytopes of the form (7.6.3) are *lattice* polytopes.

Proposition 7.6.3. *Let $G = (V, E)$ be a graph with a fixed base orientation. Then $\mathsf{P}_G(\mathbf{b})$ is a lattice polytope for every $\mathbf{b} \in \mathbb{Z}^V$.*

Proof. For most $\mathbf{b} \in \mathbb{Z}^V$, the polytope $\mathsf{P}_G(\mathbf{b})$ is empty and there is nothing to prove. So let's assume that $\mathsf{P}_G(\mathbf{b}) \neq \varnothing$. Since a vertex $f \in \mathsf{P}_G(\mathbf{b})$ is a face, there is a linear function ℓ such that f uniquely maximizes ℓ over $\mathsf{P}_G(\mathbf{b})$. We will show that if ℓ has a unique maximizer f, then f has to be integer valued. For this, we can reuse an idea that we already appealed to in the proof of Proposition 7.6.1. Assume that $f \in \mathsf{P}_G(\mathbf{b})$ maximizes ℓ and set $R := \{e \in E : 0 < f(e) < 1\}$. We claim that if $R \neq \varnothing$, then the graph $G_f = (V, R)$ contains a cycle. Sure enough, for any $v \in V$ and $b_v \in \mathbb{Z}$, the left-hand side of (7.6.1) has to feature either zero or at least two nonintegral evaluations of f. The claim now follows from Exercise 7.35.

For a cycle C in G_f, let f_C be as in (7.6.5). Then f_C is supported on the edges in R and there is a $\lambda \in \mathbb{R} \setminus \{0\}$ with $|\lambda|$ sufficiently small such that $f \pm \lambda f_C \in \mathsf{P}_G(\mathbf{b})$ and

$$\ell(f - \lambda f_C) \ \leq \ \ell(f) \ \leq \ \ell(f + \lambda f_C) \, .$$

Since we assumed that f is the unique maximizer, it follows that $R = \varnothing$ and hence $f \in \{0,1\}^E$. □

To take stock of what we have achieved so far, we note that Propositions 7.6.1 and 7.6.3 imply that $\mathsf{P}_G(\mathbf{b})$ is a $\xi(G)$-dimensional lattice polytope for any $\mathbf{b} \in \mathcal{C}(G)$. With a nod to Ehrhart–Macdonald reciprocity (Theorem 5.2.3) it follows from (7.6.2) that the number of nowhere-zero $\mathbb{Z}_n$-flows is

$$\varphi_G(n) \ = \ \sum_{\mathbf{b} \in \mathcal{C}(G)} \operatorname{ehr}_{\mathsf{P}_G^\circ(\mathbf{b})}(n) \ = \ (-1)^{\xi(G)} \sum_{\mathbf{b} \in \mathcal{C}(G)} \operatorname{ehr}_{\mathsf{P}_G(\mathbf{b})}(-n) \, . \tag{7.6.6}$$

Since $\mathsf{P}_G(\mathbf{b}) \subseteq [0,1]^E$, the lattice points of $\mathsf{P}_G(\mathbf{b})$ are exactly its vertices, and so $(-1)^{\xi(G)} \varphi_G(-1)$ equals the total number of vertices of all $\mathsf{P}_G(\mathbf{b})$ for $\mathbf{b} \in \mathcal{C}(G)$. Each vertex is of the form $\mathbf{e}_R \in \{0,1\}^E$ for a certain $R \subseteq E$. Which sets R appear is our next concern; we take a scenic detour via arrangements. We recall that an orientation of G is **totally cyclic** if every edge $e \in E$ is contained in a directed cycle.

Proposition 7.6.4. *Let $G = (V, E)$ be a graph with a fixed base orientation. This orientation is totally cyclic if and only if there is some $f \in \mathcal{F}_G$ with $f(e) > 0$ for all $e \in E$.*

Proof. Assume that the orientation of G is totally cyclic. If C is a directed cycle in G, then the corresponding flow f_C given in (7.6.5) for this orientation takes values in $\{0,1\}$. Let $f = \sum_C f_C$, where the sum is over all directed cycles of G. Since every edge is contained in a directed cycle, $f(e) > 0$ for all $e \in E$.

For the converse implication, suppose that the orientation is not totally cyclic. Then Exercise 7.36 says that there exists a coherently oriented edge cut, that is, a minimal set of edges whose removal increases the number of components of G, depicted in Figure 7.14, and so there cannot be a flow with all positive values. □

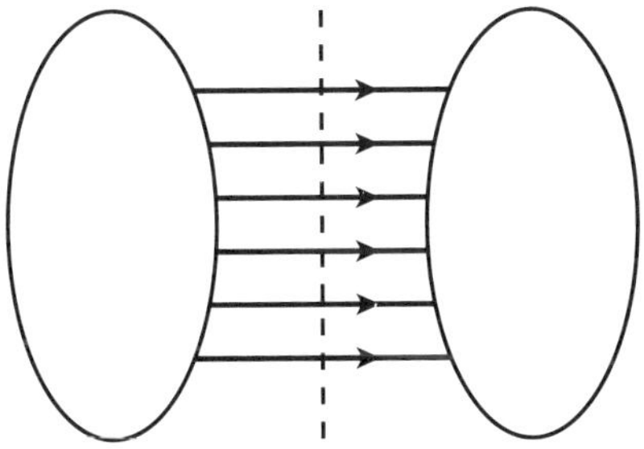

Figure 7.14. An illustration of an edge cut appearing in our proof of Proposition 7.6.4.

We defined $\mathcal{F}_G$ and $\mathcal{C}(G)$ with respect to a base orientation of G. Any other orientation of G is given by a subset $R \subseteq E$ of edges whose orientation will be reversed. On the level of flow spaces, the map $T_R : \mathbb{R}^E \to \mathbb{R}^E$ mapping f to $f' = T_R(f)$ via

$$f'(e) := \begin{cases} -f(e) & \text{if } e \in R, \\ f(e) & \text{if } e \in E \setminus R \end{cases}$$

defines a linear isomorphism between the flow spaces of G and its reorientation by R.

In Exercise 7.37 you will show that if $e \in E$ is a bridge, then $\mathcal{F}_G$ is contained in the coordinate hyperplane

$$\mathsf{H}^e := \left\{ g \in \mathbb{R}^E : g(e) = 0 \right\}.$$

Hence, if G is bridgeless, then

$$\mathcal{H}^G := \{ \mathsf{H}^e \cap \mathcal{F}_G : e \in E \}$$

is an arrangement of hyperplanes in $\mathcal{F}_G \cong \mathbb{R}^{\xi(G)}$. It is called the **cographical** hyperplane arrangement associated to G. Exercise 7.38 gives the number of distinct hyperplanes in $\mathcal{H}^G$.

The regions of the coordinate hyperplanes in $\mathbb{R}^E$ are in bijection with the different signs that a function $f : E \to \mathbb{R} \setminus \{0\}$ can have and this is inherited by $\mathcal{H}^G$. Thinking back to Section 7.2, you might venture a guess as to what the regions of $\mathcal{H}^G$ correspond to.

Theorem 7.6.5. *Let $G = (V, E)$ be a bridgeless graph and $R \subseteq E$. Then the following are equivalent:*

(a) *The set of real-valued nowhere-zero flows $f \in \mathcal{F}_G \setminus \bigcup \mathcal{H}^G$ such that $\{e : f(e) < 0\} = R$ is a region of $\mathcal{H}^G$.*

(b) *Reversing the orientation of the edges in R yields a totally cyclic orientation.*

(c) *$\mathbf{e}_R$ is a vertex of $\mathsf{P}_G(\mathbf{b})$ for some $\mathbf{b} \in \mathcal{C}(G)$.*

Proof. Each region of $\mathcal{H}^G$ is the set of all $f \in \mathcal{F}_G$ with $f(e) \neq 0$ and $f(e) < 0$ for all edges e in some fixed $R \subseteq E$. For any such f, it follows that $f' = T_R(f)$ is a strictly positive element in the cycle space corresponding to the reorientation R. The first equivalence now follows from Proposition 7.6.4.

To show that (a) implies (c), let $f \in \mathcal{F}_G$, with $f(e) \neq 0$ for all $e \in E$, be a representative of a region. The region is uniquely determined by

$$R := \{e \in E : f(e) < 0\}.$$

Since $\mathcal{H}^G$ is a central arrangement of hyperplanes, we can scale f inside its region such that $0 < |f(e)| < 1$ for all $e \in E$. We observe that $f' := \mathbf{e}_R + f$ satisfies $0 < f'(e) < 1$ for all $e \in E$ and hence $f' \in (0,1)^E$. In particular, there is a unique $\mathbf{b} \in \mathbb{Z}^V$ such that

$$f' \in \mathcal{F}_G(\mathbf{b}) = \mathbf{e}_R + \mathcal{F}_G.$$

Hence $f' \in \mathsf{P}^\circ_G(\mathbf{b})$ and $\mathbf{b} \in \mathcal{C}(G)$. Since $\mathbf{e}_R \in \mathsf{P}_G(\mathbf{b})$ as well and since $\mathbf{e}_R$ is a vertex of the cube $[0,1]^E$, it has to also be a vertex of $\mathsf{P}_G(\mathbf{b})$.

Conversely, for a given $\mathbf{b} \in \mathcal{C}(G)$, let $f \in \mathsf{P}^\circ_G(\mathbf{b})$. For any vertex $\mathbf{e}_R \in \mathsf{P}_G(\mathbf{b})$,

$$f' := f - \mathbf{e}_R \in \mathcal{F}_G \qquad \text{and} \qquad f'(e) \neq 0$$

for all $e \in E$. This determines a region of $\mathcal{H}^G$, which is independent of the choice of f. □

We are (finally!) ready to prove Theorem 1.2.5.

Proof of Theorem 1.2.5. It follows from (7.6.6) that $(-1)^{\xi(G)}\varphi_G(-n)$ is the total number of lattice points in $n\,\mathsf{P}_G(\mathbf{b})$ for all $\mathbf{b} \in \mathcal{C}(G)$. The lattice points in the relative interior of $n\,\mathsf{P}_G(\mathbf{b})$ are nowhere-zero $\mathbb{Z}_n$-flows and hence correspond to pairs $(f, \varnothing)$, i.e., f together with the original orientation. So we need to only worry about the lattice points in the boundary of $n\mathsf{P}_G(\mathbf{b})$.

For $f \in n\,\partial\mathsf{P}_G(\mathbf{b}) \cap \mathbb{Z}^E$ we set

$$Q := \{e : f(e) = 0\} \qquad \text{and} \qquad R := \{e : f(e) = n\}.$$

Note that $Q \cup R$ corresponds to the edges that get zero flow when we reduce the values of f modulo n. Choose $g \in n\,\mathsf{P}^\circ_G(\mathbf{b})$ such that $g(e) \neq f(e)$ for all $e \in E$. In particular, $h := g - f$ is contained in the flow space $\mathcal{F}_G$ and $h(e) \neq 0$ for all $e \in E$. By Theorem 7.6.5, h determines a totally cyclic reorientation of G, and in Exercise 7.39, you will show that the orientation yields a totally cyclic orientation on $G/\operatorname{supp}(f)$.

Conversely, let $\bar{f} : E \to \{0, \dots, n-1\}$ be a $\mathbb{Z}_n$-flow and let $R \subseteq E \setminus \operatorname{supp}(f)$ be a totally cyclic orientation for $G/\operatorname{supp}(f)$. Define $f : E \to \mathbb{Z}$ by setting

$$f(e) := \begin{cases} n & \text{if } e \in R, \\ \bar{f}(e) & \text{if } e \in E \setminus R. \end{cases}$$

Then f satisfies (7.6.1) for some unique $\mathbf{b} \in \mathbb{Z}^V$. To finish the proof, we need to show that $\mathbf{b} \in \mathcal{C}(G)$. Again by Exercise 7.39, we can choose a totally cyclic reorientation $R' \subseteq E$ of G such that $R \subseteq R'$. By Theorem 7.6.5, there is some $g \in \mathcal{F}_G$ with $g(e) < 0$ for $e \in R'$ and $g(e) > 0$ otherwise. Then

$$f + \epsilon g \in n(0,1)^E \cap \mathcal{F}_G(n\mathbf{b})$$

for sufficiently small $\epsilon > 0$. This shows that $\mathsf{P}^\circ_G(\mathbf{b}) \neq \varnothing$ and $\mathbf{b} \in \mathcal{C}(G)$. $\square$

The basis of $\mathcal{F}_G$ constructed in the course of our proof of Proposition 7.6.1 furnishes an inside-out view on nowhere-zero $\mathbb{Z}_n$-flows, in the sense of Section 7.3. Our proof of Proposition 7.6.1 shows that cycle bases are actually *lattice* bases. You are invited to convince yourself in Exercise 7.42.

Corollary 7.6.6. *Let G be a connected graph and T a spanning tree of G. If $f \in \mathcal{F}_G \cap \mathbb{Z}^E$ is an integral flow, then there is a unique $\mathbf{y} \in \mathbb{Z}^{E \setminus E(T)}$ such that*

$$f = \sum_{e \in E \setminus E(T)} y_e f_{C_e}. \tag{7.6.7}$$

Corollary 7.6.6 means that we can parametrize integral flows on G in terms of a cycle basis. For each $e \in E \setminus E(T)$, we can choose the orientation of the fundamental cycle C_e such that $f_{C_e}(e) = 1$. Moreover, $f_{C_e}(e') = 0$ for all $e' \in E \setminus E(T)$ and $e' \neq e$. Hence, if $f \in \mathcal{F}_G$ is given by (7.6.7), then $f(e) = y_e$ for each $e \in E \setminus E(T)$. For the remaining edges, the values are more involved to determine. Luckily, we do not have to know the exact values, just that they are not zero modulo n.

For $e \in E(T)$, let $\overline{\mathsf{H}}_{e,1}$ be the set of $\mathbf{y} \in \mathbb{R}^{E \setminus E(T)}$ such that the corresponding flow defined in (7.6.7) satisfies $f(e) = 1$. This is a hyperplane in $\mathbb{R}^{E \setminus E(T)} \cong \mathbb{R}^{\xi(G)}$. Defining the hyperplane $\overline{\mathsf{H}}_{e,0}$ for $f(e) = 0$ accordingly, we obtain a hyperplane arrangement

$$\overline{\mathcal{H}}^G := \{\overline{\mathsf{H}}_{e,0}, \overline{\mathsf{H}}_{e,1} : e \in E(T)\}$$

and the following result establishes the connection to inside-out polytopes; for more see Exercise 7.43.

Proposition 7.6.7. *Let $y \in \mathbb{Z}^{E \setminus E(T)}$ and let $f : E \to \mathbb{Z}$ be the corresponding flow given by* (7.6.7). *Then f is a nowhere-zero $\mathbb{Z}_n$-flow if and only if*

$$\frac{1}{n}\mathbf{y} \ \in \ (0,1)^{E \setminus E(T)} \setminus \bigcup_{e \in E(T)} \left(\overline{\mathsf{H}}_{e,0} \cup \overline{\mathsf{H}}_{e,1}\right).$$

Notes

Jakob Steiner [**175**] was probably among the first to consider the combinatorics of arrangements of lines (in the plane) and planes (in 3-space); see also [**78**, Ch. 18]. Studying arrangements of lines in the plane might sound easy but this subject has many deep results and still unresolved problems as can be found in Branko Grünbaum's books [**77, 79**]. Nowadays hyperplane arrangements play important roles in many fields [**129**]. The lattice of flats of an arrangement of hyperplanes was introduced by Henry Crapo and Gian-Carlo Rota [**48**]. Our lattices of flats belong to the class of *geometric lattices* which are cryptomorphisms for matroids; see, for example, [**131, 183**]. We already referenced in Chapter 3 the pioneering work of Thomas Zaslavsky on hyperplane arrangements, starting with [**187**]. Much of this chapter owes its existence to his work, including the geometric viewpoint initiated in [**25**], which introduced inside-out polytopes. Lemma 7.2.4 is due to Curtis Greene [**74, 75**]. The proof of Theorem 1.1.5 via inside-out polytopes appeared (together with Theorem 7.3.9) in [**25**]. A comprehensive introduction to the combinatorics of arrangements is in [**169**].

Corollary 7.3.3 is the Euclidean analogue of the *finite-field method* [**11, 48**] (see Exercise 7.15 for details); in fact, the proof of Corollary 7.3.3 is essentially that of [**11**].

The hyperplane arrangements in Corollaries 7.3.5 and 7.3.6 are examples of *Coxeter arrangements* or *reflection arrangements*: their hyperplanes correspond to finite reflection groups. The real braid arrangements in Corollary 7.3.5 correspond to root systems of type A, the arrangements in Corollary 7.3.6 to root systems of types B and C, and the arrangements in Exercise 7.11 correspond to root systems of type D. These arrangements basically correspond to the three infinite families of reflection groups, that is, finite groups of linear transformations generated by reflection. Much more can be said about the interplay between properties of a Coxeter arrangement and the associated reflection group; see, for example, [**34, 91**].

The infinite hyperplane arrangement that gave rise to alcoved polytopes is the affine reflection arrangement of type A. Alcoved polytopes with their triangulations into alcoves were studied by Thomas Lam and Alexander Postnikov in [**109**]. The alcoved triangulation of hypersimplices was described by Stanley's comment in [**65**] and in the context of Gröbner bases by Sturmfels [**178**]. The Lipschitz polytopes were discovered in [**149**]. Theorem 7.4.12 is due to Nan Li [**113**], where the number of *big ascents* is called the *cover* statistic. Alcoved polytopes for other types of affine reflection arrangements are studied in [**108**].

Zonotopes naturally appear in many disciplines; see [**36**] for references and the many ways to characterize zonotopes. The relation between zonotopes and hyperplane arrangements marks only the beginning of the deep theory of *oriented matroids*; see [**190**, Ch. 7] and [**35**]. Corollary 7.5.10 and Proposition 7.5.12 are due to Geoffrey Shephard [**156**]. Interestingly, Proposition 7.5.12 also holds true if one considers *all* fine zonotopal tilings, including the nonregular ones; see [**142**]. Corollary 7.5.16 was stated without proof in [**163**] and with proof in [**167**]. The expression for the leading coefficient of $\mathrm{ehr}_{\mathsf{Z}}(n)$, the volume of Z, was obtained earlier by Peter McMullen [**156**, p. 321].

The most famous graphical zonotope, namely that of a complete graph, is the *permutahedron* (see Exercise 7.30) which goes back to at least 1911 [**151**]. More geometric combinatorics surrounding the permutahedron can be found in [**138**]. General graphical zonotopes first surfaced in Michael Koren's work on degree sequences of graphs [**104**] and, independently, Zaslavsky's inception of colorings of *signed graphs* [**188**]—he calls graphical zonotopes *acyclotopes*. The first systematic study of graphical zonotopes seems to be [**167**]; see also [**71**].

The geometric perspective on nowhere-zero $\mathbb{Z}_n$-flows is from [**37**], and that of nowhere-zero integral flows from [**26**], but describing real-valued flows on graphs via polytopes has a long history. The polytopes $\mathsf{P}_G(\mathbf{b})$ are called *b-transshipments*, and the value b_v for $v \in V$ describes how much *flow* is lost or gained. For much more on flows and transshipments, see, e.g., [**153**, Chs. 10 & 11]. Properties of the cographical arrangement and, in particular, Theorem 7.6.5 were studied in [**75**].

Exercises

7.1 ⬠ Verify that the order polytopes given in (7.1.5) indeed are the maximal cells of a subdivision (as compared to just a dissection).

7.2 ⬠ Let $G = ([d], E)$ be a simple graph and $[d] = V_1 \uplus \cdots \uplus V_k$ a partition into connected components of G. Show that the lineality space of $\mathcal{H}_G$ is spanned by the characteristic vectors $\mathbf{e}_{V_i} \in \{0,1\}^d$ for $i = 1, \dots, k$.

7.3 ⬠ Let $G = (V, E)$ be a graph and $S \subseteq E$. Show that G/S is well defined, i.e., independent on the order with which we contract the edges in S.

7.4 Prove that $G/S_1 = G/S_2$ if and only if S_1 and S_2 differ only in edges that complete cycles in G.

7.5 ⬠ Prove Proposition 7.2.2: Let $G = (V, E)$ be a graph. Then $\mathcal{L}(G)$ and $\mathcal{L}(\mathcal{H}_G)$ are canonically isomorphic as posets.

7.6 ⬠ Prove Proposition 7.2.3: Let $G = (V, E)$ be a graph, $S \subseteq E$ a flat of G, and $F \in \mathcal{L}(\mathcal{H}_G)$ the corresponding flat of $\mathcal{H}_G$. Then

$$\mathcal{H}_G|_F \cong \mathcal{H}_{G/S} \quad \text{and} \quad \mathcal{L}(\mathcal{H}_G|_F) \cong \mathcal{L}(\mathcal{H}_{G/S}).$$

7.7 ⬠ Prove Proposition 7.2.5: Let $\mathcal{H}$ be a central hyperplane arrangement and H_0 a hyperplane in general position relative to $\mathcal{H}$. Then

$$\mathcal{L}(\mathcal{H}|_{\mathsf{H}_0}) \cong \mathcal{L}(\mathcal{H}) \setminus \{\text{lineal}(\mathcal{H})\}$$

via $F \mapsto F \cap \mathsf{H}_0$.

7.8 Given a graph G, let F be a flat of the associated graphical arrangement $\mathcal{H}_G$. Show that, as a function of n,

$$\left|(n+1)(0,1)^V \cap F \cap \mathbb{Z}^V\right|$$

is a polynomial.

7.9 Show that every face of a hyperplane arrangement $\mathcal{H}$ is the face of a region of $\mathcal{H}$.

7.10 Let $\mathcal{H}$ be an arrangement in $\mathbb{R}^d$ consisting of k hyperplanes in general position. Prove that

$$\chi_{\mathcal{H}}(n) = \binom{k}{0} n^d - \binom{k}{1} n^{d-1} + \binom{k}{2} n^{d-2} - \cdots + (-1)^d \binom{k}{d}.$$

(Note that this implies Exercise 3.67, by Zaslavsky's Theorem 3.6.4.)

7.11 Let $\mathcal{H} = \{\{x_j = \pm x_k\} : 1 \le j < k \le d\}$. Prove that[5]

$$\chi_{\mathcal{H}}(n) = (n-1)(n-3)\cdots(n-2d+5)(n-2d+3)(n-d+1).$$

7.12 ⬠ Let $\mathcal{H}$ be a central hyperplane arrangement. For $F, G \in \mathcal{L}(\mathcal{H})$, give an interpretation of $\mu_{\mathcal{L}(\mathcal{H})}(F, G)$.

7.13 ⬠ Prove Theorem 7.2.7 without appealing to geometry.

7.14 ⬠ Show that the hyperplane H_0 in the proof of Theorem 7.2.7 is in general position.

[5] The last factor of $\chi_{\mathcal{H}}(n)$ is not a typographical error.

7.15 This exercise gives an alternative to Corollary 7.3.3 for computing characteristic polynomials. Given a rational hyperplane arrangement $\mathcal{H}$ in $\mathbb{R}^d$, let q be a prime power such that $\mathcal{H}$, when viewed in $\mathbb{F}_q^d$, yields the same semilattice of flats as $\mathcal{H}$ viewed in $\mathbb{R}^d$. Prove that

$$\chi_{\mathcal{H}}(q) \ = \ \left|\mathbb{F}_q^d \setminus \bigcup \mathcal{H}\right| .$$

Use this to recompute some of the characteristic polynomials above.

7.16 Prove that all the coefficients of $O_{\mathsf{P},\mathcal{H}_G}(n-1)$ are nonnegative. Conclude that, for any graph G, the coefficients of $\chi_G(n)$ alternate in sign.

7.17 ☖ Let Π be a connected poset and let $\mathsf{Q} \subseteq \mathbb{R}^\Pi$ be the collection of order-preserving Lipschitz functions. Show that the lineality space of Q is given by the constant functions.

7.18 ☖ A poset Π is a **rooted tree** if Π has a unique minimal element $\hat{0}$ and for every $a \in \Pi$, the interval $[\hat{0}, a]$ is a chain. If Π is a rooted tree, then for every $i \in \Pi$, there is a unique saturated chain $\hat{0} = i_0 \prec\!\cdot\, i_1 \prec\!\cdot \cdots \prec\!\cdot\, i_k = i$. Let $T_\Pi : \mathbb{R}^d \to \mathbb{R}^d$ be the linear transformation given by

$$T_\Pi(\mathbf{y})_i \ = \ y_{i_0} + y_{i_1} + \cdots + y_{i_k} \, .$$

(a) Show that T_Π is invertible and lattice-preserving.
(b) Show that $T_\Pi([0,1]^d) = \mathsf{Lip}_\Pi$.
(c) Show that Lip_Π is linearly isomorphic to a cube if and only if Π is the disjoint union of rooted trees.

7.19 ☖ Show that the inequalities given in (7.4.5) are irredundant.

7.20 Show that Lipschitz polytopes are compressed.

7.21 Determine for which posets Π on d elements it holds that $\mathrm{DC}(\Pi) = \mathfrak{S}_d$.

7.22 ☖ Prove Lemma 7.4.10: For $\tau \in \mathfrak{S}_d$ and $\mathbf{q} \in \mathbb{Z}^d_{\geq 0}$, let

$$\begin{aligned} a(\tau, \mathbf{q}) \ &:= \ \left|\left\{ i \in \mathrm{Asc}(\tau^{-1}) \, : \, q_{\tau^{-1}(i)} < q_{\tau^{-1}(i+1)} \right\}\right| , \\ d(\tau, \mathbf{q}) \ &:= \ \left|\left\{ i \in \mathrm{Des}(\tau^{-1}) \, : \, q_{\tau^{-1}(i)} \leq q_{\tau^{-1}(i+1)} \right\}\right| . \end{aligned}$$

Then $h^*_{\mathbb{H}_{\mathbf{w}}(\mathbf{q}+\triangle^\tau)}(z) \ = \ z^{a(\tau,\mathbf{q})+d(\tau,\mathbf{q})}$.

7.23 ☖ Let $\mathbf{a}_1, \mathbf{b}_1, \dots, \mathbf{a}_m, \mathbf{b}_m \in \mathbb{R}^d$ such that $\mathbf{b}_1 - \mathbf{a}_1, \dots, \mathbf{b}_m - \mathbf{a}_m$ are linearly independent. Show that $\mathsf{Z}(\mathbf{a}_1, \dots, \mathbf{a}_m; \mathbf{b}_1, \dots, \mathbf{b}_m)$ is affinely isomorphic to $[0,1]^m$.

7.24 ☖ Let $\mathbf{u}_1, \dots, \mathbf{u}_r \in \mathbb{R}^m$ be a collection of r distinct points. For $\varepsilon > 0$, let

$$\ell_\varepsilon(\mathbf{x}) \ := \ \varepsilon x_1 + \varepsilon^2 x_2 + \cdots + \varepsilon^m x_m \, .$$

Show that there is an $\varepsilon_0 > 0$ and $1 \leq i \leq r$ such that $\ell_\varepsilon(\mathbf{u}_i) < \ell_\varepsilon(\mathbf{u}_j)$ for all $j \neq i$ and $0 < \varepsilon < \varepsilon_0$. (*Hint:* For fixed j, interpret $\ell_\varepsilon(\mathbf{u}_j)$ as a polynomial in ε.)

7.25 ○ The **lexicographic ordering** on $\mathbb{R}^m$ is given as follows. For $\mathbf{u}, \mathbf{v} \in \mathbb{R}^m$, we define $\mathbf{u} \preceq_{\text{lex}} \mathbf{v}$ if $\mathbf{u} = \mathbf{v}$ or if the smallest index $1 \le i \le m$ for which $u_i \neq v_i$ we have $v_i - u_i > 0$. Given a finite set $\mathbf{u}_1, \dots, \mathbf{u}_m \in \mathbb{R}^m$ of distinct points, show that there is a sufficiently small $\varepsilon > 0$ such that

$$\mathbf{u}_i \preceq_{\text{lex}} \mathbf{u}_j \quad \Longleftrightarrow \quad \ell_\varepsilon(\mathbf{u}_i) \le \ell_\varepsilon(\mathbf{u}_j),$$

for all $1 \le i < j \le m$.

7.26 ○ Complete our proof of Proposition 7.5.15: Let $S = \{\mathbf{z}_1, \dots, \mathbf{z}_k\} \subset \mathbb{Z}^d$ be a collection of k linearly independent lattice vectors. Then

$$\operatorname{ehr}_{\widehat{\square}(S)}(n) = n^k |\operatorname{gd}(S)|.$$

7.27 ○ Let $G = ([d], E)$ be a graph (possibly with loops and parallel edges). Show that for $E' \subseteq E$, the vectors $\{\mathbf{e}_i - \mathbf{e}_j : ij \in E'\}$ are linearly independent if and only if $G[E'] = ([d], E')$ has no cycles.

7.28 ○ Prove Theorem 7.5.5: Let Z be a zonotope with central hyperplane arrangement $\mathcal{H} = \mathcal{H}(\mathsf{Z})$. Then $\Phi(\mathsf{Z})$ and $\Phi(\mathcal{H})$ are anti-isomorphic.

7.29 ○ Let $\mathsf{Z} \subset \mathbb{R}^d$ be a zonotope with face lattice $\Phi = \Phi(\mathsf{Z})$. For faces $\mathsf{F}, \mathsf{F}' \in \Phi$, we write $\mathsf{F} \sim \mathsf{F}'$ if $\mathsf{F} = \mathsf{F}' + \mathbf{t}$ for some $\mathbf{t} \in \mathbb{R}^d$. We write $[\mathsf{F}]$ for the equivalence class of F and $\Phi/\!\sim$ for the collection of equivalence classes.

(a) Show that

$$[\mathsf{F}] \preceq [\mathsf{G}] \quad :\Longleftrightarrow \quad \mathsf{F}' \subseteq \mathsf{G}' \text{ for some } \mathsf{F}' \in [\mathsf{F}] \text{ and } \mathsf{G}' \in [\mathsf{G}]$$

defines a partial order on Φ.

(b) Let $\mathcal{H}$ be the hyperplane arrangement $\mathcal{H}$ associated to Z. Use Proposition 7.5.6 to infer that $\Phi/\!\sim$ and $\mathcal{L}(\mathcal{H})$ are anti-isomorphic as posets.

7.30 The **permutahedron** is defined as

$$\mathsf{P}_d := \operatorname{conv}\{(\pi(1), \pi(2), \dots, \pi(d)) : \pi \in S_d\},$$

that is, P_d is the convex hull of $(1, 2, \dots, d)$ and all points formed by permuting its entries. Show that it is the zonotope

$$\mathsf{P}_d = \mathbf{1} + \sum_{1 \le i < j \le d} [\mathbf{e}_i, \mathbf{e}_j]$$

and compute its Ehrhart polynomial.

7.31 ○ Prove Theorem 7.5.8: Let $\mathsf{Z} = \mathsf{Z}(\mathbf{z}_1, \dots, \mathbf{z}_m) \subset \mathbb{R}^d$ and $\boldsymbol{\delta} \in \mathbb{R}^m$ with associated zonotope $\widehat{\mathsf{Z}} \subset \mathbb{R}^{d+1}$. Then

$$\mathcal{P}(\mathsf{Z}, \boldsymbol{\delta}) := \left\{\pi(\mathsf{F}) : \mathsf{F} \text{ bounded face of } \widehat{\mathsf{Z}} + \uparrow_{\mathbb{R}}\right\}$$

is a zonotopal tiling of Z. (*Hint:* Revise the argumentation in our proof of Theorem 5.1.5.)

7.32 ⬠ Let $\mathcal{H} = \{\mathsf{H}_1, \mathsf{H}_2, \dots, \mathsf{H}_m\}$ be a central and essential hyperplane arrangement in $\mathbb{R}^d$ and $\mathcal{H}^{\boldsymbol{\delta}}$ a fixed affinization. Let $\mathbf{v}_1, \dots, \mathbf{v}_s$ be the vertices of $\mathcal{H}^{\boldsymbol{\delta}}$. For every 1-dimensional flat L in $\mathcal{L}(\mathcal{H})$, pick a nonzero element $\mathbf{r}$. This yields a set $\{\mathbf{r}_1, \dots, \mathbf{r}_t\}$ of nonzero vectors in $\mathbb{R}^d$.

(a) Let F be a face of $\mathcal{H}^{\boldsymbol{\delta}}$. Show that F is of the form

$$\mathsf{F} = \operatorname{conv}(V) + \operatorname{cone}(R)$$

for unique subsets $V \subseteq \{\mathbf{v}_1, \dots, \mathbf{v}_s\}$ and $R \subseteq \{\pm\mathbf{r}_1, \pm\mathbf{r}_2, \dots, \pm\mathbf{r}_t\}$.

(b) Show that the regions of $\mathcal{H}^{\boldsymbol{\delta}}$ are in bijection to the inclusion-maximal sets $V \subseteq \{\mathbf{v}_1, \dots, \mathbf{v}_s\}$ and $R \subseteq \{\pm\mathbf{r}_1, \pm\mathbf{r}_2, \dots, \pm\mathbf{r}_t\}$ such that

i) for any $\mathbf{v}, \mathbf{v}' \in V$, there is no i with $\mathbf{v} \in (\mathsf{H}_i^{\boldsymbol{\delta}_i})^<$ and $\mathbf{v}' \in (\mathsf{H}_i^{\boldsymbol{\delta}_i})^>$;

ii) for any $\mathbf{v} \in V$ and $\mathbf{t} \in R$, there is no i with $\mathbf{v} \in (\mathsf{H}_i^{\boldsymbol{\delta}_i})^<$ and $\mathbf{t} \in (\mathsf{H}_i)^>$.

(c) Deduce from this Lemma 7.5.20: Let $\mathcal{H}^{\boldsymbol{\delta}}$ be an essential affine hyperplane arrangement with vertices $\mathbf{v}_1, \dots, \mathbf{v}_r$ given by $\sigma_1, \dots, \sigma_r \in \{<, =, >\}^m$. Then $\Phi(\mathcal{H}^{\boldsymbol{\delta}})$ is determined by $\mathcal{H}$ and $\sigma_1, \dots, \sigma_r$.

7.33 Let $S = \{\mathbf{z}_1, \dots, \mathbf{z}_m\} \in \mathbb{R}^d$ be a spanning collection of vectors. Let

$$L := \{(\lambda_1, \dots, \lambda_m) \in \mathbb{R}^m : \lambda_1\mathbf{z}_1 + \dots + \lambda_m\mathbf{z}_m = \mathbf{0}\},$$

a linear subspace of dimension $m - d$.

(a) Let $\boldsymbol{\delta}, \boldsymbol{\delta}' \in \mathbb{R}^m$. Show that the affinizations $\mathcal{H}^{\boldsymbol{\delta}}$ and $\mathcal{H}^{\boldsymbol{\delta}'}$ differ by a translation if and only if $\boldsymbol{\delta} - \boldsymbol{\delta}' \in L$.

(b) Assume that $S \setminus \{\mathbf{z}_i\}$ is still spanning for all $i \in [m]$. Show that $\{x_i = 0\} \cap L$ is a hyperplane in L.

(c) Show that this arrangement of hyperplanes is associated to the zonotope $\widehat{\mathsf{Z}}$ of Corollary 7.5.21.

7.34 ⬠ Verify the claim made by Lemma 7.6.2: The function f_C is an integer-valued flow for any undirected cycle C of G.

7.35 ⬠ Let $G = (V, E)$ be a graph such that every node $w \in V$ has either no or at least two incident edges. Show that G contains a cycle.

7.36 ⬠ Show that, if an orientation of a given graph G is not totally cyclic, then there exists a coherently oriented edge cut (as illustrated in Figure 7.14). Conclude that there cannot be a flow with all positive values.

7.37 ⬠ Let $G = (V, E)$ be an oriented graph and $e \in E$ a bridge, i.e., $G \setminus e$ has strictly more connected components than G. Show that $f(e) = 0$ for every $f \in \mathcal{F}_G$.

7.38 Can you determine the number of hyperplanes of the cographical arrangement from G?

7.39 ⬠ Let $G = (V, E)$ be a bridgeless graph and $R \subseteq E$.

(a) Show that a totally cyclic orientation of G induces a totally cyclic orientation of the contraction G/R.
(b) Conversely, show that every totally cyclic orientation of G/R induces a totally cyclic orientation of G.

7.40 Extend Theorem 7.3.9 to the case that P is not full dimensional. (*Hint:* The only subtle case is when the affine span of P contains no lattice point; thus $I_{\mathsf{P}^\circ, \mathcal{H}}(n) = O_{\mathsf{P}, \mathcal{H}}(n) = 0$ for certain n.)

7.41 ⬠ Show that the linear transformation $T : \mathbb{R}^{d-1} \to \mathbb{R}^{d-1}$ defined in (7.4.2) satisfies that $T(\mathbf{p})$ is a lattice point if and only if $\mathbf{p}$ is a lattice point, and give an explicit inverse.

7.42 ⬠ Prove Corollary 7.6.6: Let G be a connected graph and T a spanning tree of G. If $f \in \mathcal{F}_G \cap \mathbb{Z}^E$ is an integral flow, then there is a unique $\mathbf{y} \in \mathbb{Z}^{E \setminus E(T)}$ such that

$$f = \sum_{e \in E \setminus E(T)} y_e f_{C_e} .$$

7.43 Given a graph $G = (V, E)$ together with an orientation ρ, an **integral flow** is a map $f : E \to \mathbb{Z}$ that assigns an integer $f(e)$ to each edge $e \in E$ such that there is conservation of flow at every node v:

$$\sum_{\stackrel{e}{\to} v} f(e) = \sum_{v \stackrel{e}{\to}} f(e) .$$

If $|f(e)| < n$ for all $e \in E$, we called f an n-flow in Exercise 1.14. Let

$$\psi_G(n) := |\{f \text{ nowhere-zero } n\text{-flow on } {}_\rho G\}| .$$

(a) Convince yourself that $\psi_G(n) = I_{\mathsf{P}^\circ, \mathcal{H}}(n)$, where $\mathsf{P} = [-1, 1]^E \cap \mathcal{F}_G$ and $\mathcal{H}$ is the coordinate hyperplane arrangement in $\mathbb{R}^E$, and use this to show that $\psi_G(n)$ is a polynomial in n.
(b) Now assume G is bridgeless. Prove that $(-1)^{\xi(G)} \psi_G(-n)$ counts the number of pairs (f, ρ), where f is an $(n + 1)$-flow and ρ is a totally-cyclic reorientation of $G/\operatorname{supp}(f)$. (In particular, $(-1)^{\xi(G)} \psi_G(0)$ equals the number of totally-cyclic orientations of G.)

Bibliography

[1] *The On-Line Encyclopedia of Integer Sequences*, published electronically at http://oeis.org, 2014.

[2] Karim A. Adiprasito and Raman Sanyal, *Relative Stanley-Reisner theory and upper bound theorems for Minkowski sums*, Publ. Math. Inst. Hautes Études Sci. **124** (2016), 99–163, DOI 10.1007/s10240-016-0083-7. MR3578915

[3] Edward E. Allen, *Descent monomials, P-partitions and dense Garsia-Haiman modules*, J. Algebraic Combin. **20** (2004), no. 2, 173–193, DOI 10.1023/B:JACO.0000047281.84115.b7. MR2104675

[4] George E. Andrews, *The theory of partitions*, Cambridge Mathematical Library, Cambridge University Press, Cambridge, 1998. Reprint of the 1976 original. MR1634067

[5] George E. Andrews and Kimmo Eriksson, *Integer partitions*, Cambridge University Press, Cambridge, 2004. MR2122332

[6] George E. Andrews, Peter Paule, and Axel Riese, *MacMahon's partition analysis. VIII. Plane partition diamonds, Special issue in honor of Dominique Foata's 65th birthday (Philadelphia, PA, 2000)*, Adv. in Appl. Math. **27** (2001), no. 2-3, 231–242, DOI 10.1006/aama.2001.0733. MR1868964

[7] Kenneth Appel and Wolfgang Haken, *Every planar map is four colorable. I. Discharging*, Illinois J. Math. **21** (1977), no. 3, 429–490. MR0543792

[8] Kenneth Appel, Wolfgang Haken, and John Koch, *Every planar map is four colorable. II. Reducibility*, Illinois J. Math. **21** (1977), no. 3, 491–567. MR0543793

[9] Federico Ardila, Thomas Bliem, and Dido Salazar, *Gelfand-Tsetlin polytopes and Feigin-Fourier-Littelmann-Vinberg polytopes as marked poset polytopes*, J. Combin. Theory Ser. A **118** (2011), no. 8, 2454–2462, DOI 10.1016/j.jcta.2011.06.004. MR2834187

[10] Isao Arima and Hiroyuki Tagawa, *Generalized (P, ω)-partitions and generating functions for trees*, J. Combin. Theory Ser. A **103** (2003), no. 1, 137–150, DOI 10.1016/S0097-3165(03)00091-8. MR1986835

[11] Christos A. Athanasiadis, *Characteristic polynomials of subspace arrangements and finite fields*, Adv. Math. **122** (1996), no. 2, 193–233, DOI 10.1006/aima.1996.0059. MR1409420

[12] Christos A. Athanasiadis, *Ehrhart polynomials, simplicial polytopes, magic squares and a conjecture of Stanley*, J. Reine Angew. Math. **583** (2005), 163–174, DOI 10.1515/crll.2005.2005.583.163. MR2146855

[13] Peter Barlow, *An Elementary Investigation of the Theory of Numbers*, J. Johnson & Co., London, 1811.

[14] Alexander I. Barvinok, *A polynomial time algorithm for counting integral points in polyhedra when the dimension is fixed*, Math. Oper. Res. **19** (1994), no. 4, 769–779, DOI 10.1287/moor.19.4.769. MR1304623

[15] Alexander Barvinok, *A course in convexity*, Graduate Studies in Mathematics, vol. 54, American Mathematical Society, Providence, RI, 2002. MR1940576

[16] Alexander Barvinok, *Integer points in polyhedra*, Zurich Lectures in Advanced Mathematics, European Mathematical Society (EMS), Zürich, 2008. MR2455889

[17] Alexander Barvinok and James E. Pommersheim, *An algorithmic theory of lattice points in polyhedra*, New perspectives in algebraic combinatorics (Berkeley, CA, 1996), Math. Sci. Res. Inst. Publ., vol. 38, Cambridge Univ. Press, Cambridge, 1999, pp. 91–147. MR1731815

[18] Victor V. Batyrev, *Dual polyhedra and mirror symmetry for Calabi-Yau hypersurfaces in toric varieties*, J. Algebraic Geom. **3** (1994), no. 3, 493–535. MR1269718

[19] Margaret M. Bayer and Louis J. Billera, *Generalized Dehn-Sommerville relations for polytopes, spheres and Eulerian partially ordered sets*, Invent. Math. **79** (1985), no. 1, 143–157, DOI 10.1007/BF01388660. MR774533

[20] Matthias Beck and Benjamin Braun, *Euler-Mahonian statistics via polyhedral geometry*, Adv. Math. **244** (2013), 925–954, DOI 10.1016/j.aim.2013.06.002. MR3077893

[21] Matthias Beck, Jesús A. De Loera, Mike Develin, Julian Pfeifle, and Richard P. Stanley, *Coefficients and roots of Ehrhart polynomials*, Integer points in polyhedra—geometry, number theory, algebra, optimization, Contemp. Math., vol. 374, Amer. Math. Soc., Providence, RI, 2005, pp. 15–36, DOI 10.1090/conm/374/06897. MR2134759

[22] Matthias Beck, Christian Haase, and Frank Sottile, *Formulas of Brion, Lawrence, and Varchenko on rational generating functions for cones*, Math. Intelligencer **31** (2009), no. 1, 9–17, DOI 10.1007/s00283-008-9013-y. MR2480796

[23] Matthias Beck and Neville Robbins, *Variations on a generating-function theme: enumerating compositions with parts avoiding an arithmetic sequence*, Amer. Math. Monthly **122** (2015), no. 3, 256–263, DOI 10.4169/amer.math.monthly.122.03.256. MR3327715

[24] Matthias Beck and Sinai Robins, *Computing the continuous discretely, Integer-point enumeration in polyhedra; With illustrations by David Austin*, 2nd ed., Undergraduate Texts in Mathematics, Springer, New York, 2015. MR3410115

[25] Matthias Beck and Thomas Zaslavsky, *Inside-out polytopes*, Adv. Math. **205** (2006), no. 1, 134–162, DOI 10.1016/j.aim.2005.07.006. MR2254310

[26] Matthias Beck and Thomas Zaslavsky, *The number of nowhere-zero flows on graphs and signed graphs*, J. Combin. Theory Ser. B **96** (2006), no. 6, 901–918, DOI 10.1016/j.jctb.2006.02.011. MR2274083

[27] Dale Beihoffer, Jemimah Hendry, Albert Nijenhuis, and Stan Wagon, *Faster algorithms for Frobenius numbers*, Electron. J. Combin. **12** (2005), Research Paper 27, 38. MR2156681

[28] Ulrich Betke and Peter McMullen, *Lattice points in lattice polytopes*, Monatsh. Math. **99** (1985), no. 4, 253–265, DOI 10.1007/BF01312545. MR799674

[29] Louis J. Billera and Gábor Hetyei, *Linear inequalities for flags in graded partially ordered sets*, J. Combin. Theory Ser. A **89** (2000), no. 1, 77–104, DOI 10.1006/jcta.1999.3008. MR1736134

[30] Louis J. Billera and Carl W. Lee, *A proof of the sufficiency of McMullen's conditions for f-vectors of simplicial convex polytopes*, J. Combin. Theory Ser. A **31** (1981), no. 3, 237–255, DOI 10.1016/0097-3165(81)90058-3. MR635368

[31] Garrett Birkhoff, *Lattice theory*, 3rd ed., American Mathematical Society Colloquium Publications, vol. 25, American Mathematical Society, Providence, R.I., 1979. MR598630

[32] George D. Birkhoff, *A determinant formula for the number of ways of coloring a map*, Ann. of Math. (2) **14** (1912/13), no. 1-4, 42–46, DOI 10.2307/1967597. MR1502436

[33] Anders Björner, *Topological methods*, Handbook of combinatorics, Vol. 1, 2, Elsevier Sci. B. V., Amsterdam, 1995, pp. 1819–1872. MR1373690

[34] Anders Björner and Francesco Brenti, *Combinatorics of Coxeter groups*, Graduate Texts in Mathematics, vol. 231, Springer, New York, 2005. MR2133266

[35] Anders Björner, Michel Las Vergnas, Bernd Sturmfels, Neil White, and Günter M. Ziegler, *Oriented matroids*, 2nd ed., Encyclopedia of Mathematics and its Applications, vol. 46, Cambridge University Press, Cambridge, 1999. MR1744046

[36] Ethan D. Bolker, *A class of convex bodies*, Trans. Amer. Math. Soc. **145** (1969), 323–345, DOI 10.2307/1995073. MR0256265

[37] Felix Breuer and Raman Sanyal, *Ehrhart theory, modular flow reciprocity, and the Tutte polynomial*, Math. Z. **270** (2012), no. 1-2, 1–18, DOI 10.1007/s00209-010-0782-6. MR2875820

[38] Charles J. Brianchon, *Théorème nouveau sur les polyèdres*, J. Ecole (Royale) Polytechnique **15** (1837), 317–319.

[39] Graham Brightwell and Peter Winkler, *Counting linear extensions*, Order **8** (1991), no. 3, 225–242, DOI 10.1007/BF00383444. MR1154926

[40] Michel Brion, *Points entiers dans les polyèdres convexes* (French), Ann. Sci. École Norm. Sup. (4) **21** (1988), no. 4, 653–663. MR982338

[41] Heinz Bruggesser and Peter Mani, *Shellable decompositions of cells and spheres*, Math. Scand. **29** (1971), 197–205 (1972), DOI 10.7146/math.scand.a-11045. MR0328944

[42] Winfried Bruns and Joseph Gubeladze, *Polytopes, rings, and K-theory*, Springer Monographs in Mathematics, Springer, Dordrecht, 2009. MR2508056

[43] Winfried Bruns and Tim Römer, *h-vectors of Gorenstein polytopes*, J. Combin. Theory Ser. A **114** (2007), no. 1, 65–76, DOI 10.1016/j.jcta.2006.03.003. MR2275581

[44] Thomas Brylawski and James Oxley, *The Tutte polynomial and its applications*, Matroid applications, Encyclopedia Math. Appl., vol. 40, Cambridge Univ. Press, Cambridge, 1992, pp. 123–225, DOI 10.1017/CBO9780511662041.007. MR1165543

[45] Arthur Cayley, *The collected mathematical papers. Volume 10*, Cambridge Library Collection, Cambridge University Press, Cambridge, 2009. Reprint of the 1896 original. MR2866374

[46] Anastasia Chavez and Nicole Yamzon, *The Dehn-Sommerville relations and the Catalan matroid*, Proc. Amer. Math. Soc. **145** (2017), no. 9, 4041–4047, DOI 10.1090/proc/13554. MR3665055

[47] William Y. C. Chen, Alan J. X. Guo, Peter L. Guo, Harry H. Y. Huang, and Thomas Y. H. Liu, *s-inversion sequences and P-partitions of type B*, SIAM J. Discrete Math. **30** (2016), no. 3, 1632–1643, DOI 10.1137/130942140. MR3539893

[48] Henry H. Crapo and Gian-Carlo Rota, *On the foundations of combinatorial theory: Combinatorial geometries*, Preliminary edition, The M.I.T. Press, Cambridge, Mass.-London, 1970. MR0290980

[49] Jesús A. De Loera, David Haws, Raymond Hemmecke, Peter Huggins, and Ruriko Yoshida, *A User's Guide for LattE v1.1, software package LattE*, 2004. Electronically available at `https://www.math.ucdavis.edu/~latte/`.

[50] Jesús A. De Loera, Raymond Hemmecke, Jeremiah Tauzer, and Ruriko Yoshida, *Effective lattice point counting in rational convex polytopes*, J. Symbolic Comput. **38** (2004), no. 4, 1273–1302, DOI 10.1016/j.jsc.2003.04.003. MR2094541

[51] Jesús A. De Loera, Jörg Rambau, and Francisco Santos, *Triangulations, Structures for algorithms and applications*, Algorithms and Computation in Mathematics, vol. 25, Springer-Verlag, Berlin, 2010. MR2743368

[52] Max Dehn, *Die Eulersche Formel im Zusammenhang mit dem Inhalt in der Nicht-Euklidischen Geometrie* (German), Math. Ann. **61** (1906), no. 4, 561–586, DOI 10.1007/BF01449498. MR1511363

[53] Boris N. Delaunay, *Sur la sphère vide.*, Bull. Acad. Sci. URSS **1934** (1934), no. 6, 793–800.

[54] Graham Denham, *Short generating functions for some semigroup algebras*, Electron. J. Combin. **10** (2003), Research Paper 36, 7. MR2014523

[55] The Sage Developers, *Sagemath, the Sage Mathematics Software System (Version 7.6)*, 2017, `http://www.sagemath.org`.

[56] Richard Ehrenborg and Margaret A. Readdy, *On valuations, the characteristic polynomial, and complex subspace arrangements*, Adv. Math. **134** (1998), no. 1, 32–42, DOI 10.1006/aima.1997.1693. MR1612379

[57] Eugène Ehrhart, *Sur les polyèdres rationnels homothétiques à n dimensions* (French), C. R. Acad. Sci. Paris **254** (1962), 616–618. MR0130860

[58] Eugène Ehrhart, *Sur la partition des nombres* (French), C. R. Acad. Sci. Paris **259** (1964), 3151–3153. MR0168517

[59] Eugène Ehrhart, *Sur un problème de géométrie diophantienne linéaire. I. Polyèdres et réseaux* (French), J. Reine Angew. Math. **226** (1967), 1–29, DOI 10.1515/crll.1967.226.1. MR0213320

[60] Eugène Ehrhart, *Sur un problème de géométrie diophantienne linéaire. II. Systèmes diophantiens linéaires* (French), J. Reine Angew. Math. **227** (1967), 25–49, DOI 10.1515/crll.1967.227.25. MR0217010

[61] Leonhard Euler, *Demonstatio nonnullarum insignium proprietatum, quibus solida hedris planis inclusa sunt praedita*, Novi Comm. Acad. Sci. Imp. Petropol. **4** (1752/53), 140–160.

[62] Leonhard Euler, *Elementa doctrinae solidorum*, Novi Comm. Acad. Sci. Imp. Petropol. **4** (1752/53), 109–140.

[63] William Feller, *An introduction to probability theory and its applications. Vol. I*, Third edition, John Wiley & Sons, Inc., New York-London-Sydney, 1968. MR0228020

[64] Valentin Féray and Victor Reiner, *P-partitions revisited*, J. Commut. Algebra **4** (2012), no. 1, 101–152, DOI 10.1216/JCA-2012-4-1-101. MR2913529

[65] Dominique Foata, *Distributions eulériennes et mahoniennes sur le groupe des permutations* (French), Higher combinatorics (Proc. NATO Advanced Study Inst., Berlin, 1976), NATO Adv. Study Inst. Ser., Ser. C: Math. Phys. Sci., vol. 31, Reidel, Dordrecht-Boston, Mass., 1977, pp. 27–49. With a comment by Richard P. Stanley. MR519777

[66] Hans Freudenthal, *Simplizialzerlegungen von beschränkter Flachheit* (German), Ann. of Math. (2) **43** (1942), 580–582, DOI 10.2307/1968813. MR0007105

[67] Wei Gao, Qing-Hu Hou, and Guoce Xin, *On P-partitions related to ordinal sums of posets*, European J. Combin. **30** (2009), no. 5, 1370–1381, DOI 10.1016/j.ejc.2008.10.007. MR2514659

[68] Ewgenij Gawrilow and Michael Joswig, *polymake: a framework for analyzing convex polytopes*, Polytopes—combinatorics and computation (Oberwolfach, 1997), DMV Sem., vol. 29, Birkhäuser, Basel, 2000, pp. 43–73. MR1785292

[69] Ladnor Geissinger, *The face structure of a poset polytope*, Proceedings of the Third Caribbean Conference on Combinatorics and Computing (Bridgetown, 1981), Univ. West Indies, Cave Hill Campus, Barbados, 1981, pp. 125–133. MR657196

[70] Israel M. Gelfand, Mikhail M. Kapranov, and Andrei V. Zelevinsky, *Discriminants, resultants and multidimensional determinants*, Modern Birkhäuser Classics, Birkhäuser Boston, Inc., Boston, MA, 2008. Reprint of the 1994 edition. MR2394437

[71] Laura Gellert and Raman Sanyal, *On degree sequences of undirected, directed, and bidirected graphs*, European J. Combin. **64** (2017), 113–124, DOI 10.1016/j.ejc.2017.04.002. MR3658823

[72] Ira M. Gessel, *A historical survey of P-partitions*, The mathematical legacy of Richard P. Stanley, Amer. Math. Soc., Providence, RI, 2016, pp. 169–188. MR3617222

[73] Jørgen P. Gram, *Om Rumvinklerne i et Polyeder*, Tidsskrift for Math. (Copenhagen) **4** (1874), no. 3, 161–163.

[74] Curtis Greene, *Acyclic orientations*, Higher Combinatorics (M. Aigner, ed.), NATO Adv. Study Inst. Ser., Ser. C: Math. Phys. Sci., vol. 31, Reidel, Dordrecht, 1977, pp. 65–68.

[75] Curtis Greene and Thomas Zaslavsky, *On the interpretation of Whitney numbers through arrangements of hyperplanes, zonotopes, non-Radon partitions, and orientations of graphs*, Trans. Amer. Math. Soc. **280** (1983), no. 1, 97–126, DOI 10.2307/1999604. MR712251

[76] Peter M. Gruber, *Convex and discrete geometry*, Grundlehren der Mathematischen Wissenschaften [Fundamental Principles of Mathematical Sciences], vol. 336, Springer, Berlin, 2007. MR2335496

[77] Branko Grünbaum, *Arrangements and spreads*, American Mathematical Society Providence, R.I., 1972. Conference Board of the Mathematical Sciences Regional Conference Series in Mathematics, No. 10. MR0307027

[78] Branko Grünbaum, *Convex polytopes*, 2nd ed., Graduate Texts in Mathematics, vol. 221, Springer-Verlag, New York, 2003. Prepared and with a preface by Volker Kaibel, Victor Klee and Günter M. Ziegler. MR1976856

[79] Branko Grünbaum, *Configurations of points and lines*, Graduate Studies in Mathematics, vol. 103, American Mathematical Society, Providence, RI, 2009. MR2510707

[80] Philip Hall, *The Eulerian functions of a group*, Q. J. Math. **7** (1936), 134–151.

[81] Allen Hatcher, *Algebraic topology*, Cambridge University Press, Cambridge, 2002. MR1867354

[82] Silvia Heubach and Toufik Mansour, *Combinatorics of compositions and words*, Discrete Mathematics and its Applications (Boca Raton), CRC Press, Boca Raton, FL, 2010. MR2531482

[83] Takayuki Hibi, *Algebraic combinatorics on convex polytopes*, Carslaw Publications, Glebe, 1992. MR3183743

[84] Takayuki Hibi, *Dual polytopes of rational convex polytopes*, Combinatorica **12** (1992), no. 2, 237–240, DOI 10.1007/BF01204726. MR1179260

[85] Takayuki Hibi, *Stanley's problem on (P, ω)-partitions*, Words, Languages and Combinatorics (Kyoto, 1990), World Sci. Publ., River Edge, NJ, 1992, pp. 187–201.

[86] Friedrich Hirzebruch, *Eulerian polynomials*, Münster J. Math. **1** (2008), 9–14. MR2502493

[87] Verner E. Hoggatt Jr. and D. A. Lind, *Fibonacci and binomial properties of weighted compositions*, J. Combinatorial Theory **4** (1968), 121–124. MR0218253

[88] John F. P. Hudson, *Piecewise linear topology*, University of Chicago Lecture Notes prepared with the assistance of J. L. Shaneson and J. Lees, W. A. Benjamin, Inc., New York-Amsterdam, 1969. MR0248844

[89] June Huh, *Milnor numbers of projective hypersurfaces and the chromatic polynomial of graphs*, J. Amer. Math. Soc. **25** (2012), no. 3, 907–927, DOI 10.1090/S0894-0347-2012-00731-0. MR2904577

[90] June Huh and Eric Katz, *Log-concavity of characteristic polynomials and the Bergman fan of matroids*, Math. Ann. **354** (2012), no. 3, 1103–1116, DOI 10.1007/s00208-011-0777-6. MR2983081

[91] James E. Humphreys, *Reflection groups and Coxeter groups*, Cambridge Studies in Advanced Mathematics, vol. 29, Cambridge University Press, Cambridge, 1990. MR1066460

[92] Masa-Nori Ishida, *Polyhedral Laurent series and Brion's equalities*, Internat. J. Math. **1** (1990), no. 3, 251–265, DOI 10.1142/S0129167X90000150. MR1078514

[93] François Jaeger, *On nowhere-zero flows in multigraphs*, Proceedings of the Fifth British Combinatorial Conference (Univ. Aberdeen, Aberdeen, 1975), Utilitas Math., Winnipeg, Man., 1976, pp. 373–378. Congressus Numerantium, No. XV. MR0395778

[94] François Jaeger, *Flows and generalized coloring theorems in graphs*, J. Combin. Theory Ser. B **26** (1979), no. 2, 205–216, DOI 10.1016/0095-8956(79)90057-1. MR532588

[95] Katharina Jochemko and Raman Sanyal, *Arithmetic of marked order polytopes, monotone triangle reciprocity, and partial colorings*, SIAM J. Discrete Math. **28** (2014), no. 3, 1540–1558, DOI 10.1137/130944849. MR3262594

[96] Katharina Jochemko and Raman Sanyal, *Combinatorial mixed valuations*, Adv. Math. **319** (2017), 630–652, DOI 10.1016/j.aim.2017.08.032. MR3695886

[97] Katharina Jochemko and Raman Sanyal *Combinatorial positivity of translation-invariant valuations and a discrete Hadwiger theorem*, J. Eur. Math. Soc. **20** (2018), no. 9, 2181-2208.

[98] Ravi Kannan, *Lattice translates of a polytope and the Frobenius problem*, Combinatorica **12** (1992), no. 2, 161–177, DOI 10.1007/BF01204720. MR1179254

[99] Daniel A. Klain and Gian-Carlo Rota, *Introduction to geometric probability*, Lezioni Lincee. [Lincei Lectures], Cambridge University Press, Cambridge, 1997. MR1608265

[100] Victor Klee, *A combinatorial analogue of Poincaré's duality theorem*, Canad. J. Math. **16** (1964), 517–531, DOI 10.4153/CJM-1964-053-0. MR0189039

[101] Donald E. Knuth, *A note on solid partitions*, Math. Comp. **24** (1970), 955–961, DOI 10.2307/2004628. MR0277401

[102] Matthias Köppe, *A primal Barvinok algorithm based on irrational decompositions*, SIAM J. Discrete Math. **21** (2007), no. 1, 220–236, DOI 10.1137/060664768. Software `LattE macchiato` available at `http://www.math.ucdavis.edu/~mkoeppe/latte/`. MR2299706

[103] Matthias Köppe and Sven Verdoolaege, *Computing parametric rational generating functions with a primal Barvinok algorithm*, Electron. J. Combin. **15** (2008), no. 1, Research Paper 16, 19. MR2383436

[104] Michael Koren, *Extreme degree sequences of simple graphs*, J. Combinatorial Theory Ser. B **15** (1973), 213–224. MR0329967

[105] Maximilian Kreuzer and Harald Skarke, *Classification of reflexive polyhedra in three dimensions*, Adv. Theor. Math. Phys. **2** (1998), no. 4, 853–871, DOI 10.4310/ATMP.1998.v2.n4.a5. MR1663339

[106] Maximilian Kreuzer and Harald Skarke, *Complete classification of reflexive polyhedra in four dimensions*, Adv. Theor. Math. Phys. **4** (2000), no. 6, 1209–1230, DOI 10.4310/ATMP.2000.v4.n6.a2. MR1894855

[107] Joseph P. S. Kung, Gian-Carlo Rota, and Catherine H. Yan, *Combinatorics: the Rota way*, Cambridge Mathematical Library, Cambridge University Press, Cambridge, 2009. MR2483561

[108] Thomas Lam and Alexander Postnikov, *Alcoved polytopes. II*, Preprint (`arXiv:math/1202.4015`), July 2006.

[109] Thomas Lam and Alexander Postnikov, *Alcoved polytopes. I*, Discrete Comput. Geom. **38** (2007), no. 3, 453–478, DOI 10.1007/s00454-006-1294-3. MR2352704

[110] Jim Lawrence, *Valuations and polarity*, Discrete Comput. Geom. **3** (1988), no. 4, 307–324, DOI 10.1007/BF02187915. MR947219

[111] Jim Lawrence, *Polytope volume computation*, Math. Comp. **57** (1991), no. 195, 259–271, DOI 10.2307/2938672. MR1079024

[112] Jim Lawrence, *A short proof of Euler's relation for convex polytopes*, Canad. Math. Bull. **40** (1997), no. 4, 471–474, DOI 10.4153/CMB-1997-056-4. MR1611351

[113] Nan Li, *Ehrhart h^*-vectors of hypersimplices*, Discrete Comput. Geom. **48** (2012), no. 4, 847–878, DOI 10.1007/s00454-012-9452-2. MR3000568

[114] László Lovász, *Combinatorial problems and exercises*, 2nd ed., North-Holland Publishing Co., Amsterdam, 1993. MR1265492

[115] Ian G. Macdonald, *Polynomials associated with finite cell-complexes*, J. London Math. Soc. (2) **4** (1971), 181–192, DOI 10.1112/jlms/s2-4.1.181. MR0298542

[116] Percy A. MacMahon, *Memoir on the theory of the partitions of numbers. Part V. Partitions in two-dimensional space*, Proc. Roy. Soc. London Ser. A **85** (1911), no. 578, 304–305.

[117] Percy A. MacMahon, *Combinatory analysis*, Two volumes (bound as one), Chelsea Publishing Co., New York, 1960. MR0141605

[118] Claudia Malvenuto, *P-partitions and the plactic congruence*, Graphs Combin. **9** (1993), no. 1, 63–73, DOI 10.1007/BF01195328. MR1215586

[119] P. McMullen, *The numbers of faces of simplicial polytopes*, Israel J. Math. **9** (1971), 559–570, DOI 10.1007/BF02771471. MR0278183

[120] Peter McMullen, *Lattice invariant valuations on rational polytopes*, Arch. Math. (Basel) **31** (1978/79), no. 5, 509–516, DOI 10.1007/BF01226481. MR526617

[121] Peter McMullen, *On simple polytopes*, Invent. Math. **113** (1993), no. 2, 419–444, DOI 10.1007/BF01244313. MR1228132

[122] Ezra Miller and Bernd Sturmfels, *Combinatorial commutative algebra*, Graduate Texts in Mathematics, vol. 227, Springer-Verlag, New York, 2005. MR2110098

[123] Hermann Minkowski, *Volumen und Oberfläche* (German), Math. Ann. **57** (1903), no. 4, 447–495, DOI 10.1007/BF01445180. MR1511220

[124] Hermann Minkowski, *Gesammelte Abhandlungen von Hermann Minkowski. Unter Mitwirkung von Andreas Speiser und Hermann Weyl, herausgegeben von David Hilbert. Band I, II.*, Leipzig u. Berlin: B. G. Teubner. Erster Band. Mit einem Bildnis Hermann Minkowskis und 6 Figuren im Text. xxxvi, 371 S.; Zweiter Band. Mit einem Bildnis Hermann Minkowskis, 34 Figuren in Text und einer Doppeltafel. iv, 466 S. gr. 8° (1911)., 1911.

[125] Leo Moser and E. L. Whitney, *Weighted compositions*, Canad. Math. Bull. **4** (1961), 39–43, DOI 10.4153/CMB-1961-006-0. MR0125060

[126] James R. Munkres, *Elements of algebraic topology*, Addison-Wesley Publishing Company, Menlo Park, CA, 1984. MR755006

[127] Isabella Novik and Ed Swartz, *Applications of Klee's Dehn-Sommerville relations*, Discrete Comput. Geom. **42** (2009), no. 2, 261–276, DOI 10.1007/s00454-009-9187-x. MR2519879

[128] Kathryn L. Nyman, *The peak algebra of the symmetric group*, J. Algebraic Combin. **17** (2003), no. 3, 309–322, DOI 10.1023/A:1025000905826. MR2001673

[129] Peter Orlik and Louis Solomon, *Combinatorics and topology of complements of hyperplanes*, Invent. Math. **56** (1980), no. 2, 167–189, DOI 10.1007/BF01392549. MR558866

[130] Peter Orlik and Hiroaki Terao, *Arrangements of hyperplanes*, Grundlehren der Mathematischen Wissenschaften [Fundamental Principles of Mathematical Sciences], vol. 300, Springer-Verlag, Berlin, 1992. MR1217488

[131] James Oxley, *Matroid theory*, 2nd ed., Oxford Graduate Texts in Mathematics, vol. 21, Oxford University Press, Oxford, 2011. MR2849819

[132] SeungKyung Park, *P-partitions and q-Stirling numbers*, J. Combin. Theory Ser. A **68** (1994), no. 1, 33–52, DOI 10.1016/0097-3165(94)90090-6. MR1295782

[133] Sam Payne, *Ehrhart series and lattice triangulations*, Discrete Comput. Geom. **40** (2008), no. 3, 365–376, DOI 10.1007/s00454-007-9002-5. MR2443289

[134] Micha A. Perles and Geoffrey C. Shephard, *Angle sums of convex polytopes*, Math. Scand. **21** (1967), 199–218 (1969), DOI 10.7146/math.scand.a-10860. MR0243425

[135] T. Kyle Petersen, *Enriched P-partitions and peak algebras*, Adv. Math. **209** (2007), no. 2, 561–610, DOI 10.1016/j.aim.2006.05.016. MR2296309

[136] Georg Alexander Pick, *Geometrisches zur Zahlenlehre*, Sitzenber. Lotos (Prague) **19** (1899), 311–319.

[137] Henri Poincaré, *Sur la généralisation d'un theorem d'Euler relatif aux polyèdres*, C. R. Acad. Sci. Paris (1893), 144–145.

[138] Alexander Postnikov, *Permutohedra, associahedra, and beyond*, Int. Math. Res. Not. IMRN **6** (2009), 1026–1106, DOI 10.1093/imrn/rnn153. MR2487491

[139] Jorge L. Ramírez Alfonsín, *The Diophantine Frobenius problem*, Oxford Lecture Series in Mathematics and its Applications, vol. 30, Oxford University Press, Oxford, 2005. MR2260521

[140] Ronald C. Read, *An introduction to chromatic polynomials*, J. Combinatorial Theory **4** (1968), 52–71. MR0224505

[141] Victor Reiner and Volkmar Welker, *On the Charney-Davis and Neggers-Stanley conjectures*, J. Combin. Theory Ser. A **109** (2005), no. 2, 247–280, DOI 10.1016/j.jcta.2004.09.003. MR2121026

[142] Jürgen Richter-Gebert and Günter M. Ziegler, *Zonotopal tilings and the Bohne-Dress theorem*, Jerusalem combinatorics '93, Contemp. Math., vol. 178, Amer. Math. Soc., Providence, RI, 1994, pp. 211–232, DOI 10.1090/conm/178/01902. MR1310586

[143] John Riordan, *An introduction to combinatorial analysis*, Dover Publications, Inc., Mineola, NY, 2002. Reprint of the 1958 original [Wiley, New York; MR0096594 (20 #3077)]. MR1949650

[144] Neville Robbins, *On Tribonacci numbers and 3-regular compositions*, Fibonacci Quart. **52** (2014), no. 1, 16–19. MR3181091

[145] Neville Robbins, *On r-regular compositions*, J. Combin. Math. Combin. Comput. **96** (2016), 195–199. MR3495368

[146] Gian-Carlo Rota, *On the foundations of combinatorial theory. I. Theory of Möbius functions*, Z. Wahrscheinlichkeitstheorie und Verw. Gebiete **2** (1964), 340–368 (1964), DOI 10.1007/BF00531932. MR0174487

[147] Steven V Sam, *A bijective proof for a theorem of Ehrhart*, Amer. Math. Monthly **116** (2009), no. 8, 688–701, DOI 10.4169/193009709X460813. MR2572104

[148] Francisco Santos and Günter M. Ziegler, *Unimodular triangulations of dilated 3-polytopes*, Trans. Moscow Math. Soc., posted on 2013, 293–311, DOI 10.1090/s0077-1554-2014-00220-x. MR3235802

[149] Raman Sanyal and Christian Stump, *Lipschitz polytopes of posets and permutation statistics*, J. Combin. Theory Ser. A **158** (2018), 605–620, DOI 10.1016/j.jcta.2018.04.006. MR3800139

[150] Ludwig Schläfli, *Gesammelte mathematische Abhandlungen. Band I* (German), Verlag Birkhäuser, Basel, 1950. MR0034587

[151] Pieter Hendrik Schoute, *Analytic treatment of the polytopes regularly derived from the regular polytopes*, Verhandelingen der Koninklijke Akademie von Wetenschappen te Amsterdam **11** (1911), no. 3.

[152] Alexander Schrijver, *Theory of linear and integer programming*, Wiley-Interscience Series in Discrete Mathematics, John Wiley & Sons, Ltd., Chichester, 1986. A Wiley-Interscience Publication. MR874114

[153] Alexander Schrijver, *Combinatorial optimization. Polyhedra and efficiency. Vol. C*, Algorithms and Combinatorics, vol. 24, Springer-Verlag, Berlin, 2003. Disjoint paths, hypergraphs; Chapters 70–83. MR1956926

[154] Paul D. Seymour, *Nowhere-zero 6-flows*, J. Combin. Theory Ser. B **30** (1981), no. 2, 130–135, DOI 10.1016/0095-8956(81)90058-7. MR615308

[155] Geoffrey C. Shephard, *An elementary proof of Gram's theorem for convex polytopes*, Canad. J. Math. **19** (1967), 1214–1217, DOI 10.4153/CJM-1967-110-7. MR0225228

[156] Geoffrey C. Shephard, *Combinatorial properties of associated zonotopes*, Canad. J. Math. **26** (1974), 302–321, DOI 10.4153/CJM-1974-032-5. MR0362054

[157] Andrew V. Sills, *Compositions, partitions, and Fibonacci numbers*, Fibonacci Quart. **49** (2011), no. 4, 348–354. MR2852008

[158] Duncan M. Y. Sommerville, *The relation connecting the angle-sums and volume of a polytope in space of n dimensions*, Proc. Royal Soc. Lond. Ser. A **115** (1927), 103–119.

[159] Eugene Spiegel and Christopher J. O'Donnell, *Incidence algebras*, Monographs and Textbooks in Pure and Applied Mathematics, vol. 206, Marcel Dekker, Inc., New York, 1997. MR1445562

[160] Richard P. Stanley, *Ordered structures and partitions*, American Mathematical Society, Providence, R.I., 1972. Memoirs of the American Mathematical Society, No. 119. MR0332509

[161] Richard P. Stanley, *Acyclic orientations of graphs*, Discrete Math. **5** (1973), 171–178, DOI 10.1016/0012-365X(73)90108-8. MR0317988

[162] Richard P. Stanley, *Combinatorial reciprocity theorems*, Advances in Math. **14** (1974), 194–253, DOI 10.1016/0001-8708(74)90030-9. MR0411982

[163] Richard P. Stanley, *Decompositions of rational convex polytopes*, Ann. Discrete Math. **6** (1980), 333–342. Combinatorial mathematics, optimal designs and their applications (Proc. Sympos. Combin. Math. and Optimal Design, Colorado State Univ., Fort Collins, Colo., 1978). MR593545

[164] Richard P. Stanley, *The number of faces of a simplicial convex polytope*, Adv. in Math. **35** (1980), no. 3, 236–238, DOI 10.1016/0001-8708(80)90050-X. MR563925

[165] Richard P. Stanley, *Some aspects of groups acting on finite posets*, J. Combin. Theory Ser. A **32** (1982), no. 2, 132–161, DOI 10.1016/0097-3165(82)90017-6. MR654618

[166] Richard P. Stanley, *Two poset polytopes*, Discrete Comput. Geom. **1** (1986), no. 1, 9–23, DOI 10.1007/BF02187680. MR824105

[167] Richard P. Stanley, *A zonotope associated with graphical degree sequences*, Applied geometry and discrete mathematics, DIMACS Ser. Discrete Math. Theoret. Comput. Sci., vol. 4, Amer. Math. Soc., Providence, RI, 1991, pp. 555–570. MR1116376

[168] Richard P. Stanley, *A monotonicity property of h-vectors and h^*-vectors*, European J. Combin. **14** (1993), no. 3, 251–258, DOI 10.1006/eujc.1993.1028. MR1215335

[169] Richard P. Stanley, *An introduction to hyperplane arrangements*, Geometric combinatorics, IAS/Park City Math. Ser., vol. 13, Amer. Math. Soc., Providence, RI, 2007, pp. 389–496. MR2383131

[170] Richard P. Stanley, *Enumerative combinatorics. Volume 1*, 2nd ed., Cambridge Studies in Advanced Mathematics, vol. 49, Cambridge University Press, Cambridge, 2012. MR2868112

[171] Richard P. Stanley and Jim Pitman, *A polytope related to empirical distributions, plane trees, parking functions, and the associahedron*, Discrete Comput. Geom. **27** (2002), no. 4, 603–634, DOI 10.1007/s00454-002-2776-6. MR1902680

[172] Alan Stapledon, *Weighted Ehrhart theory and orbifold cohomology*, Adv. Math. **219** (2008), no. 1, 63–88, DOI 10.1016/j.aim.2008.04.010. MR2435420

[173] Alan Stapledon, *Inequalities and Ehrhart δ-vectors*, Trans. Amer. Math. Soc. **361** (2009), no. 10, 5615–5626, DOI 10.1090/S0002-9947-09-04776-X. MR2515826

[174] Alan Stapledon, *Additive number theory and inequalities in Ehrhart theory*, Int. Math. Res. Not. IMRN **5** (2016), 1497–1540, DOI 10.1093/imrn/rnv186. MR3509934

[175] Jakob Steiner, *Einige Gesetze über die Theilung der Ebene und des Raumes* (German), J. Reine Angew. Math. **1** (1826), 349–364, DOI 10.1515/crll.1826.1.349. MR1577621

[176] Ernst Steinitz, *Polyeder und Raumeinteilungen*, Encyclopädie der mathematischen Wissenschaften, Band 3 (Geometrie), Teil 3AB12 (1922), 1–139.

[177] John R. Stembridge, *Enriched P-partitions*, Trans. Amer. Math. Soc. **349** (1997), no. 2, 763–788, DOI 10.1090/S0002-9947-97-01804-7. MR1389788

[178] Bernd Sturmfels, *Gröbner bases and convex polytopes*, University Lecture Series, vol. 8, American Mathematical Society, Providence, RI, 1996. MR1363949

[179] William T. Tutte, *A ring in graph theory*, Proc. Cambridge Philos. Soc. **43** (1947), 26–40. MR0018406

[180] Alexander N. Varchenko, *Combinatorics and topology of the arrangement of affine hyperplanes in the real space* (Russian), Funktsional. Anal. i Prilozhen. **21** (1987), no. 1, 11–22. MR888011

[181] Sven Verdoolaege, *Software package* barvinok, (2004), electronically available at http://freshmeat.net/projects/barvinok/.

[182] Hermann Weyl, *Elementare Theorie der konvexen Polyeder* (German), Comment. Math. Helv. **7** (1934), no. 1, 290–306, DOI 10.1007/BF01292722. MR1509514

[183] Neil White (ed.), *Theory of matroids*, Encyclopedia of Mathematics and its Applications, vol. 26, Cambridge University Press, Cambridge, 1986. MR849389

[184] Hassler Whitney, *A logical expansion in mathematics*, Bull. Amer. Math. Soc. **38** (1932), no. 8, 572–579, DOI 10.1090/S0002-9904-1932-05460-X. MR1562461

[185] Herbert S. Wilf, *Which polynomials are chromatic?* (English, with Italian summary), Colloquio Internazionale sulle Teorie Combinatorie (Roma, 1973), Accad. Naz. Lincei, Rome, 1976, pp. 247–256. Atti dei Convegni Lincei, No. 17. MR0453579

[186] Herbert S. Wilf, *generatingfunctionology*, 3rd ed., A K Peters, Ltd., Wellesley, MA, 2006. MR2172781

[187] Thomas Zaslavsky, *Facing up to arrangements: face-count formulas for partitions of space by hyperplanes*, Mem. Amer. Math. Soc. **1** (1975), no. issue 1, 154, vii+102. MR0357135

[188] Thomas Zaslavsky, *Signed graph coloring*, Discrete Math. **39** (1982), no. 2, 215–228, DOI 10.1016/0012-365X(82)90144-3. MR675866

[189] Doron Zeilberger, *The composition enumeration reciprocity theorem*, Personal Journal of Shalosh B. Ekhad and Doron Zeilberger (2012), http://www.math.rutgers.edu/~zeilberg/mamarim/mamarimhtml/comp.html.

[190] Günter M. Ziegler, *Lectures on polytopes*, Graduate Texts in Mathematics, vol. 152, Springer-Verlag, New York, 1995. MR1311028

Notation Index

The following table contains a list of symbols that are frequently used throughout the book. The page numbers refer to the first appearance/definition of each symbol.

Notation	Meaning	Page
$[a,b]$	an interval in a poset	12
$a \prec\!\cdot\, b$	cover relation in a poset	12
$\operatorname{aff}(S)$	affine hull of $S \subseteq \mathbb{R}^d$	58
$\operatorname{Ast}_{\mathbf{v}}(\mathsf{P})$	$\{\mathsf{F} \in \Phi(\mathsf{P}) : \mathbf{v} \notin \mathsf{F}\}$, the antistar of the vertex $\mathbf{v}$	188
$\operatorname{Asc}(\sigma)$	$\{j \in [d-1] : \sigma(j) < \sigma(j+1)\}$, the ascent set of σ	214
$\operatorname{asc}(\sigma)$	$\lvert \operatorname{Asc}(\sigma) \rvert$, the ascent number of σ	225
B_d	Boolean lattice of all subsets of $[d]$	34
$b(\mathcal{H})$	number of relatively bounded regions of $\mathcal{H}$	90
C	a polyhedral cone	55
$\mathsf{C}^\vee$	polar cone	62
$cp_{\Pi,\phi}(n)$	number of (Π,ϕ)-chain partitions of n	138
$CP_{\Pi,\phi}(n)$	generating function of (Π,ϕ)-chain partitions of n	138
$\mathbb{C}^\Pi$	vector space of functions $\Pi \to \mathbb{C}$	41
$\mathbb{C}[x]$	vector space of polynomials with complex coefficients	108
$\mathbb{C}[x]_{\le d}$	polynomials with complex coefficients of degree $\le d$	108
$\mathbb{C}[\![z]\!]$	vector space of formal power series	110
$c_A(n)$	number of compositions of n with parts in A	116
$c_\Pi(n)$	number of compositions of n that respect the poset Π	227
$\operatorname{comaj}(\sigma)$	$\sum_{j \in \operatorname{Asc}\sigma} j$, the comajor index of σ	226

Symbol	Meaning	Page
$\mathcal{J}(\Pi)$	lattice of order ideals of the poset Π	30
$(I f)(n)$	$f(n)$, the identity operator applied to $f(n)$	109
$I(\Pi)$	incidence algebra of the poset Π	30
$I_{\mathsf{P},\mathcal{H}}(t)$	Ehrhart function of inside-out polytope $(\mathsf{P},\mathcal{H})$	245
$\mathrm{JH}(\Pi)$	$\{\tau \in \mathfrak{S}_d : \tau^{-1} \in \mathrm{Lin}(\Pi)\}$, Jordan–Hölder set of Π	211
$[k]$	set $\{1, 2, \ldots, k\}$	ix
K_d	complete graph on d nodes	24
K_Π	order cone of the poset Π	203
$\mathsf{K}_1 + \mathsf{K}_2$	Minkowski sum of $\mathsf{K}_1, \mathsf{K}_2 \subseteq \mathbb{R}^d$	64
$l_\Pi(x,y)$	length of a maximal chain in $[x,y]$ in the poset Π	38
$\mathrm{lineal}(\mathsf{Q})$	lineality space of the polyhedron Q	57
$\mathrm{Lin}(\Pi)$	set of linear extensions of the poset Π	206
Lip_Π	Lipschitz polytope of the poset Π	255
$\mathcal{L}(G)$	flats of the graph G partially ordered by inclusion	42
$\mathcal{L}(\mathcal{H})$	intersection poset of the hyperplane arrangement $\mathcal{H}$	88
$\mathrm{maj}(\sigma)$	$\sum_{j \in \mathrm{Des}\,\sigma} j$, the major index of σ	221
μ_Π	Möbius function of the poset Π	33
$\binom{n}{d}$	binomial coefficient	x
$[n]_q$	$1 + q + \cdots + q^{n-1}$, a q-integer	221
$\mathcal{N}(\Pi, \preceq)$	poset of refinements of the poset $(\Pi, \preceq)$	210
O_Π	order polytope of the poset Π	214
$\Omega_\Pi(n)$	order polynomial of the poset Π	14
$\Omega^\circ_\Pi(n)$	strict order polynomial of the poset Π	13
P, Q	a polyhedron or polytope	16
P°	relative interior of the polyhedron P	16
$\partial\mathsf{P}$	relative boundary of the polyhedron P	59
PC_d	collection of polyconvex sets in $\mathbb{R}^d$	72
$\mathsf{PC}(\mathcal{H})$	collection of $\mathcal{H}$-polyconvex sets	74
$(\mathsf{P}, \mathcal{H})$	an inside-out polytope	246
$[\mathbf{p}, \mathbf{q}]$	line segment with endpoints $\mathbf{p}$ and $\mathbf{q}$	60
Π	a poset	12
$p_\Pi(n)$	number of Π-partitions of the integer n	228
$p^\circ_\Pi(n)$	number of strict Π-partitions of the integer n	228
$P_\Pi(z)$	$\sum_{t \geq 0} p_\Pi(t)\, z^t$	228
$p_A(n)$	restricted partition function for A	120
$pl(n)$	number of plane partitions of n	117
$\mathrm{Pull}(\mathsf{P})$	pulling triangulation of a polytope P	189

Symbol	Meaning	Page
$r(\mathcal{H})$	number of regions of the arrangement $\mathcal{H}$	90
$\mathrm{rk}_\Pi(x)$	the rank of $x \in \Pi$	48
$\mathrm{rec}(\mathsf{Q})$	recession cone of the polyhedron Q	55
$[S]$	indicator function of the set S	91
$\|\mathcal{S}\|$	support of the polyhedral complex $\mathcal{S}$	156
$S(d,r)$	Stirling number of the second kind	14
$c(d,r)$	Stirling number of the first kind	48
$s(d,k)$	Eulerian number	192
$(S\,f)(n)$	$f(n+1)$, the shift operator applied to $f(n)$	109
$\mathrm{supp}(f)$	support of a flow (or vector) f	7
$\binom{S}{d}$	$\{A \subseteq S : \|A\| = d\}$	x
$\sigma_S(\mathbf{z})$	integer-point transform of S	125
$\mathfrak{S}_d$	set of bijections/permutations of $[d]$	49
$\mathcal{T}$	a triangulation	18
$\mathrm{T}_{\mathbf{q}}(\mathsf{Q})$	tangent cone of the polyhedron Q at the point $\mathbf{q}$	82
$\mathrm{T}_{\mathsf{F}}(\mathsf{Q})$	tangent cone of the polyhedron Q at the face F	83
$\mathbf{v} * \mathsf{P}$	pyramid with apex $\mathbf{v}$ and base P	71
$\mathrm{vert}(\mathsf{P})$	vertex set of the polytope P	61
$\mathrm{vol}(S)$	(relative) volume of S	152
$\mathrm{Vis}_{\mathbf{p}}(\mathsf{P})$	complex of faces of P visible from $\mathbf{p}$	91
$\mathrm{Vis}_{\mathbf{p}}(\mathcal{S})$	subcomplex of cells of $\mathcal{S}$ visible from $\mathbf{p}$	168
$\xi(G)$	cyclotomic number of the graph G	12
ζ_Π	zeta function of the poset Π	31
$Z_\Pi(n)$	zeta polynomial of the poset Π	36
$\mathsf{Z}(\mathbf{z}_1, \dots, \mathbf{z}_m)$	a zonotope	261
$\hat{0}$	minimum of a poset	32
$\hat{1}$	maximum of a poset	32
$x \vee y$	join of elements in a poset	37
$x \wedge y$	join of elements in a poset	37
$\preceq$, $\preceq_\Pi$	partial order relation (of a poset Π)	12
$\Box$	(half-open) parallelpiped	127
$\hat{\Box}$, $\check{\Box}$	fundamental parallelpipeds	134
$\pentagon$	an exercise used in the text	xii

Index

Selected Published Titles in This Series

195 **Matthias Beck and Raman Sanyal,** Combinatorial Reciprocity Theorems, 2018
193 **Martin Lorenz,** A Tour of Representation Theory, 2018
192 **Tai-Peng Tsai,** Lectures on Navier-Stokes Equations, 2018
191 **Theo Bühler and Dietmar A. Salamon,** Functional Analysis, 2018
190 **Xiang-dong Hou,** Lectures on Finite Fields, 2018
189 **I. Martin Isaacs,** Characters of Solvable Groups, 2018
188 **Steven Dale Cutkosky,** Introduction to Algebraic Geometry, 2018
187 **John Douglas Moore,** Introduction to Global Analysis, 2017
186 **Bjorn Poonen,** Rational Points on Varieties, 2017
185 **Douglas J. LaFountain and William W. Menasco,** Braid Foliations in Low-Dimensional Topology, 2017
184 **Harm Derksen and Jerzy Weyman,** An Introduction to Quiver Representations, 2017
183 **Timothy J. Ford,** Separable Algebras, 2017
182 **Guido Schneider and Hannes Uecker,** Nonlinear PDEs, 2017
181 **Giovanni Leoni,** A First Course in Sobolev Spaces, Second Edition, 2017
180 **Joseph J. Rotman,** Advanced Modern Algebra: Third Edition, Part 2, 2017
179 **Henri Cohen and Fredrik Strömberg,** Modular Forms, 2017
178 **Jeanne N. Clelland,** From Frenet to Cartan: The Method of Moving Frames, 2017
177 **Jacques Sauloy,** Differential Galois Theory through Riemann-Hilbert Correspondence, 2016
176 **Adam Clay and Dale Rolfsen,** Ordered Groups and Topology, 2016
175 **Thomas A. Ivey and Joseph M. Landsberg,** Cartan for Beginners: Differential Geometry via Moving Frames and Exterior Differential Systems, Second Edition, 2016
174 **Alexander Kirillov Jr.,** Quiver Representations and Quiver Varieties, 2016
173 **Lan Wen,** Differentiable Dynamical Systems, 2016
172 **Jinho Baik, Percy Deift, and Toufic Suidan,** Combinatorics and Random Matrix Theory, 2016
171 **Qing Han,** Nonlinear Elliptic Equations of the Second Order, 2016
170 **Donald Yau,** Colored Operads, 2016
169 **András Vasy,** Partial Differential Equations, 2015
168 **Michael Aizenman and Simone Warzel,** Random Operators, 2015
167 **John C. Neu,** Singular Perturbation in the Physical Sciences, 2015
166 **Alberto Torchinsky,** Problems in Real and Functional Analysis, 2015
165 **Joseph J. Rotman,** Advanced Modern Algebra: Third Edition, Part 1, 2015
164 **Terence Tao,** Expansion in Finite Simple Groups of Lie Type, 2015
163 **Gérald Tenenbaum,** Introduction to Analytic and Probabilistic Number Theory, Third Edition, 2015
162 **Firas Rassoul-Agha and Timo Seppäläinen,** A Course on Large Deviations with an Introduction to Gibbs Measures, 2015
161 **Diane Maclagan and Bernd Sturmfels,** Introduction to Tropical Geometry, 2015
160 **Marius Overholt,** A Course in Analytic Number Theory, 2014
159 **John R. Faulkner,** The Role of Nonassociative Algebra in Projective Geometry, 2014
158 **Fritz Colonius and Wolfgang Kliemann,** Dynamical Systems and Linear Algebra, 2014
157 **Gerald Teschl,** Mathematical Methods in Quantum Mechanics: With Applications to Schrödinger Operators, Second Edition, 2014